# Lecture Notes in Computer Science 2791

Edited by G. Goos, J. Hartmanis, and J. van Leeuwen

**Springer**
*Berlin*
*Heidelberg*
*New York*
*Hong Kong*
*London*
*Milan*
*Paris*
*Tokyo*

Kim G. Larsen   Peter Niebert (Eds.)

# Formal Modeling
# and Analysis
# of Timed Systems

First International Workshop, FORMATS 2003
Marseille, France, September 6-7, 2003
Revised Papers

 Springer

Series Editors

Gerhard Goos, Karlsruhe University, Germany
Juris Hartmanis, Cornell University, NY, USA
Jan van Leeuwen, Utrecht University, The Netherlands

Volume Editors

Kim G. Larsen
Aalborg University, Department of Computer Science
Fr. Bajersvej 7E, 9220 Aalborg East, Denmark
E-mail: kgl@cs.auc.dk

Peter Niebert
Université de Provence, Laboratoire d'Informatique Fondamentale, CMI
39, Rue Joliot-Curie, 13453 Marseille Cedex 13, France
E-mail: peter.niebert@lif.univ-mrs.fr

Library of Congress Control Number: 2004103614

CR Subject Classification (1998): F.3, D.2, D.3, C.3

ISSN 0302-9743
ISBN 3-540-21671-5 Springer-Verlag Berlin Heidelberg New York

Springer-Verlag is a part of Springer Science+Business Media

springeronline.com

© Springer-Verlag Berlin Heidelberg 2004
Printed in Germany

Typesetting: Camera-ready by author, data conversion by DA-TeX Gerd Blumenstein
Printed on acid-free paper      SPIN: 10931844      06/3142      5 4 3 2 1 0

# Timed Automata and Timed Languages
# Challenges and Open Problems[*]

Eugene Asarin

VERIMAG, Centre Equation
2 ave de Vignate, 38610 Gières
France
Eugene.Asarin@imag.fr

**Abstract.** The first years of research in the area of timed systems were marked by a spectacular progress, but also by many natural and important problems left behind without solutions. Some of those are really hard, some have been completely overlooked, some are known only to small groups of researchers but have never been really attacked by the community.

The aim of this talk is to present several open problems and research directions in the domain of timed systems which seem important to the author. In particular we will consider variants of timed automata, theory of timed languages, timed games etc.

---

[*] Partially supported by the European community project IST-2001-35304 Ametist

# Towards Efficient Partition Refinement for Checking Reachability in Timed Automata[*]

Agata Półrola[1], Wojciech Penczek[2,3], and Maciej Szreter[2]

[1] Faculty of Mathematics
University of Lodz
Banacha 22, 90-238 Lodz, Poland
polrola@math.uni.lodz.pl
[2] Institiute of Computer Science
PAS
Ordona 21, 01-237 Warsaw, Poland
{penczek,mszreter}@ipipan.waw.pl
[3] Institute of Informatics
Podlasie Academy
Sienkiewicza 51, 08-110 Siedlce, Poland

**Abstract.** The paper presents a new method for building abstract models for Timed Automata, enabling on-the-fly reachability analysis. Our pseudo-simulating models, generated by a modified partitioning algorithm, are in many cases much smaller than forward-reachability graphs commonly applied for this kind of verification. A theoretical description of the method is supported by some preliminary experimental results.

## 1 Introduction

Model checking is an approach commonly applied for automated verification of *reachability properties*. Given a system and a property $p$, reachability model checking consists in an exploration of the (reachable) state space of the system, testing whether there exists a state where $p$ holds. The main problem of this approach is caused by the size of the state space, which in many cases, in particular for timed systems, can be very large (even infinite). One of the solutions to this problem consists in applying finite *abstract models* of systems, preserving reachability properties. To this aim, *forward-reachability graphs* are most commonly used [6, 8, 14]. Reachability analysis on these models is usually performed *on-the-fly*, while generating a model, i.e., given a property $p$, newly obtained states of the model are examined, and the generation of the model is finished as soon as a state satisfying $p$ is found [6]. An alternative solution are *symbolic methods*, one of which, very intensively investigated recently, consists in exploiting SAT-based Bounded Model Checking (BMC) [3, 21]. In the BMC approach, satisfiability of a formula encoding reachability of a state satisfying $p$

---

[*] Partly supported by the State Committee for Scientific Research under the grant No. 8T11C 01419

is tested, using a symbolic path of a bounded length encoding the unfolding of a transition relation. Since the length of this path affects dramatically the size of its propositional encoding, the BMC methods are mainly applicable for proving reachability, but can become ineffective when no state satisfying $p$ can be found (see the discussion in the section on experimental results). Therefore, verification methods based on building (small) abstract models of systems still have a practical importance, and developing efficient algorithms for generating such models remains an important subject of research.

Our paper presents a new method for generating abstract models of Timed Automata using a modified minimization (partitioning) algorithm [5]. The very first motivation for our approach has been taken from [20], where the authors claim that minimal bisimulating models (b-models, for short) for Timed Automata could often be smaller than the corresponding forward-reachability ones (fr-models, for short). Since *simulating* (*s-*) *models* [17] are usually smaller than minimal b-models, they could be used instead of the latter. However, it is clear that there should exist abstract models preserving reachability properties that are even smaller than the minimal s-models, as the latter preserve the whole language of ACTL. To define these models we relax the requirement on the transition relation of the s-models, formulated for all the predecessors of each state, such that it applies to one of them only, and call the new class of models pseudo-simulating ones (ps-models, for short). The models can be generated using a modification of the partitioning algorithm for s-models [10]. Moreover, the method can be used in an on-the-fly manner for reachability verification.

The rest of the paper is organised as follows: Section 2 presents the related work. In Section 3, we introduce Timed Automata and their concrete and abstract models usually considered in the literature. Then, in Sections 4 - 6 we provide a definition, an algorithm, and an implementation of ps-models for Timed Automata. Sections 7 and 8 contain experimental results and final remarks.

## 2   Related Work

Different aspects of the reachability analysis for Timed Automata have been usually studied on fr-models [8, 14, 16]. In [8], some abstractions allowing to reduce their sizes are proposed, while in [14], data structures for effective verification are shown. Alternative methods of reachability verification consist in exploiting SAT-solvers [3, 21], BDDs (a solution for closed automata shown in [4]), untimed histories of states and a bisimulation relation [13], or partitioning to obtain pseudo-b-models [19]. Partitioning-based reachability analysis was studied also for other kinds of systems [7, 15]. Moreover, the paper [12] presents various reachability-preserving equivalence relations. We provide a comparison with models generated by these relations in the full version of this work [18].

Minimization algorithms for b-models were introduced in [5, 15]. The first of them was applied to s-models in [10]. Implementations for Timed Automata and b-models can be found in [1, 2, 20, 22], and for s-models - in [11]. The paper [20] contains some examples showing that b-models can be smaller than the corresponding fr- ones.

## 3   Timed Automata

Let $\mathbb{R}$ ($\mathbb{R}_+$) denote the set of (non-negative) reals, and $\mathbb{N}$ - the set of natural numbers. Let $\mathcal{X} = \{x_1, \ldots, x_n\}$ be a finite set of variables, called *clocks*. A *valuation* on $\mathcal{X}$ is a $n$-tuple $v = (v_1, \ldots, v_n) \in \mathbb{R}^n_+$, where $v_i$ is the value of the clock $x_i$ in $v$. For a valuation $v$ and $\delta \in \mathbb{R}$, $v + \delta$ denotes the valuation $v'$ s.t. for all $x_i \in \mathcal{X}$, $v'_i = v_i + \delta$. For a valuation $v$ and a subset of clocks $X \subseteq \mathcal{X}$, $v[X := 0]$ denotes the valuation $v'$ such that for all $x_i \in X$, $v'_i = 0$ and for all $x_i \in \mathcal{X} \setminus X$, $v'_i = v_i$. By an *atomic constraint* for $\mathcal{X}$ we mean an expression of the form $x_i \sim c$ or $x_i - x_j \sim c$, where $x_i, x_j \in \mathcal{X}$, $\sim \in \{\leq, <, >, \geq\}$ and $c \in \mathbb{N}$. A valuation $v$ *satisfies* an atomic constraint $x_i \sim c$ ($x_i - x_j \sim c$) if $v_i \sim c$ ($v_i - v_j \sim c$, respectively). A (*time*) *zone* of $\mathcal{X}$ is a convex polyhedron in $\mathbb{R}^n_+$ defined by a finite set of atomic constraints, i.e., the set of all the valuations satisfying all these constraints. The set of all the time zones of $\mathcal{X}$ is denoted by $Z(n)$.

**Definition 1.** *A timed automaton $\mathcal{A}$ is a tuple $(\Sigma, S, \mathcal{X}, s^0, E, \mathcal{I})$, where $\Sigma$ is a finite set of actions, $\mathcal{X} = \{x_1, \ldots, x_n\}$ is a finite set of clocks, $E \subseteq S \times \Sigma \times Z(n) \times 2^{\mathcal{X}} \times S$ is a transition relation. Each element $e$ of $E$ is denoted by $s \xrightarrow{a, z, Y} s'$, which represents a transition from location $s$ to $s'$, performing an action $a$, with the set $Y \subseteq \mathcal{X}$ of clocks to be reset, and with a zone $z$ defining the enabling condition for $e$. The function $\mathcal{I} : S \longrightarrow Z(n)$, called a location invariant, assigns to each location a zone defining the conditions under which $\mathcal{A}$ can be in this location.*

A *concrete state* of $\mathcal{A}$ is a pair $q = (s, v)$, where $s \in S$ and $v \in \mathbb{R}^n_+$ is a valuation such that $v \in \mathcal{I}(s)$. The set of all the concrete states is denoted by $Q$. The initial state of $\mathcal{A}$ is $i = 1, \ldots, n$. The states of $\mathcal{A}$ can change as a result of passing some time or performing an action as follows: the automaton can change from $(s, v)$ to $(s', v')$ on $e \in E$ (denoted by $(s, v) \xrightarrow{e}_d (s', v')$) iff $e : s \xrightarrow{a, z, Y} s'$, $v \in z$, and $v' = v[Y := 0] \in \mathcal{I}(s')$; and can change from $(s, v)$ to $(s', v')$ by passing some time $\delta \in \mathbb{R}_+$ (denoted by $(s, v) \xrightarrow{\delta}_d (s', v')$) iff $s = s'$ and $v' = v + \delta \in \mathcal{I}(s)$. The structure $F_d = (Q, q^0, \rightarrow_d)$ is the *concrete dense state space* of $\mathcal{A}$.

Besides the relation $\rightarrow_d$ defined above, other kinds of transition relations can also be introduced. For our purposes, we define the *concrete (discrete) successor relation* $\rightarrow \in Q \times E \times Q$ as follows: for $q, q' \in Q$ and $e \in E$, let $q \xrightarrow{e} q'$ denote that $q'$ is obtained from $q$ by passing some time, performing the transition $e \in E$, and then passing some time again. Formally, $q \xrightarrow{e} q'$ iff $(\exists q_1, q_2 \in Q)(\exists \delta_1, \delta_2 \in \mathbb{R}_+)$ $q \xrightarrow{\delta_1}_d q_1 \xrightarrow{e}_d q_2 \xrightarrow{\delta_2}_d q'$. The state $q'$ is called a *successor* of $q$, whereas the structure $F_c = (Q, q^0, \rightarrow)$ is called the *concrete (discrete) state space* of $\mathcal{A}$.

Let $q \in Q$. A $q$-run of $\mathcal{A}$ is a finite sequence of concrete states $q_0 \xrightarrow{e_0} q_1 \xrightarrow{e_1} q_2 \xrightarrow{e_2} \ldots \xrightarrow{e_{n-1}} q_n$, where $q_0 = q$ and $e_i \in E$ for each $i < n$. A state $q' \in Q$ is *reachable* if there exists a $q^0$-run and $i \in \mathbb{N}$ such that $q' = q_i$. The set of all the reachable states of $\mathcal{A}$ will be denoted by $Reach_{\mathcal{A}}$.

## 3.1   Models for Timed Automata

Let $PV$ be a set of propositional variables, and let $V_c : Q \rightarrow 2^{PV}$ be a valuation function, which assigns the same propositions to the states with the same location, i.e., $V_c((s, v)) = V_c((s', v'))$ for all $s = s'$.

**Definition 2.** *Let $F_c = (Q, q^0, \rightarrow)$ be the concrete (discrete) state space of a timed automaton $\mathcal{A}$. A structure $M_c = (F_c, V_c)$ is called a* concrete *(discrete) model of $\mathcal{A}$.*

Since concrete state spaces (and therefore concrete models) of Timed Automata are usually infinite, they cannot be directly applied to model checking. Therefore, in order to reduce their sizes we define finite abstractions, preserving properties to be verified. The idea is to combine into classes (sets) the concrete states that are indistinguishable w.r.t. these properties.

**Definition 3.** *Let $M_c = (F_c, V_c)$ be a concrete model for $\mathcal{A}$. A structure $M = (G, V)$, where $G = (W, w_0, \rightarrow)$ is a directed, rooted, edge-labelled [1] graph with a node set $W$, $w_0 \in W$ is the initial node, and $V : W \rightarrow 2^{PV}$ is a valuation function, is called an* abstract *(discrete) model for $\mathcal{A}$ if the following conditions are satisfied:*

 − *each node $w \in W$ is a set of states of $Q$ and $q^0 \in w_0$;*
 − *for each $w \in W$ and $q \in w$ we have $V_c(q) = V(w)$;*
 − *$(\forall w_1, w_2 \in Reach(W))(\forall e \in E)$ $w_1 \xrightarrow{e} w_2$ iff $(\exists q_1 \in w_1)(\exists q_2 \in w_2)$ $q_1 \xrightarrow{e} q_2$, where $Reach(W) = \{w \in W \mid w \cap Reach_{\mathcal{A}} \neq \emptyset\}$.*

*The graph $G$ is called an* abstract state space *of $\mathcal{A}$, whereas its nodes are called* abstract states. *The abstract model $M$ is* complete *iff $(\forall q \in Q)(\exists w \in W)$ $q \in w$.*

In what follows, we consider complete abstract models only.

   In the literature, abstract models generated for a *dense semantics* (i.e., derived from the concrete state space $F_d$) are usually considered. One of them are *surjective models*. Below, we provide their definition adapted for the discrete case:

**Definition 4.** *A model $M = (G, V)$ for $\mathcal{A}$, where $G = (W, w_0, \rightarrow)$, is called* surjective *iff*
*$(\forall w_1, w_2 \in Reach(W))(\forall e \in E)$ if $w_1 \xrightarrow{e} w_2$ then $(\forall q_2 \in w_2)(\exists q_1 \in w_1)$ $q_1 \xrightarrow{e} q_2$.*

   An example of surjective models are forward reachability (fr-) models, commonly applied for reachability verification [6, 8, 14]. Reachability analysis on these models is usually performed *on-the-fly*, together with their generation [6]

---

[1] The edges are labelled with the names of transitions in $E$.

(notice that in the worst case the whole model must be generated). The models can be further improved by applying various abstractions [8, 14].

Another class of abstract models considered in the literature are *bisimulating* (*b-*) *models*. These models are usually generated for the dense semantics [1, 20], but again their definition can be easily adapted also for the discrete one:

**Definition 5.** *A model* $M = (G, V)$ *for* $\mathcal{A}$, *where* $G = (W, w_0, \rightarrow)$, *is* bisimulating *iff*
$(\forall w_1, w_2 \in Reach(W))(\forall e \in E)$ *if* $w_1 \xrightarrow{e} w_2$ *then* $(\forall q_1 \in w_1)(\exists q_2 \in w_2)\, q_1 \xrightarrow{e} q_2$.

Moreover, in [17], the following *simulating* (*s-*) *models* were introduced:

**Definition 6.** *A model* $M = (G, V)$ *for* $\mathcal{A}$, *where* $G = (W, w_0, \rightarrow)$, *is* simulating *iff for each* $w \in W$ *there exists a non-empty* $w^{cor} \subseteq w$ *such that* $q^0 \in w_0^{cor}$ *and* $(\forall w_1, w_2 \in Reach(W))(\forall e \in E)$ *if* $w_1 \xrightarrow{e} w_2$ *then* $(\forall q_1 \in w_1^{cor})(\exists q_2 \in w_2^{cor})\, q_1 \xrightarrow{e} q_2$.

Both b- and s- models preserve reachability properties.

## 3.2   Zones and Regions

Finite abstract models built for Timed Automata use *regions* as states.

**Definition 7.** *Given a timed automaton* $\mathcal{A}$, *let* $s \in S$, *and* $Z \in Z(n)$. *A region* $R \subseteq S \times \mathbb{R}_+^n$ *is a set of states* $R = \{(s, v) \mid v \in Z\}$, *denoted by* $(s, Z)$. *The region* $(s, \emptyset)$ *is identified with the empty region.*

Let $v, v' \in \mathbb{R}_+^n$, $Z, Z' \in Z(n)$, and $R, R' \in S \times Z(n)$. We define the following operations on zones and regions:

- $v \leq v'$ iff $\exists \delta \in \mathbb{R}_+$ such that $v' = v + \delta$;
- $Z \setminus Z'$ is a set of disjoint zones s.t. $\{Z'\} \cup (Z \setminus Z')$ is a partition of $Z$;
- $R \setminus R' = \{(s, Z'') \mid Z'' \in Z \setminus Z'\}$ for regions $R = (s, Z)$ and $R' = (s, Z')$;
- $Z[Y := 0] = \{v[Y := 0] \mid v \in Z\}$;   $[Y := 0]Z = \{v \mid v[Y := 0] \in Z\}$;
- $Z \nearrow := \{v' \in \mathbb{R}^n \mid (\exists v \in Z)\, v \leq v'\}$;   $Z \swarrow := \{v' \in \mathbb{R}^n \mid (\exists v \in Z)\, v' \leq v\}$.

Notice that the operations $\cap$, $\nearrow$, $\swarrow$, $Z[Y := 0]$ and $[Y := 0]Z$ preserve zones. These results together with the implementation of $Z \setminus Z'$ can be found in [1, 20].

## 4   Pseudo-simulating Models

In [20], the authors claim that minimal b-models, generated for the dense semantics, are often smaller than the corresponding fr- ones. Since s-models are usually smaller than the former, they could be better for reachability verification. However, in order to test reachability even more effectively, we introduce *pseudo-simulating* (*ps-*) *models* (which are never bigger than s- ones), and provide an algorithm for an on-the-fly reachability verification. The idea behind the definition of ps-models consists in relaxing the requirement on the transition

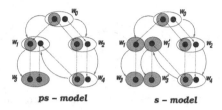

**ps – model**          **s – model**

**Fig. 1.** A ps- and s-model generated for the same case

relation of the s-models, formulated for all the predecessors of each state (see Def. 6), such that it applies to one of them only. The selected predecessor needs to be reachable from the beginning state in the minimal number of steps.

Before we give the definition, we need some auxiliary notions. For two nodes $w, w'$ of $G$, let $w \rightarrow w'$ denote that there exists $e \in E$ s.t. $w \xrightarrow{e} w'$. A *path* $\pi$ in $G$ is a finite sequence of nodes and edges of the form $\pi = w_1 \xrightarrow{e_1} w_2 \xrightarrow{e_2} \ldots \xrightarrow{e_{k-1}} w_k$, with $e_i \in E$ for all $i < k$ (labels on the edges can be then omitted). We say that $\pi$ is from $w_1$ to $w_k$. A path is of length $k$ if it contains $k$ edges. For a node $w \in W$, the *depth* of $w$, denoted by $dpt(w)$, is the length of a shortest path from $w_0$ to $w$ in $G$ if there is such, otherwise, the depth of $w$ is assumed to be infinite.

**Definition 8.** *A model $M = (G, V)$ for $\mathcal{A}$, where $G = (W, w_0, \rightarrow)$, is* pseudo-simulating *iff for each $w \in W$ there exists a non-empty $w^{cor} \subseteq w$ such that $q^0 \in w_0^{cor}$, and*
$(\forall w_1, w_2 \in Reach(W))(\forall e \in E)$ *if $w_1 \xrightarrow{e} w_2$, then there exists $w \in Reach(W)$ and $h \in E$ such that $w \xrightarrow{h} w_2$ and $dpt(w)$ is minimal in the set $\{dpt(w') \mid w' \xrightarrow{h'} w_2$, for some $h' \in E\}$, and (\*) $(\forall q_1 \in w^{cor}) (\exists q_2 \in w_2^{cor}) q_1 \xrightarrow{h} q_2$.*

The following example shows a difference between s- and ps- models:

*Example 1.* Fig. 1 presents a ps- and s-model generated for the same case. The *cors* of the classes are coloured; circles and straight lines are used for drawing the concrete model, while ellipses and arcs - for abstract ones. In the ps-model, the state of $w_4^{cor}$ does not need to have successors in $w_1^{cor}$. This, however, is required in the s-model, which results in creating two additional nodes $w_1'$ and $w_3'$.

Let $M = (G, V)$, where $G = (W, w_0, \rightarrow)$, be a ps-model for $\mathcal{A}$. A run $\rho = q_0 \xrightarrow{e_1} q_1 \xrightarrow{e_2} \ldots \xrightarrow{e_n} q_n$ of $\mathcal{A}$ is said to be *inscribed* in a path $\pi = w_0 \xrightarrow{e_1} w_1 \xrightarrow{e_2} \ldots \xrightarrow{e_n} w_n$ in $G$, if $q_i \in w_i$ for all $i = 0, \ldots, n$.

Denote all the edges $w \xrightarrow{e} w_2$ in $G$ satisfying the condition (\*) of Def. 8 by $w \xRightarrow{e} w'$. Moreover, let $w \Rightarrow w'$ denote that there exists $e \in E$ s.t. $w \xRightarrow{e} w'$. Next, we characterise ps-models:

**Theorem 1.** *The following conditions hold:*

a) *Each $q^0$-run of $\mathcal{A}$ is inscribed in a path of $G$,*
b) *For each $w \in Reach(W)$, there is $n \in \mathbb{N}$ and $\pi = w_0 \Rightarrow w_1 \Rightarrow \ldots \Rightarrow w_n$ in $G$ s.t. $w = w_n$,*

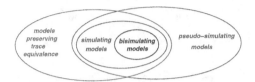

**Fig. 2.**  Relations between various kinds of models

*c)  For each path $\pi = w_0 \Rightarrow w_1 \Rightarrow \ldots \Rightarrow w_n$, there exists a $q^0$-run $\rho = q^0 \xrightarrow{e_0} q_1 \xrightarrow{e_1} \ldots \xrightarrow{e_{n-1}} q_n$ of $\mathcal{A}$ inscribed in $\pi$ and such that $q_i \in w_i^{cor}$ for each $i \leq n$.*

A proof can be found in [18]. It it easy to see from the above theorem that the ps-models preserve reachability.

Fig. 2 shows the relations between ps-models and some other well-known classes of models considered in the literature. A proof and an extended comparison, including also other classes of models, can be found in the full version of this paper [18].

## 5    A Minimization Algorithm for Ps-Models

Ps-models can be generated using a modification of the well-known *minimization (partitioning) algorithm* [5]. In order to give the algorithm, we introduce the following notions:

By a *partition* $\Pi \subseteq 2^Q$ of the set of concrete states $Q$ of $\mathcal{A}$ we mean a set of disjoint *classes* $X \subseteq Q$ the union of which equals $Q$. For a given partition $\Pi$ of $Q$, $X, Y \in \Pi$ and $e \in E$ we introduce the functions:

- $pre_e(X, Y) = \{x \in X \mid \exists y \in Y : x \xrightarrow{e} y\}$;
- $post_e(X, Y) = \{y \in Y \mid \exists x \in X : x \xrightarrow{e} y\}$.

In order to generate ps-models, instead of a partition of $Q$, we use a *d-cor-partition* $\Pi \subseteq 2^Q \times 2^Q \times (\mathbb{N} \cup \{\infty\})$, defined as a set of triples of the form $\widehat{X} = (X, X^{cor}, dpt(X))$, where $\Pi|_1$ (i.e., the projection of $\Pi$ on the first component) is a partition of $Q$, and $X^{cor} \subseteq X$. By $q \in \widehat{X}$ we mean that $q \in X$. Define $\widehat{X} \xrightarrow{e} \widehat{Y}$ iff $X \xrightarrow{e} Y$. Moreover, we introduce

- $Pre_\Pi^e(\widehat{X}) = \{\widehat{Y} \in \Pi \mid pre_e(Y, X) \neq \emptyset\}$, $Pre_\Pi(\widehat{X}) = \bigcup_{e \in E} Pre_\Pi^e(\widehat{X})$,
- $Post_\Pi^e(\widehat{X}) = \{\widehat{Y} \in \Pi \mid post_e(X, Y) \neq \emptyset\}$, $Post_\Pi(\widehat{X}) = \bigcup_{e \in E} Post_\Pi^e(\widehat{X})$.

A class $\widehat{X}$ is *reachable* if there is a concrete state $q \in X$ which is reachable.

Below, we introduce the notion of *ps-unstability*. Intuitively, a class is *ps-unstable* w.r.t. its successor $\widehat{Y}$ in $\Pi$ if there is no predecessor of $\widehat{Y}$ with a minimal depth such that its *cor* contains only states with successors in $Y^{cor}$ (see also Fig. 3).

**Definition 9.** *Let $\Pi$ be a given d-cor-partition, and $\widehat{X}, \widehat{Y} \in \Pi$.*

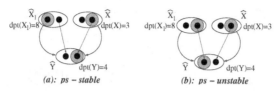

**Fig. 3.** $Ps$-stability and $ps$-unstability

- The class $\widehat{X}$ is $ps$-unstable w.r.t. $\widehat{Y}$ iff for some $e \in E$ we have $pre_e(X, Y) \neq \emptyset$, and for all $h \in E$ and all $\widehat{X_1} \in \Pi$ such that $\widehat{X_1} \xrightarrow{h} \widehat{Y}$ and $dpt(X_1)$ is minimal in $\{dpt(X'_1) \mid \widehat{X'_1} \in Pre_\Pi(\widehat{Y})\}$ we have $pre_h(X_1^{cor}, Y^{cor}) \neq X_1^{cor}$;
- $\Pi$ is $ps$-stable iff $(G \mid_1, V \mid_1)$, where $G \mid_1 = (\Pi \mid_1, [q^0], \rightarrow)$, is a $ps$-model with $X^{cor}$ and $dpt(X)$ satisfying Def. 8 w.r.t. $X$ for each $\widehat{X} \in \Pi$.

*Example 2.* Fig. 3 illustrates the notions of $ps$-stability and $ps$-unstability. Consider classes $\widehat{X}, \widehat{X_1}, \widehat{Y}$ of a partition $\Pi$ with the components $dpt$ as shown in the figure. In the part (a), both the classes $\widehat{X}$ and $\widehat{X_1}$ are $ps$-stable w.r.t. $\widehat{Y}$, since all the states of $X^{cor}$ have successors in $Y^{cor}$, and $\widehat{X}$ is the predecessor of $\widehat{Y}$ with the minimal depth. In contrary, in (b) both the classes are $ps$-unstable w.r.t. $\widehat{Y}$, since its predecessor $\widehat{X}$ does not satisfy the required condition.

The minimization algorithm for ps-models is a modification of the algorithm for s-models [10]. It starts from an initial $d$-$cor$-partition $\Pi_0$, in which the component $dpt$ of the class containing $q^0$ is equal to 0 and its $cor$ is the singleton $\{q^0\}$, whereas for all the other classes $\widehat{X} \in \Pi_0$, $dpt(X) = \infty$ and $X^{cor} = X$. Then, it constructs a minimal model $M_{min}^{ps} = (G_{min}^{ps}, V)$, where $G_{min}^{ps} = (\Pi^{st}, ([q^0], \{q^0\}, 0), \rightarrow)$, $\Pi^{st}$ is the reachable part of a $ps$-stable partition $\Pi$ obtained by a refinement of $\Pi_0$, $\Pi \mid_1$ is *compatible* with $\Pi_0 \mid_1$ (i.e., each class of $\Pi_0 \mid_1$ is a union of classes of $\Pi \mid_1$), and $q^0 \in ([q^0], \{q^0\}, 0)$. The algorithm is parameterised by a non-deterministic function $Split(\widehat{X}, \Pi)$, defined for the classes $\widehat{X} \in \Pi$ with $dpt(X) \neq \infty$ (the explanation for considering these classes only will be given later). The function refines $\Pi$ by choosing a class $\widehat{Y} \in \Pi$ w.r.t. which $\widehat{X}$ is $ps$-unstable, and then splitting either $\widehat{X}$, or a class $\widehat{X_1} \in \Pi$ s.t. $dpt(X_1)$ is minimal in the set $\{dpt(X'_1) \mid \widehat{X'_1} \in Pre_\Pi(\widehat{Y})\}$, in order to make $\widehat{X}$ $ps$-stable w.r.t. $\widehat{Y}$. Before defining the above function, we introduce another function $dpt_\Pi(X)$, defined for $X \in \Pi \mid_1$, which is used for computing the component $dpt(X)$ when a new class $\widehat{X}$ is created. The function returns a value which is a possible depth of $\widehat{X}$, determined by the analysis of the classes $\widehat{Y} \in \Pi$ for which there is $e \in E$ s.t. $pre_e(Y, X) \neq \emptyset$ (notice that the components $dpt$ of the classes in a given step of the algorithm can differ from their depths in the model obtained when the algorithm terminates). More precisely, $dpt_\Pi([q^0]) = 0$, $dpt_\Pi(X) = 1 + min\{dpt(V) \mid \widehat{V} \in \Pi \ \wedge \ pre_e(V, X) \neq \emptyset$ for some $e \in E\}$ if there exists $\widehat{V} \in \Pi$ and $e \in E$ such that $pre_e(V, X) \neq \emptyset$ and $dpt(V) \neq \infty$, and $dpt_\Pi(X) = \infty$, otherwise. For $\widehat{X}, \widehat{Y} \in \Pi$ s.t. $pre_e(X, Y) \neq \emptyset$ and $pre_e(X^{cor}, Y^{cor}) \neq X^{cor}$ for some $e \in E$,

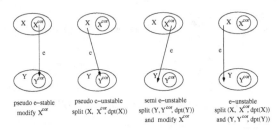

pseudo e-stable     pseudo e-unstable     semi e-unstable     e-unstable

**Fig. 4.** The four cases of the function $Sp$

we define also an auxiliary function $Sp(\widehat{X}, \widehat{Y}, e, \Pi)$, which splits $\widehat{X}$ w.r.t. $\widehat{Y}$ as follows (see also Fig. 4):

1. $Sp(\widehat{X}, \widehat{Y}, e, \Pi) = \{(X, pre_e(X^{cor}, Y^{cor}), dpt_\Pi(X))\}$
   if $\widehat{X}$ is *pseudo e-stable* w.r.t. $(Y, Y^{cor}, dpt(Y))$, i.e., $pre_e(X^{cor}, Y^{cor}) \neq \emptyset$;

2. $Sp(\widehat{X}, \widehat{Y}, e, \Pi) =$
   $\{(X \setminus X^{cor}, pre_e(X, Y^{cor}), dpt_\Pi(X \setminus X^{cor})), (X^{cor}, X^{cor}, dpt_\Pi(X^{cor}))\}$ if $\widehat{X}$
   is *pseudo e-unstable* w.r.t. $\widehat{Y}$, i.e., $pre_e(X^{cor}, Y^{cor}) = \emptyset \wedge pre_e(X, Y^{cor}) \neq \emptyset$;

3. $Sp(\widehat{X}, \widehat{Y}, e, \Pi) = \{(X, pre_e(X^{cor}, Y), dpt_\Pi(X)), (Y^{cor}, Y^{cor}, dpt_\Pi(Y^{cor})),$
   $(Y \setminus Y^{cor}, Y \setminus Y^{cor}, dpt_\Pi(Y \setminus Y^{cor})))\}$ if $\widehat{X}$ is *semi e-unstable* w.r.t. $\widehat{Y}$, i.e.,
   $pre_e(X^{cor}, Y^{cor}) = pre_e(X, Y^{cor}) = \emptyset \wedge pre_e(X^{cor}, Y) \neq \emptyset$;

4. $Sp(\widehat{X}, \widehat{Y}, e, \Pi) = \{(pre_e(X, Y), pre_e(X, Y), dpt_\Pi(pre_e(X, Y))),$
   $(X \setminus pre_e(X, Y), X^{cor}, dpt_\Pi(X \setminus pre_e(X, Y))), (Y^{cor}, Y^{cor}, dpt_\Pi(Y^{cor})), (Y \setminus$
   $Y^{cor}, Y \setminus Y^{cor}, dpt_\Pi(Y \setminus Y^{cor})))\}$ if $\widehat{X}$ is *e-unstable* w.r.t. $\widehat{Y}$,
   i.e., $pre_e(X^{cor}, Y^{cor}) = pre_e(X, Y^{cor}) = pre_e(X^{cor}, Y) = \emptyset$.

Then, we define

- $Split(\widehat{X}, \Pi) = \{\widehat{X}\}$ if $\widehat{X}$ is *ps-stable* w.r.t. all $\widehat{Y}$ in $\Pi$.

Otherwise, a class $\widehat{Y}$ and a transition $e \in E$ are chosen, for which $pre_e(X, Y) \neq \emptyset$ and $\widehat{X}$ is *ps-unstable* w.r.t. $\widehat{Y}$, and then

a) if $dpt(Y) \geq dpt(X) + 1$, then $Split(\widehat{X}, \Pi) = Sp(\widehat{X}, \widehat{Y}, e, \Pi)$;
b) if $dpt(Y) < dpt(X) + 1$, then we choose a class $\widehat{X_1}$ s.t. for some $h \in E$ we
   have $pre_h(X_1, Y) \neq \emptyset$, $pre_h(X_1^{cor}, Y) \neq X_1^{cor}$ and $dpt(X_1) = min\{dpt(X_1') \mid$
   $\widehat{X_1'} \in Pre_\Pi(\widehat{Y})\}$, and $Split(\widehat{X}, \Pi) = Sp(\widehat{X_1}, \widehat{Y}, h, \Pi)$.

Intuitively, if $\widehat{X}$ is *ps-unstable* w.r.t. $\widehat{Y}$ and from the analysis of $\Pi$ of a given step we can assume that in the model obtained when the algorithm terminates $\widehat{X}$ will be the predecessor of $\widehat{Y}$ of the minimal depth, then we apply to these classes the appropriate case of the function $Sp$. Otherwise, i.e., if $dpt(Y)$ indicates that $\widehat{Y}$ has another predecessor with a depth smaller than $dpt(X)$, we apply the function $Sp$ to $\widehat{Y}$ and to its predecessor of a smallest value of $dpt$ (see also Fig. 5).

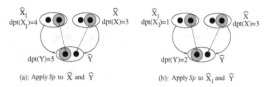

(a): Apply $Sp$ to $\widehat{X}$ and $\widehat{Y}$          (b): Apply $Sp$ to $\widehat{X_1}$ and $\widehat{Y}$

**Fig. 5.** Two cases of the function $Split(\widehat{X}, \Pi)$

*Example 3.* Fig. 5 presents two cases of the function $Split(\widehat{X}, \Pi)$. Consider a class $\widehat{Y}$ and its two predecessors $\widehat{X}$ and $\widehat{X_1}$ shown in the figure, and a step of the algorithm in which the components $dpt$ of these classes are equal to the ones given in the picture. If in this step stability of $\widehat{X}$ is checked, then the appropriate case of the function $Sp$ will be applied to $\widehat{X}$ and $\widehat{Y}$ in the case (a), and to $\widehat{X_1}$ and $\widehat{Y}$ in the case (b).

Notice that while applying the function $Split$ to $([q^0], \{q^0\}, 0)$, $cor$ of the class containing $q^0$ remains unchanged, which ensures that $q^0 \in w_0^{cor}$ always holds.

The minimization algorithm for s-models maintains two sets **stable** and **reachable**, which contain the stable and reachable classes of $\Pi$ of a given step, respectively. Classes to be split are chosen from **reachable**. The algorithm terminates when the sets **reachable** and **stable** are equal. The algorithm for ps-models is similar. However, in this case, the modifications made to **reachable** in a step of the algorithm result in changing the components $dpt$ of some classes. More precisely, if a class $\widehat{X}$ is ps-stable w.r.t. all its successors, then it is added to the set **stable**, and all its successors $\widehat{Y}$ - to **reachable**. Before adding $\widehat{Y}$ to **reachable** we set $dpt(Y) = min\{dpt(X)+1, dpt(Y)\}$. Then, the set **reachable** contains only classes with the components $dpt$ different than $\infty$ (this explains why it is sufficient to have the function $Split$ defined for such classes only). Moreover, if a class $\widehat{Y}$ is removed from **reachable**, then $dpt(Y)$ is set to $\infty$.

In order to generate ps-models more effectively, the set **reachable** can be replaced by a **list reachable** sorted w.r.t. the depths of the classes. This makes the algorithm work in a BFS-like mode, and the case b) of the function $Split$ never occurs. (More precisely, in a BFS like behaviour of the algorithm, the case b), i.e., ps-unstability of a class $\widehat{X}$ w.r.t. a class $\widehat{Y}$ with $dpt(Y) < dpt(X) + 1$, can occur only when $\widehat{Y}$ is the initial class. This, however, can be avoided as well, which is explained later).

The models obtained this way are usually smaller than the ones we get when classes to be split can be chosen in an arbitrary order.

*Example 4.* Fig. 6 shows an influence of the order in which classes to be split are chosen from $\Pi$, on the size of the generated model. Consider $\widehat{X}, \widehat{X_1} \in \Pi$ and a step of the algorithm in which components $dpt$ of the classes are equal to ones given in the picture. In the case when ps-stability of the class $\widehat{X_1}$ is checked before adding $\widehat{X}$ to the list **reachable**, the class is found ps-unstable w.r.t. $\widehat{Y}$, and due to $dpt(X) = \infty$, the function $Sp$ is applied to $\widehat{X_1}$ and $\widehat{Y}$. In a further

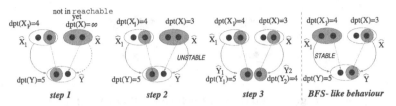

**Fig. 6.** A difference between ps-models obtained when classes to be split are chosen in two different orders

step, $ps$-stability of $\widehat{X}$ is checked, and since at this step $dpt(X)$ satisfies the condition $dpt(X) < dpt(X_1)$, the function $Sp$ is applied to $\widehat{X}$ and $\widehat{Y}$. The above process is shown on the left-hand side of the picture. On the other hand, if the algorithm works in a BFS-like mode, $ps$-stability of the predecessor of $\widehat{Y}$ with the minimal depth is always checked first. This prevents unnecessary partitionings.

Notice that in spite of the algorithm operating in the BFS-like mode, in the case when $ps$-stability of a class $\widehat{X}$ w.r.t. its successor $\widehat{Y} = ([q^0], \{q^0\}, 0)$ is checked, it can be impossible to find a predecessor $\widehat{X}_1$ of $\widehat{Y}$ which satisfies the condition $pre(X_1^{cor}, Y^{cor}) = X_1^{cor}$, and the case b) of the function $Split$ can occur. However, it is easy to see that partitionings in this case are not necessary, since they do not influence the reachability information. The problem can be easily solved by adding a fictitious self-loop $\widehat{Y} \Rightarrow \widehat{Y}$, which makes the initial class its own predecessor of the minimal depth satisfying the condition (*) of Def. 8.

Due to the BFS-like behaviour of the algorithm, on-the-fly reachability analysis is possible. The process of generating a model can be stopped as soon as a class $\widehat{X}$ satisfying a tested property is added to `reachable`, since in this case we have a finite path $\pi := ([q^0], \{q^0\}, 0) \Rightarrow \widehat{X}_1 \Rightarrow \ldots \Rightarrow \widehat{X}$, which proves reachability of a state $q \in X$.

The pseudo-code of the algorithm for generating ps-models, enabling on-the-fly reachability analysis, is presented in Fig. 7. The termination of the algorithm follows from the termination of the algorithm for s-models.

## 6   Implementation for Timed Automata

In order to implement the above algorithm for Timed Automata, we have to define an initial $d$-$cor$-partition $\Pi_0$ such that $\Pi_0|_1$ is a partition of the concrete state space $Q$ for a given automaton $\mathcal{A}$, and to implement the functions $pre_e$, $Pre$ and $Post$. Since abstract models for Timed Automata usually use regions as states, we need to define $\Pi_0$ and the above functions to satisfy this requirement. Therefore, as an initial $d$-$cor$-partition of $Q$ we can assume a set of classes whose first components are regions corresponding to locations of the automaton and invariants associated with them. $Cor$s of all the classes besides the initial one are equal to their first components, and their depths are set to $\infty$. $Cor$ of the initial

1.   $\Pi := \Pi_0$; $reachable := \{([q^0], \{q^0\}, 0)\}$; $stable := \emptyset$;
2.   while $(\exists \widehat{X} \in reachable \setminus stable)$ do
3.       begin
4.           $C_{\widehat{X}} := Split(\widehat{X}, \Pi)$;
5.           if $(C_{\widehat{X}} = \{\widehat{X}\})$ then
6.               begin
7.                   $stable := stable \cup \{\widehat{X}\}$;
8.                   for $\widehat{Y} \in Post_\Pi(\widehat{X})$ do $dpt(Y) := min\{dpt(X) + 1, dpt(Y)\}$;
9.                   $reachable := reachable \cup Post_\Pi(\widehat{X})$;
10.                  if $(\exists \widehat{Y} \in Post_\Pi(\widehat{X}))$ s.t. $Y \models p$ then return "YES";
11.              end;
12.          else
13.              begin
14.                  $Y_X := \{\widehat{Y} \in \Pi \mid Y$ has been split or $Y^{cor}$ changed $\}$;
15.                  for $\widehat{Z} \in \{\widehat{Y} \in C_{\widehat{X}} \mid q^0 \notin Y\}$ do $dpt(Z) := \infty$;
16.                  $stable := stable \setminus Pre_\Pi(Y_X) \setminus Y_X$;
17.                  $reachable := (reachable \setminus Y_X) \cup \{\widehat{Y} \in C_{\widehat{X}} \mid q^0 \in Y\}$;
18.                  $\Pi := (\Pi \setminus Y_X) \cup C_{\widehat{X}}$;
19.              end;
20.      end;
21.  return "NO";

**Fig. 7.** A minimization algorithm for an on-the-fly reachability analysis on ps-models

class is a singleton $\{q^0\}$, and its depth is equal to 0. Formally, we assume $\Pi_0 = \{((s^0, \mathbb{R}_+^n \cap \mathcal{I}(s^0)), \{q^0\}, 0)\} \cup \{((s, Z), (s, Z), \infty) \mid s \neq s^0 \wedge Z = \mathbb{R}_+^n \cap \mathcal{I}(s)\}$. In order to deal with the algorithm, for regions $R, R'$ we define $(R \setminus R', R \setminus R') = \{(R'', R'') \mid R'' \in R \setminus R'\}$. Then, for a given $d$-cor-partition $\Pi$, $(s, Z), (s', Z') \in \Pi|_1$ and a transition $e : s \xrightarrow{a,z,Y} s' \in E$ we introduce [11]:

- $pre_e((s, Z), (s', Z')) = (s, Z \cap (([Y := 0](Z' \swarrow \cap \mathcal{I}(s')) \cap z \cap \mathcal{I}(s)) \swarrow))$;
- for $e : s' \xrightarrow{a,z,Y} s$, $Pre_\Pi^e(((s, Z), (s, Z_1)))$
  $= \{((s', Z'), (s', Z_1')) \in \Pi \mid Z' \cap (([Y := 0](Z \swarrow \cap \mathcal{I}(s))$
  $\cap z \cap \mathcal{I}(s')) \swarrow) \neq \emptyset\}$; for $e : s \xrightarrow{a,z,Y} s'$, $Post_\Pi^e(((s, Z), (s, Z_1)))$
  $= \{((s', Z'), (s', Z_1')) \in \Pi \mid ((Z \nearrow \cap \mathcal{I}(s) \cap z)[Y := 0]) \nearrow \cap Z' \neq \emptyset\}$.

Notice that while computing $Z \nearrow$ or $Z \swarrow$ for the zone $Z$ of a region $(s, Z)$, we need to ensure that the invariant for the location $s$ is satisfied. Due to that, in the above operations zones are intersected with invariants for appropriate locations.

# 7   Experimental Results

We have implemented our algorithm (as a component of the tool VerIcs [9]) in the C++ programming language, and run it on the machine equipped with 772

| | $\mathcal{A}$ | | forw | | -ai-ax | | b-model | | ps-model | |
|---|---|---|---|---|---|---|---|---|---|---|
| | stats | edg. | stats | edges | stats | edges | stats | edges | stats | edges |
| RCS, K=700 | 24 | 37 | 14 | 15 | 12 | 13 | 18 | 24 | 11 | 16 |
| RCS, K=800 | 24 | 37 | 14 | 15 | 12 | 13 | 18 | 24 | 9 | 11 |
| RCS, K=900 | 24 | 37 | 14 | 15 | 12 | 13 | 18 | 24 | 9 | 11 |
| RCS, K=1000 | 24 | 37 | 14 | 15 | 12 | 13 | 18 | 24 | 9 | 11 |
| mutex 2 proc. | 28 | 36 | 196601 | 229386 | 34 | 44 | 39 | 57 | 25 | 37 |
| mutex 3 proc. | 152 | 240 | ⋆ | ⋆ | 431 | 622 | 225 | 469 | 145 | 472 |
| mutex 4 proc. | 752 | 1408 | ⋆ | ⋆ | 7336 | 11490 | 1391 | 4047 | 848 | 5369 |
| mutex 5 proc. | 3552 | 7680 | ⋆ | ⋆ | 156794 | 263650 | 9933 | 38761 | 5855 | 65283 |
| CSMA/CD 2 senders | 15 | 26 | 156 | 366 | 35 | 49 | 49 | 52 | 17 | 30 |
| CSMA/CD 3 senders | 46 | 131 | 9242 | 37647 | 289 | 579 | 373 | 920 | 100 | 273 |
| CSMA/CD 4 senders | 126 | 500 | ⋆ | ⋆ | 2727 | 9826 | 2390 | 12590 | 554 | 2246 |
| CSMA/CD 5 senders | 324 | 1777 | ⋆ | ⋆ | ⋆ | ⋆ | 21327 | 238705 | 4959 | 32400 |

**Fig. 8.** Experimental results for various examples

MB of main memory and the Linux operating system. We have considered three examples: the well-known Fischer's mutual exclusion protocol, the CSMA/CD protocol used by networking machines to control the use of a common bus, and the railroad crossing system (RCS) [1, 13], consisting of three automata representing its components, and an additional automaton of a specification for the property "whenever the gate is down, it is moved back up within $K$ seconds for some $K$". The property is violated when an "erroneous" state of the specification is reachable, which holds for $K < 700$. The sizes of our models are compared with the sizes of fr- and b-models, obtained using the tool Kronos [23]. We provide the results only for the cases in which the whole model must be generated. In the table in Fig. 8, the column $\mathcal{A}$ gives the size of the automaton, $forw$ shows the size of the fr-model generated without any abstractions, and -ai-ax - with the inclusion and extrapolation abstractions added [8]. The cases when memory was exhausted are denoted by ⋆.

The preliminary results show that the reduction in the size of the model can be substantial. Although the total number of edges of ps-models can be relatively large (our transition relation implies that for a given class all the classes which can be reached by passing some time, performing an action and then passing some time again are considered as its successors, whereas in b-models, for instance, only *immediate time-successors* are taken into account), the number of states of ps-models is smaller than of corresponding fr- and b- ones, which is most important for checking reachability. However, the requirement of convexity often results in generating too many classes while partitioning, since computing complementations (differences) of classes is exponential in the number of clocks. No solution which allows to avoid this operation has been found so far for s-models as well as for the discrete semantics. Our experiments show that in many cases the efficiency of the method is able to overcome the above problem, but definitely there are also examples in which this drawback is superior. Therefore, the present approach is expected to be even more efficient as soon as the problem of avoiding complementations is solved.

The comparison of our approach with other methods of reachability analysis for Timed Automata shows that our result can be considered as quite important for proving correctness (i.e., for showing that no state satisfying a tested property is reachable, which requires exploring the whole model). The paper [3], which compares several results of proving reachability for a mutual exclusion protocol by various tools (exploiting both symbolic and non-symbolic methods), shows the superiority of the BMC approach. Unfortunately, that paper contains no results for proving correctness, but the implementation of [21], which seems to be even more efficient than the one of [3], is able to deal in this case with 2 processes only, whereas it is possible to generate abstract models for more. Since our ps-models can be smaller than other abstract models shown in the literature, it seems they can be better for this kind of verification.

# 8 Conclusions and Further Work

The idea of using b-models for testing reachability comes from the paper [20], in which it is said that minimal b-models could often be smaller than the corresponding fr- ones. Since our ps-models are usually smaller than b- ones, we use them instead of the latter, combining this approach with a discrete semantics (a generalization of the semantics used for fr-models), which is sufficient for checking reachability, and can lead to better results. We provide the modification of the partitioning algorithm, enabling on-the-fly reachability analysis. Moreover, in many cases the simple "yes/no" answer for a reachability (or safety) question is not sufficient, but also the sequence of transitions leading to a state is of our interest. Due to a BFS-like behaviour of the algorithm, our method allows to answer the above question as well.

The preliminary experimental results show that our method can be very efficient. However, the main drawback consists in complementation of classes during partitioning. Therefore, our further research will concentrate around the above problem.

# References

[1] R. Alur, C. Courcoubetis, D. Dill, N. Halbwachs, and H. Wong-Toi. An implementation of three algorithms for timing verification based on automata emptiness. In Proc. of the 13th IEEE Real-Time Systems Symposium (RTSS '92), pages 157–166. IEEE Comp. Soc. Press, 1992. 3, 6, 14

[2] R. Alur, C. Courcoubetis, D. Dill, N. Halbwachs, and H. Wong-Toi. Minimization of timed transition systems. In Proc. of the Int. Conf. on Concurrency Theory (CONCUR '92), volume 630 of LNCS, pages 340–354. Springer-Verlag, 1992. 3

[3] G. Audemard, A. Cimatti, A. Kornilowicz, and R. Sebastiani. Bounded model checking for timed systems. Technical Report 0201-05, ITC-IRST, Sommarive 16, 38050 Povo, Trento, Italy, January 2002. 2, 3, 15

[4] D. Beyer. Improvements in BDD-based reachability analysis of Timed Automata. In Proc. of Int. Symp. of Formal Methods Europe (FME '01), volume 2021 of LNCS, pages 318–343. Springer-Verlag, 2002. 3

[5] A. Bouajjani, J-C. Fernandez, N. Halbwachs, P. Raymond, and C. Ratel. Minimal state graph generation. Science of Computer Programming, 18:247–269, 1992. 3, 8

[6] A. Bouajjani, S. Tripakis, and S. Yovine. On-the-fly symbolic model checking for real-time systems. In Proc. of the 18th IEEE Real-Time Systems Symposium (RTSS'97), pages 232–243. IEEE Comp. Soc. Press, 1997. 2, 5

[7] P. D'Argenio, B. Jeannet, H. Jensen, and K. Larsen. Reachability analysis of probabilistic systems by successive refinements. In Proc. of Int. Workshop on Process Algebra and Probabilistic Methods, Performance Modeling and Verification (PAPM -PROBMIV), volume 2165 of LNCS, pages 39–56. Springer-Verlag, 2001. 3

[8] C. Daws and S. Tripakis. Model checking of real time reachability properties using abstractions. In Proc. of the 4th Int. Conf. on Tools and Algorithms for the Construction and Analysis of Systems (TACAS'98), volume 1384 of LNCS, pages 313–329. Springer-Verlag, 1998. 2, 3, 5, 6, 14

[9] P. Dembiński, A. Janowska, P. Janowski, W. Penczek, A. Półrola, M. Szreter, B. Woźna, and A. Zbrzezny. VerICS: A tool for verifying Timed Automata and Estelle specifications. In Proc. of the 9th Int. Conf. on Tools and Algorithms for the Construction and Analysis of Systems (TACAS'03), volume 2619 of LNCS, pages 278–283. Springer-Verlag, 2003. 13

[10] P. Dembiński, W. Penczek, and A. Półrola. Automated verification of infinite state concurrent systems: an improvement in model generation. In Proc. of the 4th Int. Conf. on Parallel Processing and Applied Mathematics (PPAM '01), volume 2328 of LNCS, pages 247–255. Springer-Verlag, 2002. 3, 9

[11] P. Dembiński, W. Penczek, and A. Półrola. Verification of Timed Automata based on similarity. Fundamenta Informaticae, 51(1-2):59–89, 2002. 3, 13

[12] T. Henzinger and R. Majumdar. A classification of symbolic transition systems. In Proc. of the 17th Int. Symp. on Theoretical Aspects of Computer Science (STACS'00), volume 1770 of LNCS, pages 13–34. Springer-Verlag, 2000. 3

[13] I. Kang and I. Lee. An efficient state space generation for the analysis of real-time systems. In Proc. of Int. Symposium on Software Testing and Analysis, 1996. 3, 14

[14] K. G. Larsen, F. Larsson, P. Pettersson, and W. Yi. Efficient verification of real-time systems: Compact data structures and state-space reduction. In Proc. of the 18th IEEE Real-Time System Symposium (RTSS'97), pages 14–24. IEEE Comp. Soc. Press, 1997. 2, 3, 5, 6

[15] D. Lee and M. Yannakakis. On-line minimization of transition systems. In Proc. of the 24th ACM Symp. on the Theory of Computing, pages 264–274, May 1992. 3

[16] P. Niebert, S. Tripakis, and S. Yovine. Minimum-time reachability for Timed Automata. In Proc. of the 8th IEEE Mediterranean Conf. on Control and Automation (MED '2000), Patros, Greece, July 2000. IEEE Comp. Soc. Press. 3

[17] W. Penczek. Partial order reductions for checking branching properties of Time Petri Nets. In Proc. of the Int. Workshop on Concurrency, Specification and Programming (CS&P '00), volume 140 of Informatik-Berichte, pages 189–202. Humboldt University, 2000. 3, 6

[18] A. Półrola, W. Penczek, and M. Szreter. Towards efficient partition refinement for checking reachability in Timed Automata. An electronic version of this paper, available at http://www.math.uni.lodz.pl/~polrola. 3, 8

[19] A. Półrola, W. Penczek, and M. Szreter. Reachability analysis for Timed Automata based on partitioning. Technical Report 961, ICS PAS, Ordona 21, 01-237 Warsaw, June 2003. 3

[20] S. Tripakis and S. Yovine. Analysis of timed systems using time-abstracting bisimulations. Formal Methods in System Design, 18(1):25–68, 2001. 3, 4, 6, 15

[21] B. Woźna, W. Penczek, and A. Zbrzezny. Checking reachability properties for Timed Automata via SAT. Technical Report 949, ICS PAS, Ordona 21, 01 - 237 Warsaw, October 2002. 2, 3, 15

[22] M. Yannakakis and D. Lee. An efficient algorithm for minimizing real-time transition systems. In Proc. of the 5th Int. Conf. on Computer Aided Verification (CAV '93), volume 697 of LNCS, pages 210–224. Springer-Verlag, 1993. 3

[23] S. Yovine. KRONOS: A verification tool for real-time systems. Springer International Journal of Software Tools for Technology Transfer, 1(1/2):123–133, 1997. 14

# Checking ACTL* Properties
# of Discrete Timed Automata
# via Bounded Model Checking*

Bożena Woźna and Andrzej Zbrzezny

Institute of Mathematics and Computer Science
PU of Częstochowa
Armii Krajowej 13/15, 42-200 Częstochowa, Poland
{b.wozna,a.zbrzezny}@wsp.czest.pl

**Abstract.** The main contribution of the paper consists in showing that the BMC method is feasible for ACTL* (the universal fragment of CTL*) which subsumes both ACTL and LTL. The extension to ACTL* is obtained by redefining the function returning the sufficient number of executions over which an ACTL* formula is checked, and then by combining two known translations to SAT for ACTL and LTL formulas. The proposed translation of ACTL* formulas is essentially different from the existing translations of both ACTL and LTL formulas. Moreover, ACTL* seems to be the largest set of temporal properties which can be verified by means of BMC. We have implemented our new BMC algorithm for discrete timed automata and we have presented a preliminary experimental results, which prove the efficiency of the method. The formal treatment is the basis for the implementation of the technique in the symbolic model checker √erics.

## 1 Introduction

Model checking is considered as one of the most spectacular practical applications of the theoretical computer science in verification of concurrent systems. The main idea of model checking consists in representing a program as a labeled transition system (model), representing a specification as a temporal formula, and checking automatically whether the formula holds in the model [10]. Unfortunately, the practical applicability of model checking is strongly restricted by the state explosion problem, which is mainly caused by representing concurrency of operations by their interleaving. Therefore, many different reduction techniques have been introduced in order to alleviate the state explosion. The major methods include application of partial order reductions [4, 21, 22, 28], symmetry reductions [15], abstraction techniques [12, 11], BDD-based symbolic storage methods [19], and SAT-related algorithms [2, 6, 9, 14, 18, 20, 23, 24, 25, 26, 29].

---

* Partly supported by the State Committee for Scientific Research under the grant No. 8T11C 01419.

Bounded model checking (BMC) based on SAT (satisfiability checking) methods has been introduced as a complementary technique to BDD-based symbolic model checking for LTL [6, 7]. The main idea of bounded model checking for LTL is to look for an execution of the system of some length $k$, which is a counterexample for a tested property. If no counterexample of length $k$ can be found, then $k$ is increased by one. The efficiency of this method is based upon an observation that if a system is faulty, then often only a (small) fragment of its state space is sufficient for finding an error. The above observation has been experimentally proved [6, 7, 23, 25].

The main contribution of our paper is an extension of the method BMC to verification of the branching time properties expressible in ACTL* (the universal fragment of CTL*) [10], which subsumes both ACTL and LTL. ACTL* seems to be the largest set of temporal properties which can be verified by means of BMC. Moreover, we have implemented our new BMC algorithm for Discrete Timed Automata [8] and proved its efficiency by performing several experiments for the standard mutual exclusion protocol. The proposed BMC algorithm for ACTL* is going to be a new module of the tool √erics [13].

The main idea of our new BMC method for ACTL* consists in combining a translation of a model $M$ to several symbolic paths, which can start at arbitrary states of the model, with a translation of the negation of an ACTL* formula $\varphi$. The latter translation is obtained by redefining the function $f_k$ of [23] returning the sufficient number of executions over which $\varphi$ is checked, and then by combining two known translations for ACTL [23] and LTL [6]. This is obtained by applying the LTL translation for all the LTL subformulas of $\varphi$, and the ACTL translation for all the state subformulas of $\varphi$, i.e., the formulas which begin with a path quantifier.

The rest of the paper is organized as follows. The next section contains the discussion of the related work. Then, in section 3 the bounded model checking for ACTL* is presented. The implementation of BMC for Discrete Timed Automata is described in section 4. Experimental results are presented in section 5. The last section contains final remarks.

## 2   Related Work

Our paper shows for the first time an extension of the BMC method based on SAT procedures to verification of all the properties expressible in ACTL*. It builds upon the results of [23], where an approach to applying BMC for ACTL was described. The idea of BMC for a temporal logic is taken from [6, 7]. The BMC method has been also applied for LTL model checking of 1-safe Petri Nets [18] and Timed Automata [2], for TACTL model checking of Timed Automata [25, 24], for checking reachability of Petri Nets [17] and Timed Automata [20, 26, 29], for past LTL model checking of digital circuits [3]. A motivation for considering the universal fragment of CTL* can be found in [16, 22]. The discrete timed automata were considered by several authors [8, 5] because the model checking of such automata is a very challenging and important task.

# 3   Bounded Model Checking

In this section we describe our techniques for ACTL* bounded model checking. First, we give some background and notational conventions that are used in the rest of the paper. Then, we describe the bounded semantics for ECTL*. Finally, we show the reduction of the BMC method to propositional satisfiability for ACTL* formulas.

Since the paper is an extended abstract, the intuitive explanations of the introduced definitions and proofs are omitted, but they can be found in [30].

## 3.1   Background

The specification of a system is expressed in ACTL* (the universal fragment of CTL* ) [10]. ACTL* is defined as the subset of CTL* formulas [10] that are in negation normal form (NNF)[1] and contain the universal path quantifier (A) only. ECTL* is defined in the same way, but only the existential path quantifier (E) are allowed. We consider the following operators: *the next state* (X), the *eventually* (F), the *always* (G), the *until* (U), and the *release* (R, dual to U).

The implementation of a system is described as a *Kripke structure* $M = (S, \rightarrow, s^0, \mathcal{V})$, where $S$ is a finite set of states, $\rightarrow \subseteq S \times S$ is a total binary (successor) relation on $S$ (i.e., each state has at least one $\rightarrow$-successor), $s^0$ is an initial state and $\mathcal{V} : S \longrightarrow 2^{\mathcal{PV}}$ is a valuation function such that $true \in \mathcal{V}(s)$ for all $s \in S$, where $\mathcal{PV}$ is a set of propositional variables containing the symbol $true$.

We use Kripke structures as models in order to give the semantics of the logic CTL*. For the rest of the paper we consider only Kripke structures for which we have a *Boolean encoding*. We require that $S \subseteq \{0, 1\}^n$, for $n = \lceil log_2(|S|) \rceil$, and that each state can be represented by a vector of *state variables* $\mathbf{w} = (\mathbf{w}[1], \ldots, \mathbf{w}[n])$, where $\mathbf{w}[i]$ are propositional variables for $i = 1, \ldots, n$. Moreover, we define the following propositional formulas: $I_s(\mathbf{w}) := \bigwedge_{i=1}^{n} lit(s[i], \mathbf{w}[i])^2$ for $s = (s[1], \ldots, s[n]) \in S$, $T(\mathbf{w}, \mathbf{w}')$ such that for every interpretation of states variables $Val \in \{0, 1\}^{\mathcal{SV}}$: $Val$ satisfies $T(\mathbf{w}, \mathbf{w}')$ iff $(\mathbf{Val}(\mathbf{w}), \mathbf{Val}(\mathbf{w}')) \in \rightarrow ^3$, $p(\mathbf{w})$ such that for every interpretation of states variables $Val \in \{0, 1\}^{\mathcal{SV}}$: $Val$ satisfies $p(\mathbf{w})$ iff $p \in \mathcal{V}(\mathbf{Val}(\mathbf{w}))$ for $p \in \mathcal{PV}$, $H(\mathbf{w}, \mathbf{w}')$ $:= \bigwedge_{i=1}^{n}(\mathbf{w}[i] \leftrightarrow \mathbf{w}'[i])$, $L_{k,j}(l) := T(\mathbf{w}_{k,j}, \mathbf{w}_{l,j})$. For an infinite sequence of states $\pi = (s_0, s_1, \ldots)$ we define $\pi(i) = s_i$ and $\pi^i = (s_i, s_{i+1}, \ldots)$ for $i \in \mathbb{N}$. An infinite sequence of states $\pi$ is a *path* if $(s_i, s_{i+1}) \in \rightarrow$ for all $i \in \mathbb{N}$.

---

[1] A CTL* formula is in negation normal form if the negations can occur in front of the propositional variables only.

[2] Let $\mathcal{SV}$ be a set of the state variables containing the symbols $true$ and $false$, and let $\mathcal{SF}$ be a set of propositional formulas built over $\mathcal{SV}$. $lit : \{0, 1\} \times \mathcal{SV} \to \mathcal{SF}$ is a function defined as follows: $lit(0, p) = \neg p$ and $lit(1, p) = p$.

[3] $Val : \mathcal{SV} \to \{0, 1\}$ is an interpretation for state variables and $\mathbf{Val} : \mathcal{SV}^n \to \{0, 1\}^n$ is its extension for vectors of state variables such that $\mathbf{Val}(\mathbf{w}[1], \ldots, \mathbf{w}[n]) = (Val(\mathbf{w}[1]), \ldots, Val(\mathbf{w}[n]))$.

**Definition 1.** *Let $M$ be a model, $s \in S$ be a state, $\pi$ be a path, $\alpha_1$, $\alpha_2$ be state formulas and $\beta_1$, $\beta_2$ be path formulas. $M, s \models \alpha_1$ denotes that $\alpha_1$ is true at the state $s$ in the model $M$. $M, \pi \models \beta_1$ denotes that $\beta_1$ is true along path $\pi$ in the model $M$. $M$ is omitted, if it is implicitly understood. The relation $\models$ is defined inductively as follows:*

$$s \models p \quad \text{iff } p \in \mathcal{V}(s), \qquad\qquad \pi \models \beta_1 \vee \beta_2 \text{ iff } \pi \models \beta_1 \text{ or } \pi \models \beta_2,$$

$$s \models \neg\alpha_1 \quad \text{iff } s \not\models \alpha_1, \qquad\qquad \pi \models \beta_1 \wedge \beta_2 \text{ iff } \pi \models \beta_1 \text{ and } \pi \models \beta_2,$$

$$s \models \mathrm{E}\beta_1 \quad \text{iff } \exists \pi \, (\pi(0) = s \text{ and } \pi \models \beta_1), \quad \pi \models \mathrm{X}\beta_1 \quad \text{iff } \pi^1 \models \beta_1,$$

$$s \models \alpha_1 \vee \alpha_2 \text{ iff } s \models \alpha_1 \text{ or } s \models \alpha_2, \qquad\qquad \pi \models \mathrm{F}\beta_1 \quad \text{iff } (\exists m \geq 0)\pi^m \models \beta_1,$$

$$s \models \alpha_1 \wedge \alpha_2 \text{ iff } s \models \alpha_1 \text{ and } s \models \alpha_2, \qquad\qquad \pi \models \mathrm{G}\beta_1 \quad \text{iff } (\forall m \geq 0) \, \pi^m \models \beta_1,$$

$$\pi \models \alpha_1 \quad \text{iff } \pi(0) \models \alpha_1,$$

$$\pi \models \beta_1 \mathrm{U}\beta_2 \text{ iff } (\exists m \geq 0) \left(\pi^m \models \beta_2 \text{ and } (\forall j < m) \, \pi^j \models \beta_1\right),$$

$$\pi \models \beta_1 \mathrm{R}\beta_2 \text{ iff } (\forall m \geq 0) \left(\pi^m \models \beta_2 \text{ or } (\exists j < m) \, \pi^j \models \beta_1\right),$$

**Definition 2 (Validity).** *A CTL* formula $\varphi$ is valid in $M = ((S, \rightarrow, s^0), \mathcal{V})$ (denoted $M \models \varphi$) iff $M, s^0 \models \varphi$, i.e., $\varphi$ is true at the initial state of the model $M$.*

Let $M$ be a given model and $\psi$ be a given ACTL* formula. Our aim is to show that $\psi$ does not hold in $M$ (i.e., $M \not\models \psi$), which means to show that $M \models \neg\psi$ (notice that $\neg\psi$ is an ECTL* formula). To solve this problem we use the bounded model checking method. In order to deal with the bounded model checking method for ACTL*, we have to define the bounded semantics for ECTL*, which allows us to interpret the formulas over a fragment of the considered model only.

## 3.2   Bounded Semantics of ECTL*

In order to define the bounded semantics for ECTL* we have to introduce the notations of $k$−*paths*, *loops*, and $k$−*models*.

Let $M = (S, \rightarrow, s^0, \mathcal{V})$ be a model and $k \in \mathbb{N}_+$[4]. A $k$−*path* $\pi_k = (s_0, \ldots, s_k)$ is a finite sequence of states such that $(s_i, s_{i+1}) \in \rightarrow$ for each $s_i \in S$ and $0 \leq i < k$. For a $k$−path $\pi_k = (s_0, \ldots, s_k)$ let $\pi_{i,k} = (s_i, \ldots, s_k)$ for each $i \in \{0, \ldots, k\}$. Though a $k$−path is finite, it still can represent an infinite path if there is a loop from the last state of the $k$−path to any of the previous states. A $k$−path $\pi_k$ is a *loop* if $(\pi_k(k), \pi_k(l)) \in \rightarrow$ for some $0 \leq l \leq k$. The $k$−*model* for $M$ is a tuple $M_k = (S, Path_k, s^0, \mathcal{V})$, where $Path_k$ is a set of all the $k$−paths of $M$. Note that, the set of all the $k$−paths determines the transition relation of $M$ in an unambiguous manner. Moreover, the set $Path_k$ is finite.

The bounded semantics is defined over a $k$−model $M_k$. The definition of the bounded semantics of the temporal operators depends on whether a considered $k$−path of $M_k$ is a loop or not. In order to distinguish which of the $k$−paths are loops we define the following auxiliary function.

---

[4] $\mathbb{N}_+ = \{1, 2, \ldots\}$ is the set of positive natural numbers.

**Definition 3.** *Let* $M = (S, \rightarrow, s^0, \mathcal{V})$ *be a model,* $M_k = (S, Path_k, s^0, \mathcal{V})$ *be a* $k-model$ *for* $M$ *and* $\pi_k \in Path_k$ *be a* $k-path$*. A function loop* $: Path_k \rightarrow 2^{\{0,\ldots,k\}}$ *is defined as:* $loop(\pi_k) = \{l \mid l \leq k \text{ and } (\pi_k(k), \pi_k(l)) \in \rightarrow\}$.

**Definition 4 (Bounded Semantics).** *Let* $M_k$ *be a* $k-model$, $\alpha_1$, $\alpha_2$ *be state formulas and* $\beta_1$, $\beta_2$ *be path formulas.* $M_k, s \models \alpha$ *denotes that* $\alpha$ *is true at the state* $s$ *of* $M_k$. $M_k, \pi_{0,k} \models \beta$ *denotes that* $\beta$ *is true along* $k-path$ $\pi_k$ *of* $M_k$. $M_k$ *is omitted if it is implicitly understood. The relation* $\models$ *is defined inductively as follows:*

$s \models p$ iff $p \in \mathcal{V}(s)$,   $s \models \alpha_1 \vee \alpha_2$ iff $s \models \alpha_1$ or $s \models \alpha_2$,

$s \models \neg p$ iff $p \notin \mathcal{V}(s)$,   $s \models \alpha_1 \wedge \alpha_2$ iff $s \models \alpha_1$ and $s \models \alpha_2$,

$s \models E\beta_1$   iff $\exists \pi_k \in Path_k$ $(\pi_k(0) = s$ and $\pi_{0,k} \models \beta_1)$,

$\pi_{m,k} \models \alpha_1$ iff $\pi_k(m) \models \alpha_1$,

$\pi_{m,k} \models \beta_1 \vee \beta_2$ iff $\pi_{m,k} \models \beta_1$ or $\pi_{m,k} \models \beta_2$,

$\pi_{m,k} \models \beta_1 \wedge \beta_2$ iff $\pi_{m,k} \models \beta_1$ and $\pi_{m,k} \models \beta_2$,

$\pi_{m,k} \models X\beta_1$ iff $(\pi_{m+1,k} \models \beta_1$ and $m < k)$ or $(\exists_{l \in loop(\pi_k)} \pi_{l,k} \models \beta_1$ and $m = k)$,

$\pi_{m,k} \models F\beta_1$ iff $(\exists_{m \leq i \leq k} \pi_{i,k} \models \beta_1)$ or $(\exists_{l \in loop(\pi_k)} \exists_{l \leq i < m} \pi_{i,k} \models \beta_1)$,

$\pi_{m,k} \models G\beta_1$ iff $\exists_{l \in loop(\pi_k)} \forall_{min(m,l) \leq i \leq k} \pi_{i,k} \models \beta_1$,

$\pi_{m,k} \models \beta_1 U\beta_2$ iff $\exists_{m \leq i \leq k} \big(\pi_{i,k} \models \beta_2$ and $\forall_{m \leq j < i} \pi_{j,k} \models \beta_1\big)$ or $\exists_{l \in loop(\pi_k)}$

$\qquad \big(\exists_{l \leq i < m}(\pi_{i,k} \models \beta_2$ and $\forall_{m \leq j \leq k} \pi_{j,k} \models \beta_1$ and $\forall_{l \leq j < i} \pi_{j,k} \models \beta_1)\big)$,

$\pi_{m,k} \models \beta_1 R\beta_2$ iff $\big(\exists_{l \in loop(\pi_k)} \forall_{min(m,l) \leq i \leq k} \pi_{i,k} \models \beta_2\big)$ or

$\qquad \big(\exists_{m \leq i \leq k} (\pi_{i,k} \models \beta_1$ and $\forall_{m \leq j \leq i} \pi_{j,k} \models \beta_2)\big)$ or $\exists_{l \in loop(\pi_k)}$

$\qquad \big(\exists_{l \leq i < m}(\pi_{i,k} \models \beta_1$ and $\forall_{m \leq j \leq k} \pi_{j,k} \models \beta_2$ and $\forall_{l \leq j \leq i} \pi_{j,k} \models \beta_2)\big)$.

**Definition 5 (Validity for Bounded Semantics).** *An* ECTL* *formula* $\varphi$ *is valid in a* $k-model$ $M_k$ *(denoted* $M \models_k \varphi$*) iff* $M_k, s^0 \models \varphi$.

The main theorem of this section states that the bounded semantics is equivalent to the unbounded one, which means that the model checking problem $(M \models \varphi)$ can be reduced to the bounded model checking problem $(M \models_k \varphi)$.

**Theorem 1 ([30]).** *Let* $M = (S, \rightarrow, s^0, \mathcal{V})$ *be a model,* $\varphi$ *be an* ECTL* *formula. Then for some* $k \leq (|M| \cdot 2^{|\varphi|})^2$, $M \models \varphi$ *iff* $M \models_k \varphi$.

The above theorem guarantees the completeness of our BMC method. Despite of the fact that the bound from the theorem is very large and therefore quite impractical, it is often the case that some essentially smaller bounds are sufficient. This is, in fact, the reason that the BMC method is sometimes very efficient.

## 3.3  Translation

In this subsection we show the reduction of bounded model checking for ACTL* to propositional satisfiability. We use the same general algorithm BMC as for ACTL and TACTL [23, 25]. We start with introducing a definition of a function $f_k$ determining the number of $k$-paths, which is sufficient for checking an ECTL* formula in a $k-$model $M_k$.

**Definition 6.** *Define a function* $f_k : \mathcal{FORM}^* \to \mathbb{N}$ *as follows:*

$$f_k(p) = f_k(\neg p) = 0, \text{ where } p \in \mathcal{PV}, \quad f_k(\mathrm{F}\alpha) = f_k(\alpha),$$

$$f_k(\alpha \vee \beta) = max\{f_k(\alpha), f_k(\beta)\}, \qquad f_k(\mathrm{G}\alpha) = (k+1) \cdot f_k(\alpha),$$

$$f_k(\alpha \wedge \beta) = f_k(\alpha) + f_k(\beta), \qquad f_k(\alpha \mathrm{U}\beta) = k \cdot f_k(\alpha) + f_k(\beta),$$

$$f_k(\mathrm{E}\alpha) = f_k(\alpha) + 1, \qquad\qquad f_k(\mathrm{X}\alpha) = f_k(\alpha),$$

$$f_k(\alpha \mathrm{R}\beta) = (k+1) \cdot f_k(\beta) + f_k(\alpha).$$

In order to handle an arbitrary ECTL* formula $\varphi$ the function $f_k$ is successively applied to the subformulas of $\varphi$. It is easy to see that the value of $f_k$ depends on the number of the existential quantifiers, and the temporal operators U and G appearing in $\varphi$.

Although there exist formulas for which the value of $f_k$ is exponential w.r.t. their size, in practice one uses the formulas for which the number of sufficient $k$-paths is reasonably small. Moreover, it is very important to mention that the actual depth reachable in the model is obtained by a combination of the considered $k-$paths, and it is at most equal to $k * f_k(\varphi)$, but, more importantly, it is never smaller than $k * n$, where $n$ is the number of the existential quantifiers followed by G, U or R occuring in $\varphi$. Notice also that the function $f_k$ returns 1 for all the LTL formulas, which means that all LTL formulas are checked over one symbolic path[5].

Given a model $M$, an ECTL* formula $\varphi$ and a bound $k$, we shall construct a propositional formula $[M, \varphi]_k$ which is satisfiable iff $M \models_k \varphi$. We define a *symbolic $k-$path* as a finite sequence $(\mathbf{w}_0, \ldots, \mathbf{w}_k)$ of vectors of state variables. To construct $[M, \varphi]_k$, we first define a propositional formula $[M^{\varphi,s^0}]_k$ that constrains $f_k(\varphi)$ symbolic $k$-paths to be valid $k$-paths of $M_k$. Then, we translate the ECTL* formula $\varphi$ to a propositional formula that constrains the sets of $f_k(\varphi)$ symbolic $k$-paths to satisfy $\varphi$.

**Definition 7. (Unfolding of the Transition Relation)**
*Let* $M = (S, \to, s^0, \mathcal{V})$ *be a model,* $s \in S$ *be a state,* $k$ *be a bound and* $\varphi$ *be an* ECTL* *formula. The propositional formula* $[M^{\varphi,s}]_k$ *is defined as follows:*

$$[M^{\varphi,s}]_k := I_s(\mathbf{w}_{0,0}) \wedge \bigwedge_{j=1}^{f_k(\varphi)} \bigwedge_{i=0}^{k-1} T(\mathbf{w}_{i,j}, \mathbf{w}_{i+1,j})$$

*where* $\mathbf{w}_{0,0}$ *and* $\mathbf{w}_{i,j}$ *are vectors of state variables for* $i = 0, \ldots, k$.

---

[5] LTL formulas are formulas of the form A$\alpha$, where $\alpha$ does not contain any path quantifiers.

The translation of an ECTL* formula $\varphi$ into the propositional formula $[\varphi]_{M_k}$ differs w.r.t. $k$-paths that are and that are not loops. In order to distinguish which of the $k$-paths are loops we use the propositional formulas $L_{k,j}(l)$. Furthermore, at each state $\mathbf{w}_{m,n}$ within a $k$-path of index $n$, all the state subformulas of a formula being translated to the $k$-path $n$ of the form $E\beta$ (where $\beta$ is a path formula) are translated to the $k$-paths that start at that state, i.e., beginning with $\mathbf{w}_{0,i} = \mathbf{w}_{m,n}$ for all $i \in \{1, \ldots, f_k(\varphi)\}$.

We use $[\varphi]_k^{[m,n]}$, where $0 \le m \le k$ and $0 \le n \le f_k(\varphi)$, to denote the translation of an ECTL* formula $\varphi$ at $\mathbf{w}_{m,n}$ to a propositional formula.

**Translation of an ECTL* formula.** Let $\alpha$, $\beta$ be path formulas.

$$[p]_k^{[m,n]} := p(\mathbf{w}_{m,n}), \qquad [\alpha \wedge \beta]_k^{[m,n]} := [\alpha]_k^{[m,n]} \wedge [\beta]_k^{[m,n]},$$

$$[\neg p]_k^{[m,n]} := \neg p(\mathbf{w}_{m,n}), \qquad [\alpha \vee \beta]_k^{[m,n]} := [\alpha]_k^{[m,n]} \vee [\beta]_k^{[m,n]},$$

$$[E(\alpha)]_k^{[m,n]} := \bigvee_{i \in F_k^n(\varphi)} \left( H(\mathbf{w}_{m,n}, \mathbf{w}_{0,i}) \wedge [\alpha]_k^{[0,i]} \right) \vee [\alpha]_k^{[m,n]}, where$$

$$F_k^n(\varphi) = \{1, \ldots, f_k(\varphi)\} \setminus \{n\}$$

$$[X\alpha]_k^{[m,n]} := \begin{cases} [\alpha]_k^{[m+1,n]}, & if \ m < k \\ \bigvee_{l=0}^k (L_{k,n}(l) \wedge [\alpha]_k^{[l,n]}), & otherwise \end{cases}$$

$$[F\alpha]_k^{[m,n]} := \bigvee_{i=m}^k [\alpha]_k^{[i,n]} \vee \bigvee_{l=0}^k (L_{k,n}(l) \wedge \bigvee_{i=l}^m [\alpha]_k^{[i,n]}),$$

$$[G\alpha]_k^{[m,n]} := \bigvee_{l=0}^k (L_{k,n}(l) \wedge \bigwedge_{i=min(l,m)}^k [\alpha]_k^{[i,n]}),$$

$$[\alpha U \beta]_k^{[m,n]} := \bigvee_{i=m}^k \left( [\beta]_k^{[i,n]} \wedge \bigwedge_{j=m}^{i-1} [\alpha]_k^{[j,n]} \right) \vee \bigvee_{l=0}^k \left( L_{k,n}(l) \wedge \right.$$
$$\left. \bigvee_{i=l}^m ([\beta]_k^{[i,n]} \wedge \bigwedge_{j=m}^k [\alpha]_k^{[j,n]} \wedge \bigwedge_{j=l}^{i-1} [\alpha]_k^{[j,n]}) \right),$$

$$[\alpha R \beta]_k^{[m,n]} := \bigvee_{l=0}^k (L_{k,n}(l) \wedge \bigwedge_{i=min(l,m)}^k [\beta]_k^{[i,n]}) \vee$$
$$\bigvee_{i=m}^k \left( [\alpha]_k^{[i,n]} \wedge \bigwedge_{j=m}^i [\beta]_k^{[j,n]} \right) \vee \bigvee_{l=0}^k \left( L_{k,n}(l) \wedge \right.$$
$$\left. \bigvee_{i=l}^m ([\alpha]_k^{[i,n]} \wedge \bigwedge_{j=m}^k [\beta]_k^{[j,n]} \wedge \bigwedge_{j=l}^i [\beta]_k^{[j,n]}) \right).$$

We define $[\varphi]_{M_k}$ as $[\varphi]_k^{[0,0]}$, and $[M, \varphi]_k$ as $[\varphi]_{M_k} \wedge [M^{\varphi, s^0}]_k$.

Correctness of our translation is guaranteed by the following theorem.

**Theorem 2** ([30]). *Let $M$ be a model, $k$ be a bound, and $\varphi$ be an ECTL* formula. Then, $M \models_k \varphi$ iff $[M, \varphi]_k$ is satisfiable.*

# 4    Implementation of BMC for Discrete Timed Automata

In this section we show how BMC can be applied to verification of *Discrete Timed Automata*, that are used for representing concurrent systems.

## 4.1   Discrete Timed Automata

This subsection gives a definition of *discrete timed automata* and introduce two bisimilar models for such automata, a *concrete model* and an *abstract model*.

We start with introducing some auxiliary notation. Let $\mathbb{X}$ be a finite set of variables, called clocks. A *clock valuation* is a function $v : \mathbb{X} \to \mathbb{N}$, assigning to each clock $x$ a natural value $v(x)$. The clock valuation which assigns the value $0$ to all the clocks is denoted by $v^0$. The set of all the valuations is denoted by $\mathbb{N}^n$, where $n$ is the number of clocks. For a subset $Y$ of $\mathbb{X}$ by $v[Y := 0]$ we mean the valuation $v'$ such that $\forall x \in Y$, $v'(x) = 0$ and $\forall x \in \mathbb{X} \setminus Y$, $v'(x) = v(x)$. For $\delta \in \mathbb{N}$, $v + \delta$ denotes the valuation $v''$ such that $\forall x \in \mathbb{X}, v''(x) = v(x) + \delta$.

The set $\Psi_{\mathbb{X}}$ of *clock constraints* over the set of clocks $\mathbb{X}$, for $x \in \mathbb{X}$, $c \in \mathbb{N}$, and $\sim \in \{\leq, <, =, >, \geq\}$, is defined inductively as follows: $\psi := x \sim c \mid \psi \wedge \psi$.

A clock valuation $v$ *satisfies* the clock constraint $\psi \in \Psi_{\mathbb{X}}$, if

$$v \models x \sim c \text{ iff } v(x) \sim c, \qquad v \models \psi \wedge \psi' \text{ iff } v \models \psi \text{ and } v \models \psi'$$

For each $\psi \in \Psi_{\mathbb{X}}$ by $\mathbf{p}(\psi)$ we denote the set of all the clock valuations satisfying $\psi$, i.e., $\mathbf{p}(\psi) = \{v \in \mathbb{N}^n \mid v \models \psi\}$. Now, we are ready to define a discrete timed automaton.

**Definition 8.** *A* discrete timed automaton $\mathcal{A}$ *is a 6-tuple* $(\Sigma, L, l^0, E, \mathbb{X}, \mathbb{I})$, *where* $\Sigma$ *is a finite set of actions,* $L$ *is a finite set of locations,* $l^0 \in L$ *is an initial location,* $E \subseteq L \times \Sigma \times \Psi_{\mathbb{X}} \times 2^{\mathbb{X}} \times L$ *is a transition relation,* $\mathbb{X}$ *is a finite set of clocks, and* $\mathbb{I} : L \longrightarrow \Psi_{\mathbb{X}}$ *is a state invariant function.*

*Each element* $e$ *of* $E$ *is denoted by* $e := l \xrightarrow{\sigma, \psi, Y} l'$. *This represents a transition from the location* $l$ *to the location* $l'$ *on the input action* $\sigma$. $Y \subseteq \mathbb{X}$ *is the set of all the clocks to be reset with this transition, whereas* $\psi \in \Psi_{\mathbb{X}}$ *is the enabling condition for* $e$.

Given a transition $e := l \xrightarrow{\sigma, \psi, Y} l'$, we write $source(e)$, $target(e)$, $action(e)$, $guard(e)$ and $reset(e)$ for $s$, $s'$, $\sigma$, $\psi$ and $Y$, respectively.

The semantics of the discrete timed automaton is defined by associating a transition system with it.

**Definition 9 (Concrete Model).** *A* concrete model *for the discrete timed automaton* $\mathcal{A} = (\Sigma, L, l^0, E, \mathbb{X}, \mathbb{I})$ *is a pair* $M_{\mathcal{A}} = ((Q, \to_c, q^0), \mathcal{V})$, *where* $Q = L \times \mathbb{N}^n$ *is a set of the concrete states of* $\mathcal{A}$, $q^0 = (l^0, v^0)$ *is the initial state of* $\mathcal{A}$, $\to_c \subseteq Q \times Q$ *is a total binary (successor) relation on* $Q$ *defined by action- and time-successors as follows. Let* $\sigma \in \Sigma$ *and* $\delta \in \mathbb{N}$.

1. $(l, v) \xrightarrow{\sigma}_c (l', v')$ *iff there is a transition* $l \xrightarrow{\sigma, \psi, Y} l' \in E$ *such that* $v \models \psi$ *and* $v' = v[Y := 0]$ *and* $v' \models \mathbb{I}(l')$,
2. $(l, v) \xrightarrow{\delta}_c (l', v')$ *iff* $l = l'$ *and* $v' = v + \delta$ *and* $v' \models \mathbb{I}(l')$.

$\mathcal{V} : Q \longrightarrow 2^{\mathcal{PV}}$ *is a valuation function such that* $true \in \mathcal{V}(q)$ *for all* $q \in Q$.

Notice that the set of all the concrete states of $\mathcal{A}$ is infinite. Since our aim is to translate the *model checking problem* for discrete timed automata and properties expressed in ACTL* logic to the SAT-problem, we have to define a finite abstraction of the concrete model which preserves ACTL*, namely an abstract model based on the region graph [1].

Before we give a definition of an abstract model, we introduce some auxiliary notations. Let $\Psi \subseteq \Psi_{\mathbb{X}}$ be a non-empty set of clock constraints over $\mathbb{X}$, $c_{max}$ be the largest constant appearing in a constraint of any enabling condition used in the transition relation $E$ or in a state invariant of $\mathcal{A}$. Moreover, let $n$ be the number of the clocks of $\mathcal{A}$.

**Definition 10 (Equivalence of Clock Valuations).** *For two clock valuations $v$ and $v'$ in $\mathbb{N}^n$, $v \simeq v'$ iff for all $x \in \mathbb{X}$ the following condition is met: $(v(x) > c_{max}$ and $v'(x) > c_{max})$ or $(v(x) \leq c_{max}$ and $v'(x) \leq c_{max}$ and $v(x) = v'(x))$.*

It is easy to see that the relation $\simeq$ is an equivalence relation on the set of all the clock valuations for a given discrete timed automaton.

The equivalence classes of the relation $\simeq$ are called *zones* and are denoted by $Z$ and $Z'$. The set of all the zones is denoted by $Z(n)$. The zone $Z^0 = \{v \mid (\forall x \in \mathbb{X})\ v(x) = 0\}$ is called initial. A zone $Z$ is *final* iff $v(x) > c_{max}$ for all $v \in Z$ and $x \in \mathbb{X}$. A zone $Z$ is *open* if there is a clock $x \in \mathbb{X}$ such that $v(x) > c_{max}$ for all $v \in Z$. A zone $Z$ *satisfies* a clock constraint $\psi \in \Psi_{\mathbb{X}}$ (written $Z \models \psi$) iff $\forall v \in Z$, $v \models \psi$. Define the following operation on zones: $Z[Y := 0] = \{v[Y := 0] \mid v \in Z\}$.

Note that all the zones which are not open consist of one point only.

**Definition 11 (Time Successor).** *Let $Z$ and $Z'$ be two distinct zones. If $Z$ is not final, then the zone $Z'$ is the* time successor *of $Z$ iff for each $v \in Z$ there exists $\delta \in \mathbb{N}$ such that $v + \delta \in Z'$ and $v + \delta' \in Z \cup Z'$ for all $\delta' \leq \delta$. If $Z$ is the final zone, then the* time successor *of $Z$ is the same zone $Z$. The time successor of $Z$ is denoted by $\tau(Z)$.*

**Definition 12 (Action Successor).** *The zone $Z'$ is said to be the* action successor *of $Z$ by a transition $e : l \xrightarrow{\sigma, \psi, Y} l' \in E$ iff $Z \models \psi$ and $Z' = Z[Y := 0]$. The action successor of $Z$ is denoted by $e(Z)$.*

A *region* is a pair $(l, Z)$, where $l \in L$ and $Z \in Z(n)$. Note that the set of all the regions is finite.

Now, we are ready to define an abstract model.

**Definition 13 (Abstract Model).** *An* abstract model *for the discrete timed automaton $\mathcal{A} = (\Sigma, L, l^0, E, \mathbb{X}, \mathbb{I})$ is a pair $M = (S, \rightarrow, s^0, \mathcal{V})$, where $S = L \times Z(n)$, $s^0 = (l^0, Z^0)$ and $\rightarrow \subseteq S \times (E \cup \{\tau\}) \times S$ is defined as follows:*

1. *$(l, Z) \xrightarrow{e} (l', Z')$ iff $Z' = e(Z)$, $l = source(e)$, $l' = target(e)$, and $Z' \models \mathbb{I}(l')$, for $e \in E$,*
2. *$(l, Z) \xrightarrow{\tau} (l, Z')$ iff $Z' \models \mathbb{I}(l)$ and $Z' = \tau(Z)$.*

*$\mathcal{V} : S \longrightarrow 2^{\mathcal{PV}}$ is a valuation function such that $true \in \mathcal{V}(s)$ for all $s \in S$.*

## 4.2 Distinguishing Power of ACTL*

Let $M = (S, \to, s^0, \mathcal{V})$ and $M' = (S', \to', s'^0, \mathcal{V}')$ be two models.

**Definition 14 (Simulation [16]).** *A relation $\leadsto_s \subseteq S \times S'$ is a* simulation *from $M$ to $M'$ if the following conditions hold:*

1. $s^0 \leadsto_s s'^0$,
2. *if $s \leadsto_s s'$, then $\mathcal{V}'(s') = \mathcal{V}(s)$ and for every $s'_1$ such that $(s', s'_1) \in \to'$, there is $s_1$ such that $(s, s_1) \in \to$ and $s_1 \leadsto_s s'_1$.*

*Model $M$ simulates model $M'$ ($M \leadsto_s M'$) if there is a simulation from $M$ to $M'$. Two models $M$ and $M'$ are called* bisimilar *if $M \leadsto_s M'$ and $M' \leadsto_s^{-1} M$.*

**Theorem 3 ([16]).** *If $M$ simulates $M'$, then*

- $M, s^0 \models \varphi$ *implies* $M', s'^0 \models \varphi$, *for any ACTL\* formula $\varphi$ over $\mathcal{PV}$.*
- $M', s'^0 \models \varphi$ *implies* $M, s^0 \models \varphi$, *for any ECTL\* formula $\varphi$ over $\mathcal{PV}$.*

**Lemma 1.** *Let $\mathcal{A}$ be a discrete timed automaton, and let $M$ be the concrete model for $\mathcal{A}$ and $M'$ be the abstract model for $\mathcal{A}$. Then, the models $M$ and $M'$ are bisimilar.*

*Proof.* Define the relation $\leadsto_s \subseteq Q \times S$ as follows $(l, v) \leadsto_s (l', Z)$ iff $l = l'$ and $v \in Z$. It is easy to check that the relation $\leadsto_s$ is a simulation from $M$ to $M'$ and the relation $\leadsto_s^{-1}$ is a simulation from $M'$ to $M$.

## 4.3 Implementation

To implement our new BMC method for Discrete Timed Automata we have to encode both the transition relation of the abstract model of a considered automaton and an ECTL* formula by corresponding propositional formulas. The encoding of the formula ECTL* was discussed in the section 3. This section shows the encoding of the transition relation.

The method is based on the *discretization* scheme which consists in representing each region of the abstract model of $\mathcal{A}$ by one or more appropriately chosen representative states.

Let $\mathcal{A} = (\Sigma, L, l^0, E, \mathbb{X}, \mathbb{I})$ be a timed automaton with $n$ clocks, and let $c_{max}$ be the largest constant appearing in a constraint of any enabling condition used in the transition relation $E$ or in a state invariant of $\mathcal{A}$. The discretized clock space is $C^n = \{0, 1, \ldots, c_{max} + 1\}^n$. For any zone $Z$, its discretization is defined as $\widetilde{Z} = Z \cap C^n$. A discretized zone is called a $d-zone$. Note that each zone is represented by only one representative.

Let $v \in C^n$ and $\delta \in \mathbb{N}$. Define $v' = v \oplus \delta$ as: $v'(x) = min(v(x) + \delta, c_{max} + 1)$. The operation $\oplus$ is defined in order to deal with discretizations of open zones.

The *Discrete Time Successor* of $\widetilde{Z}$, is the restriction of $\tau(Z)$ to points in $C^n$.

**Definition 15 (Discrete Time Successor).** *Let $\widetilde{Z}$, $\widetilde{Z}'$ be two distinct d-zones. If $\widetilde{Z}$ is not final, then the d-zone $\widetilde{Z}'$ is the discrete time successor of $\widetilde{Z}$ iff for each $v \in \widetilde{Z}$ there exists $\delta \in \mathbb{N}$ such that $v \oplus \delta \in \widetilde{Z}'$ and $v \oplus \delta' \in \widetilde{Z} \cup \widetilde{Z}'$ for all $\delta' \leq \delta$. If $\widetilde{Z}$ is the final d-zone, then the discrete time successor of $\widetilde{Z}$ is the same d-zone $\widetilde{Z}$. The Discrete Time Successor of $\widetilde{Z}$ is denoted by $\tau(\widetilde{Z})$.*

Before we give the definition of the discrete action successors, we introduce the following operation on discretized zones: $\widetilde{Z}[Y := 0] = \{v[Y := 0] \mid v \in \widetilde{Z}\}$.

**Definition 16 (Discrete Action Successor).** *Let $\widetilde{Z}$, $\widetilde{Z}'$ be two d-zones. The d-zone $\widetilde{Z}'$ is said to be the discrete action successor of d-zone $\widetilde{Z}$ by transition $e : s \xrightarrow{\sigma, \psi, Y} s' \in E$ iff $\widetilde{Z} \subseteq \mathbf{p}(\psi) \cap C^n$ and $\widetilde{Z}' = \widetilde{Z}[Y := 0]$. The action successor of $\widetilde{Z}$ is denoted by $e(\widetilde{Z})$.*

It is easy to see that the discretization preserves the time successor and the action successor.

Now, we are ready to define a region graph for a Discrete Timed Automaton. This structure enables us to implement the bounded model checking problem for Discrete Timed Automata.

**Definition 17.** *A region graph for the discrete timed automaton $\mathcal{A} = (\Sigma, L, l^0, E, \mathbb{X}, \mathbb{I})$ is a finite structure $\mathcal{RG}(\mathcal{A}) = (S, \rightarrow, s^0)$, where $S = \{(l, \widetilde{Z}) \mid (l, Z) \in L \times Z(n)\}$, $s^0 = (l^0, \widetilde{Z^0})$ and $\rightarrow \subseteq S \times (E \cup \{\tau\}) \times S$ is defined as follows:*

- $(l, \widetilde{Z}) \xrightarrow{e} (l', \widetilde{Z}')$ iff $\widetilde{Z}' = e(\widetilde{Z})$, $l = source(e)$, $l' = target(e)$, and $\widetilde{Z}' \subseteq \mathbf{p}(\mathbb{I}(l')) \cap C^n$, for $e \in E$,
- $(l, \widetilde{Z}) \xrightarrow{\tau} (l, \widetilde{Z}')$ iff $\widetilde{Z}' \subseteq \mathbf{p}(\mathbb{I}(l)) \cap C^n$ and $\widetilde{Z}' = \tau(\widetilde{Z})$.

*The discretized model based on the region graph of $\mathcal{A}$ is defined as $\widetilde{M} = (\mathcal{RG}(\mathcal{A}), \widetilde{\mathcal{V}})$, where $\widetilde{\mathcal{V}} : S \longrightarrow 2^{\mathcal{PV}}$ with $\widetilde{\mathcal{V}}((l, \widetilde{Z})) = \mathcal{V}((l, Z))$, where $\mathcal{V}$ is the valuation function for $\mathcal{A}$.*

Since the discretized model $\widetilde{M}$ is isomorphic with the discrete model $M$ for $\mathcal{A}$, we have $\widetilde{M} \models_k \varphi$ iff $M \models_k \varphi$, for ECTL* formula $\varphi$.

Now, we can construct a propositional formula $[\widetilde{M}, \varphi]_k$ that is satisfiable iff $\widetilde{M} \models_k \varphi$. An implementation of $[\widetilde{M}, \varphi]_k$ can be found in [30].

## 5    Experimental Results

We provide experimental results for a well-known example, Fischer's Mutual Exclusion Protocol (MUTEX) [27]. The components of the system modeled as timed automata are presented in Figure 1. The correctness of the protocol depends on the time constraints involved, but it is independent on the number of the processes. In particular, the following holds: "*Fischer's protocol ensures mutual exclusion iff $\Delta < \delta$*".

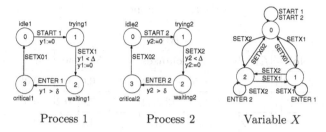

Process 1                 Process 2                 Variable $X$

**Fig. 1.** Fischer's Mutual Exclusion Protocol for two processes

As the main contribution of the paper is the ACTL* translation, we have chosen the formulas for which both the number and the length of the paths grow with the number of components. Therefore, we have tested MUTEX for the following properties:    $\psi_1$    =    $\mathrm{A}(\mathrm{F}critical_1$    →    $\mathrm{XX}(\mathrm{AF}\neg trying_1))$, $\psi_2 = \mathrm{A}$ $(\mathrm{F}critical_1$ → $\underbrace{\mathrm{X}\ldots\mathrm{X}}_{n}$ $(\mathrm{AF}(\bigvee_{i=2}^{n}\neg trying_i)$ ∨ $\mathrm{AF}\neg trying_1))$,

$\psi_3 = \mathrm{A}(\mathrm{F}(\bigwedge_{i=1}^{n}trying_i) \to \underbrace{\mathrm{X}\ldots\mathrm{X}}_{n}(\bigvee_{i=1}^{n}\mathrm{AF}\neg trying_i))$.

The first formula expresses that for each path if the process 1 reaches eventually the critical section, then in the next two steps for each path the process 1 will not eventually reach the trying section. The second formula expresses that for each path if the process 1 reaches eventually its critical section, then in the next $n$ steps for each path at least one of the processes: $2, \ldots, n$ will not eventually reach its trying sections or at none of the paths the process 1 will eventually reach its trying section. The third formula expresses that for each path if all the processes reach eventually their trying sections, then in the next $n$ steps at least one of all the processes will not eventually reach its trying section for each path.

It is easy to calculate (according to the definition of the function $f_k$) that, we need 2, 3 and $n + 1$ (n is the number of processes) symbolic $k-$paths in order to find counterexamples for $\psi_1$, $\psi_2$ and $\psi_3$, respectively, i.e., to find witnesses for $\neg\psi_1 = \mathrm{E}(\mathrm{F}critical_1 \wedge \mathrm{XX}(\mathrm{EG}trying_1))$, $\neg\psi_2 = \mathrm{E}(\mathrm{F}critical_1 \wedge \underbrace{\mathrm{X}\ldots\mathrm{X}}_{n}(\mathrm{EG}(\bigwedge_{i=2}^{n}trying_i)\wedge \mathrm{EG}trying_1))$ and $\neg\psi_3 = \mathrm{E}(\mathrm{F}(\bigwedge_{i=1}^{n}trying_i)\wedge\underbrace{\mathrm{X}\ldots\mathrm{X}}_{n}(\bigwedge_{i=1}^{n}\mathrm{EG}trying_i))$.

Note that none of the formulas can be expressed in LTL or ACTL language. Since no other experimental results of SAT-related methods are available in the literature for the Discrete Timed Automata and properties expressed in ACTL*, in this paper we show only how our method works.

We have performed our experiments on the IBM PC compatible computer equipped with the processor AMD Athlon XP 1800 (1544 MHz), 768 MB main memory and the operating system Red Hat Linux 9.0.

In Tables from 1 to 3 we show the experimental results for MUTEX system and the properties $\psi_1$, $\psi_2$ and $\psi_3$, respectively. The first column of these tables gives the number of components. The next two show the length and the number

**Table 1.** The property $\psi_1$

|       |   |          | BBMC      |          |       |       | BerkMin |       |
| ----- | - | -------- | --------- | -------- | ----- | ----- | ------- | ----- |
| NoP   | k | $f_k(\psi)$ | variables | clauses  | sec   | MB    | sec     | MB    |
| 2     | 7 | 2        | 4885      | 13884    | 0.2   | 0.0   | 0.0     | 0.0   |
| 10    | 7 | 2        | 45155     | 131044   | 2.7   | 7.6   | 0.2     | 0.0   |
| 20    | 7 | 2        | 76523     | 222735   | 4.2   | 14.2  | 0.4     | 0.2   |
| 50    | 7 | 2        | 316237    | 932008   | 21.5  | 63.5  | 1.5     | 43.0  |
| 100   | 7 | 2        | 1051863   | 3122477  | 92.0  | 217.7 | 5.7     | 152.6 |
| 150   | 7 | 2        | 2208513   | 6576018  | 232.1 | 469.3 | 14.8    | 300.2 |
| 200   | 7 | 2        | 3783013   | 11283168 | 422.0 | 618.2 | 30.0    | 579.6 |
| 220   | 7 | 2        | 4530413   | 13518828 | 502.2 | 804.7 | 38.4    | 683.1 |

of the symbolic $k$−paths. The 4th and the 5th column show the numbers of propositional variables and clauses generated by BBMC[6], respectively. The 6th and the 7th show the time and the memory consumed by BBMC to generate the set of clauses. The next two columns give the time and the memory consumed by the SAT-solver BerkMin.

**Table 2.** The property $\psi_2$

|     |    |          | BBMC      |         |      |      | BerkMin |      |
| --- | -- | -------- | --------- | ------- | ---- | ---- | ------- | ---- |
| NoP | k  | $f_k(\psi)$ | variables | clauses | sec  | MB   | sec     | MB   |
| 2   | 7  | 3        | 7721      | 22192   | 0.3  | 1.5  | 0.0     | 0.3  |
| 3   | 8  | 3        | 13098     | 37739   | 0.6  | 1.3  | 0.1     | 20.4 |
| 4   | 9  | 3        | 20141     | 58091   | 1.0  | 2.6  | 0.2     | 19.8 |
| 5   | 10 | 3        | 28365     | 81966   | 1.6  | 4.3  | 0.3     | 21.2 |
| 6   | 11 | 3        | 38159     | 110449  | 2.2  | 6.4  | 0.7     | 23.7 |
| 7   | 12 | 3        | 49631     | 143864  | 2.8  | 8.5  | 2.7     | 25.7 |
| 8   | 13 | 3        | 63493     | 184220  | 3.9  | 11.2 | 142.8   | 28.4 |
| 9   | 14 | 3        | 78688     | 228591  | 4.9  | 14.3 | 528.4   | 39.2 |

---

[6] BBMC is our tool which generates a set of clauses encoding the model of a tested system and a tested property.

**Table 3.** The property $\psi_3$

| NoP | $f_k(\psi)$ | k | BBMC variables | clauses | sec | MB | BerkMin sec | MB |
|---|---|---|---|---|---|---|---|---|
| 2 | 2 | 3 | 2525 | 7204 | 0.1 | 0.2 | 0.1 | 0.4 |
| 3 | 3 | 4 | 7301 | 21111 | 0.4 | 1.0 | 0.0 | 0.0 |
| 4 | 4 | 5 | 16530 | 48103 | 1.2 | 1.9 | 0.1 | 0.2 |
| 5 | 5 | 6 | 31381 | 91740 | 2.9 | 5.3 | 0.3 | 0.2 |
| 6 | 6 | 7 | 53804 | 157789 | 6.1 | 9.4 | 1.0 | 25.8 |
| 7 | 7 | 8 | 85755 | 252076 | 13.2 | 15.8 | 8.1 | 31.2 |
| 8 | 8 | 9 | 130623 | 384590 | 26.1 | 24.9 | 169.6 | 39.7 |
| 9 | 9 | 10 | 188517 | 555863 | 44.7 | 38.1 | 3013.2 | 131.9 |

## 6   Conclusion

We have shown that the BMC method is feasible for the branching time logic ACTL*. Then, we have implemented our new algorithm for discrete timed automata and presented preliminary experimental results, which prove the efficiency of the method.

The solution presented in this paper differs from these of [23] and [6], which could only be applied to ACTL and LTL, respectively. Our method deals with the full language of ACTL*, which subsumes both the languages of ACTL and LTL. Moreover, ACTL* seems to be the largest set of temporal properties, which can be verified by means of BMC. The present translation is based on the refined function $f_k$, returning the number of paths, which is necessary to check ACTL* formulas. The formal treatment is the basis for the implementation of the technique in the symbolic model checker $\sqrt{}$erics.

## Acknowledgements

The authors wish to thank prof. Wojciech Penczek for many useful comments and suggestions.

## References

[1] R. Alur, C. Courcoubetis, and D. Dill. Model checking in dense real-time. Information and Computation, 104(1):2–34, 1993. 26

[2] G. Audemard, A. Cimatti, A. Kornilowicz, and R. Sebastiani. Bounded model checking for timed systems. In Proc. of RT-TOOLS'02, 2002. 18, 19

[3] M. Benedetti and Alessandro Cimatti. Bounded Model checking for Past LTL. In Proc. of TACAS'03, vol. 2619 of LNCS, Springer-Verlag, 2003. 19

[4]  J. Bengtsson, B. Jonsson, J. Lilius, and W. Yi. Partial order reductions for timed systems. In Proc. of CONCUR '98, vol. 1466 of LNCS. Springer-Verlag, 1998. 18

[5]  D. Beyer. Improvements in BDD-based reachability analysis of Timed Automata. In Proc. of FME '01, vol. 2021 of LNCS. Springer-Verlag, 2002. 19

[6]  A. Biere, A. Cimatti, E. Clarke, M.Fujita, and Y. Zhu. Symbolic model checking using SAT procedures instead of BDDs. In Proc. of DAC '99, 1999. 18, 19, 31

[7]  A. Biere, A. Cimatti, E. Clarke, and Y. Zhu. Symbolic model checking without BDDs. In Proc. of TACAS '99, vol. 1579 of LNCS. Springer-Verlag, 1999. 19

[8]  M. Bozga, O. Maler, and S. Tripakis. Efficient verification of Timed Automata using dense and discrete time semantics. In Proc. of CHARME '99, 1999. 19

[9]  E. Clarke, A. Biere, R. Raimi, and Y. Zhu. Bounded model checking using satisfiability solving. Formal Methods in System Design, 19(1):7–34, 2001. 18

[10] E. M. Clarke, O. Grumberg, and D. Peled. Model Checking. MIT Press, 1999. 18, 19, 20

[11] D. Dams, O. Grumberg, and R. Gerth. Abstract interpretation of reactive systems: Abstractions preserving ACTL*, ECTL* and CTL*. In Proceedings of PROCOMET '94. Elsevier Science Publishers, 1994. 18

[12] C. Daws and S. Tripakis. Model checking of real-time reachability properties using abstractions. In Proc. of TACAS '98, vol. 1384 of LNCS. Springer-Verlag, 1998. 18

[13] P. Dembiński, A. Janowska, P. Janowski, W. Penczek, A. Półrola, M. Szreter, B. Woźna and A. Zbrzezny. √erics: A Tool for Verifying Timed Automata and Estelle Specifications. In Proc. of TACAS '03, vol. 2619 of LNCS. Springer-Verlag, 2003. 19

[14] L. de Moura, H. Rueß, and M. Sorea. Lazy theorem proving for bounded model checking over infinite domains. In Proc. of CADE '02, vol. 2392 of LNCS. Springer-Verlag, 2002. 18

[15] E. A. Emerson and A. P. Sistla. Symmetry and model checking. Formal Methods in System Design, 9:105–131, 1995. 18

[16] O. Grumberg and D. E. Long. Model checking and modular verification. In Proc. of CONCUR '91, vol. 527 of LNCS. Springer-Verlag, 1991. 19, 27

[17] K. Heljanko. Bounded reachability checking with process semantics. In Proc. of CONCUR '01, vol. 2154 of LNCS. Springer-Verlag, 2001. 19

[18] K. Heljanko and I. Niemelä. Bounded LTL model checking with stable models. In Proc. of LPNMR '2001, vol. 2173 of LNCS. Springer-Verlag, 2001. 18, 19

[19] K. L. McMillan. Symbolic Model Checking: An Approach to the State Explosion Problem. Kluwer Academic Publishers, 1993. 18

[20] P. Niebert, M. Mahfoudh, E. Asarin, M. Bozga, O. Maler, and N. Jain. Verification of Timed Automata via Satisfiability Checking. In Proc. of FTRTFT '02, vol. 2469 of LNCS. Springer-Verlag, 2002. 18, 19

[21] D. Peled. Partial order reduction: Linear and branching temporal logics and process algebras. In Proc. of POMIV '96, vol. 29 of ACM/AMS DIMACS Series. Amer. Math. Soc., 1996. 18

[22] W. Penczek, M. Szreter, R. Gerth, and R. Kuiper. Improving partial order reductions for universal branching time properties. Fundamenta Informaticae, 43:245–267, 2000. 18, 19

[23] W. Penczek, B. Woźna, and A. Zbrzezny. Bounded model checking for the universal fragment of CTL. Fundamenta Informaticae, 51(1-2):135–156, June 2002. 18, 19, 23, 31

[24] W. Penczek, B. Woźna, and A. Zbrzezny. SAT-Based Bounded Model Checking for the Universal Fragment of TCTL. Technical Report 947, ICS PAS, 2002. 18, 19

[25] W. Penczek, B. Woźna, and A. Zbrzezny. Towards bounded model checking for the universal fragment of TCTL. In Proc. of FTRTFT '02, vol. 2469 of LNCS. Springer-Verlag, 2002. 18, 19, 23

[26] Maria Sorea. Bounded model checking for timed automata. In Proc. of MTCS '02, vol. 68(5) of ENTCS. Elsevier Science Publishers, 2002. 18, 19

[27] S. Tripakis and S. Yovine. Analysis of timed systems using time-abstracting bisimulations. Formal Methods in System Design, 18(1):25–68, 2001. 28

[28] P. Wolper and P. Godefroid. Partial-order methods for temporal verification. In Proc. of CONCUR '93, vol. 715 of LNCS. Springer-Verlag, 1993. 18

[29] B. Woźna, W. Penczek, and A. Zbrzezny. Reachability for timed systems based on SAT-solvers. In Proc. of CS&P '02, vol. II of Informatik-Berichte Nr 161. Humboldt University, 2002. 18, 19

[30] B. Woźna and A. Zbrzezny. Reaching the limits for Bounded Model Checking. Technical Report 958, ICS PAS, 2003. 20, 22, 24, 28

# Removing Irrelevant Atomic Formulas for Checking Timed Automata Efficiently*

Jianhua Zhao, Xuandong Li, Tao Zheng, and Guoliang Zheng

State Key Laboratory of Novel Software Technology
Dept. of Computer Sci. and Tech. Nanjing University
Nanjing, Jiangsu, P.R.China 210093
zhaojh@nju.edu.cn

**Abstract.** Reachability analysis for timed automata can be done by enumeration of time zones, which are conjunctions of atomic formulas of the form $x - y \leq (<)n$. This paper shows that some of the atomic formulas in a generated time zone can be removed while the reachability analysis algorithm generates the same set of reachable locations. We call such formulas irrelevant ones. By removing the irrelevant formulas, the number of symbolic states associated with each location is reduced. We present two methods to detect irrelevant formulas. Case studies show that, for some kind of timed automata, these methods may significantly reduce the space requirement for reachability analysis.

## 1 Introduction

Model checking is a formal technique for validating whether a system model holds for a specific property. The basic method of model checking is exhaustive state space exploration. However, the state space increases explosively when the size of the model increases. This problem is known as 'state-space explosion'. As to the model-checking for real-time system, the state space explosion problem is even more severe because of the clock variables introduced in the system.

In the literature, most model checking tools for timed automata explore the state space by enumeration of symbolic states [1][2][3]. Generally, the reachability analysis is performed as follows. Starting from the initial symbolic state, the algorithm keeps on generating the successors of the states already generated. The algorithm terminates when either it can not generate more unexplored successors, or it reaches the destination symbolic state. Many techniques have been proposed to attack the 'state-space explosion' problem when performing reachability analysis on timed automata or parallel composition of timed automata. These techniques include compact data structure[5], partial order techniques[6][7], inactive clock reduction[8], and so on.

---

* This paper is supported by the National Natural Science Foundation of China (No.60203009, No.60233020, and No.60073031), the National 863 High-Tech Programme of China (No.2001AA113203), and by National Grand Fundamental Research 973 Program of China (No.2002CB312001).

Each symbolic state generated by the algorithm is a tuple of a location and a time zone. A time zone is a conjunction of atomic formulas of the form $x - y \leq (<)n$, where $x$, $y$ are clocks or the constant 0, $n$ is an integer. In this paper, we will try to attack the 'state-space explosion' problem caused by the clock variables by reducing the number of generated symbolic states associated with each location. We found that some atomic formulas of the time zones are irrelevant to the evolution of timed automata. The reachability algorithm can remove these formulas while generating same set of reachability locations. There are two benefits of removing irrelevant formulas. (I)The zones which have same set of relevant formulas will be reduced into one. (II) After removing irrelevant formulas, a time zone may contain other ones that were originally not contained. Thus the number of states associated with each location generated by the analysis algorithm is reduced.

The 'inactive clock reduction' technique can remove the irrelevant atomic formulas associated with inactive clocks. A clock is inactive at a location if the clock is not tested before it is reset for each path leaving this location. For the symbolic states at this location, the atomic formulas about this clock are irrelevant formulas and can be removed.

In this paper, we present two methods to detect and remove more irrelevant formulas which are not associated with inactive clocks. The case studies show that these methods reduce the space consumption significantly in some cases.

This paper is organized as follows. The second section briefly describes the timed automata and the reachability-analysis problems. The third section describes the basic idea and two methods to detect irrelevant formulas. Two basic theorems are also presented in this section. The section 4 presents the improved reachability analysis algorithm. Case studies are presented in the section 5. The last section concludes this paper.

## 2  Background

This section informally describes the timed automata and reachability analysis.

### 2.1  Timed Automata

We use $\mathcal{B}(C)$ ranged over by $D, D_1, D_2, \ldots$ to stand for the set of conjunctions of atomic formulas of the form $x - y \sim n$ for $x, y \in C \cup \{0\}$, $\sim \in \{\leq, <\}$ and $n$ being an integer. Elements of $\mathcal{B}(C)$ are called time zones over $C$.

We use $\mathcal{G}(C)$ ranged over by $g, g_1, g_2, \ldots$, to stand for the set of conjunctions of atomic formulas of the form $x \sim n$ for $x \in C$, $\sim \in \{\leq, <, >, \geq\}$ and $n$ being an integer. Elements of $\mathcal{G}(C)$ are called time guards over $C$. For any clock set $C$, we have $\mathcal{G}(C) \subseteq \mathcal{B}(C)$.

We define a connection operator $\bullet$ over atomic formulas as follows. Given two atomic formulas $x - y \sim_1 n_1$ and $y - z \sim_2 n_2$, $(x - y \sim_1 n_1) \bullet (y - z \sim_2 n_2) \overset{def}{=} (x - z \sim_3 n_1 + n2)$, where $\sim_1, \sim_2 \in \{\leq, <\}$, $\sim_3$ is either $\leq$ if $\sim_1$ and $\sim_2$ are both $\leq$, or $<$ otherwise.

A zone is *canonical* if the following condition holds. For any three clocks $x$, $y$ and $z$ ($x, y, z$ may be the constant 0.), if $x - y \sim_1 n_1$, and $y - z \sim_2 n_2$ are atomic formulas in $D$, then there is an atomic formula $x - z \sim_3 n_3$ in $D$ satisfying that $(x - z \sim_3 n_3) \Rightarrow (x - y \sim_1 n_1) \bullet (y - z \sim_2 n_2)$. Here, $\sim_1, \sim_2, \sim_3 \in \{\leq, <\}$

A timed automaton $A$ is a tuple $\ll N, l^0, C, E, I \gg$, where $N$ is a finite set of locations, $l^0 \in N$ is the start location; $C$ is a finite set of clocks; $E \subseteq N \times \mathcal{G}(C) \times 2^C \times N$ is a set of transitions; $I$ assigns each location an invariant in $\mathcal{G}(C)$. All the atomic formulas in a location invariant are of the form $x \leq (<)n$.

A timed automaton can be viewed as a conventional finite state automaton adding some clocks and time constraints. The real-number values of all the clocks increase as the time passes on. A transition of the automaton can take place if its time guard is satisfied. When a transition takes place, it can reset the values of some clocks to 0. The automaton may stay at a location as long as the location invariant is satisfied.

A symbolic state of the timed automaton $A$ is a tuple $(l, D)$, where $l$ is a location, and $D$ is in $\mathcal{B}(C)$. The symbolic state space of a time automaton can be divided into finite number of equivalence classes.

## 2.2   Reachability Analysis

The operations over the time zones can be performed efficiently[4]. So most of the model checking algorithms in the literature use enumeration of time zones to explore the state space.

The timed automata evolve by either time delay, or moving to another location through a transition. The symbolic successor operator $sp$ is as follows.

- For time delay, $sp(\delta)(l, D) \stackrel{def}{=} (l, D^\uparrow \wedge I(l))$
- For a transition $e = (l, g, r, l')$ , $sp(e)(l, D) \stackrel{def}{=} (l', r(g \wedge D) \wedge I(l'))$

We also define an operator $sp_\delta$ as $sp_\delta(e)(l, D) \stackrel{def}{=} sp(e)(sp(\delta)(l, D))$. We say $sp_\delta(e)(l, D)$ is the direct successor of $(l, D)$ w.r.t. the transition $e$. The time zone of the direct successor can be calculated as $r(D^\uparrow \wedge I(l) \wedge g) \wedge I(l')$. Let $\bar{x}$ be a clock valuation over the clock set $C$ satisfying $D^\uparrow \wedge I(l) \wedge g$. Let $\bar{x}'$ be the new valuation derived by setting the values of clocks in $r$ to 0. The valuation $\bar{x}'$ satisfies $r(D^\uparrow \wedge I(l) \wedge g) \wedge I(l')$ if and only if all the values of clocks in $C - r$ satisfies the formulas about them in $I(l')$. So the time zone can also be calculated as $r(D^\uparrow \wedge I(l) \wedge g \wedge I')$, where $I'$ is the time guard derived by removing all atomic formulas about clocks in $r$ from $I(l')$.

The basic reachability analysis algorithm depicted in Fig 1 checks whether the location $l'$ is reachable from $(l_0, D_0)$. This algorithm, or its variants, is widely used in different model checking tools[1][2][3].

## 2.3   Representing Time Zones by Weighted Directed Graphs

A timed zone is a conjunction of atomic formulas of the form $x - y \sim n$ for $x, y \in C \cup \{0\}$, $\sim \in \{\leq, <\}$ and $n$ being an integer. For convenience of reasoning in this paper, we can also represent a time zone by a weighted directed graph.

```
PASSED  := {}
WAITING := {(l_0, D_0)}
repeat begin
        get a state (l, D) from WAITING and remove it from WAITING
        if D ⊄ D' for all (l, D') ∈ PASSED then
           begin
                add (l, D) to PASSED.
                Succ := {sp_δ(e)(l, D) | e is a transition leaving l}
                if there is a non empty zone (l_1, D') ∈ Succ
                           such that l' = l_1
                    return YES
                WAITING := WAITING ∪ {z|z is non-empty ∧ z ∈ Succ}
        end
     end
until WAITING = {}
return NO.
```

**Fig. 1.** The basic reachability analysis algorithm

Given a time zone $D$ over a set of clocks $C$, the corresponding graph $G_D$ is defined as follows. Each node of the graph represents a clock or the constant 0. There are two kind of edges: equational and non-equational edges. For each atomic formula $x - y \leq n$ (or $x - y < n$), there is an equational (or non-equational) edge from node $x$ to node $y$ weighted $n$. A path $p$ is a sequence of consecutive edges. The path $p$ is an equational one if all the edges are equational ones. It's non-equational if $p$ is not equational.

The operators over time zones can be performed through weighted directed graph. Let $D$, $D_1$ and $D_2$ be three time zones.

1. $D$ is empty if and only if there is a cycle path in $G_D$ such that either the length of that cycle is negative, or the cycle is non-equational and 0-length.
2. Each edge from $x$ to $y$ in $G_{D_1 \wedge D_2}$ is the shorter one of the edges (if exist) from $x$ to $y$ in $G_{D_1}$ and $G_{D_2}$.
3. The graph $G_{D'}$ of the equivalent canonical zone of $D$ can be derived as follows. For any two nodes $x$ and $y$, there is an edge from $x$ to $y$ in $G_{D'}$ if there are paths from $x$ to $y$ in $G_D$. The edge and the shortest path are of the same length. The edge is equational if and only if the shortest path is equational.

## 3   Detecting Irrelevant Atomic Formulas

This section presents the basic idea of irrelevant formulas. The methods to detect irrelevant atomic formulas are also presented.

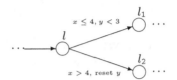

**Fig. 2.** An example for irrelevant formulas

## 3.1 Basic Idea about Irrelevant Formulas

A time zone is a conjunction of atomic formulas of the form $x - y \sim c$. Generally speaking, given a symbolic state $(l, D)$, removing an atomic formula $d$ from $D$ may result in a bigger state $(l, D')$ $(D \subseteq D')$. However, we found that under certain conditions, all the reachable locations from $(l, D')$ are still reachable location of the timed automaton. Thus, if $(l, D)$ is a symbolic state generated by reachability analysis algorithm, it can be replaced by $(l, D')$ during state generation without generating un-reachable locations. We call $d$ an irrelevant atomic formula of $(l, D)$. Here we use a timed automaton fragment in Fig 2 as an example to show that removing irrelevant atomic formulas can result in memory requirement reduction. Let $(l, -x < 4 \wedge x - y < 3)$ and $(l, -x < 4 \wedge x - y \geq 3)$ be two symbolic states of the automaton. The only direct successor of these two states is $(l_2, -x < 4 \wedge y - x < 4 \wedge y = 0)$. We can find that the constraints $x - y < 3$ and $x - y \geq 3$ are irrelevant atomic formulas. These two states are reduced to one state $(l, -x < 4)$ if the irrelevant formulas are removed.

**Definition 1.** *Let $(l, D)$ be a reachable symbolic state of a timed automaton $A$. Let $d$ be an atomic formula in $D$. Let $D'$ be the zone derived by removing $d$ from $D$. The atomic formula $d$ is an irrelevant formula of $D$ if each location reachable from $(l, D')$ is also a reachable location of the automaton.*

From the above definition, the model-checker can remove irrelevant atomic formulas from the generated states during the state-space exploration. We will present two methods to detect irrelevant formulas in subsections 3.3 and 3.4.

## 3.2 Two Basic Theorems about Zone Intersection Operator

We will first present two theorems about zone intersection. These theorems will be used to reason about the methods.

**Theorem 1.** *Let $D_1$ be a non-empty canonical zone. Let $g$ be a time guard in $\mathcal{G}(C)$. If $g \wedge D_1 = \emptyset$, there must be an atomic formula $d$ in $D_1$ such that $g \wedge d = \emptyset$. Furthermore, one of the following conclusions holds. (I) $d$ is of the form $x \sim n$ in $D$, there is a formula $x \not\sim n'$ in $g$ such that $d \wedge (x \not\sim n')$ is false. (II) $d$ is of the form $x - y \leq (<)n, (x, y \neq 0)$, there are two formulas $0 - x \leq (<)n_1$ and $y - 0 \leq (<)n_2$ in $g$ such that $d \wedge (0 - x \leq (<)n_1) \wedge (y - 0 \leq (<)n_2)$ is false.*

*Proof.* Because $g \wedge D_1$ is $\emptyset$, there must be a cycle path $p$ in $G_{g \wedge D_1}$ of which the length is negative. (We also call a non-equational 0-length path a negative one.) Because $D_1$ is not empty, each cycle in $G_{D_1}$ is not negative. So $p$ contains at lease one edge from $G_g$.

Notice that each edge in $G_g$ either leaves or arrives at the node 0, the cycle path $p$ must pass through 0 at lease once. If $p$ passes 0 more than once, we can get a negative sub-cycle $p'$ of $p$ such that $p'$ passes 0 only once. There is one or two edges in $p'$ from $G_g$ because each edge in $G_g$ either leaves or arrives at 0. The other edges in $p'$ are from $G_{D_1}$ and consecutive. Because $D_1$ is canonical, the other edges can be replaced by one edge from $G_{D_1}$. This edge, together with the edges from $G_g$, can form a negative-length cycle. Let $d$ be the atomic formula corresponding to the edge in $G_{D_1}$, we have $d \wedge g = \emptyset$.

The last cycle is composed of one edge from $G_D$ and 1 or 2 edges from $G_g$. If only one edge in the cycle is from $G_g$, the conclusion (I) holds. If two edges are from $G_g$, the conclusion (II) holds. □

**Theorem 2.** *Let $D_1$ be a canonical zone. Let $g$ be a time guard and $D_1 \wedge g \neq \emptyset$. Let $D_2'$ be the canonical zone equivalent to $g \wedge D_1$. For each atomic formula $d$ in $D_2'$, it is in one of the following categories.*

1. **Category 1** *The original atomic formula set of $D_1$.*
2. **Category 2** *The original atomic formula set of $g$.*
3. **Category 3** *The atomic formula $d$ is of the form $x \sim c$ and there are atomic formulas $d_1 \in D_1$ and $d_2 \in g$ such that $d = d_1 \bullet d_2$ or $d = d_2 \bullet d_1$.*
4. **Category 4** *The atomic formula $d$ is of the form $x - y \sim c$ $(x, y \neq 0)$ and there are atomic formulas $d_1$ and $d_2$ respectively of the form $x \leq (<)c_1$ and $0 - y \leq (<)c_2$ in one of the above three categories satisfying that $d = d_1 \bullet d_2$.*

*Proof.* Because $G_{g \wedge D_1}$ is not empty, each cycle in the graph is non-negative. If there are circles in a path $p$, we can remove them from $p$ to get a shorter path. So if there is a path from $x$ to $y$ $(x \neq y)$, there is an acyclic shortest one from $x$ to $y$.

If an edge in $G_{D_2'}$ is also in $G_{g \wedge D_1}$, the corresponding formula is in $D_1$ or $g$. So the formula is in the category 1 or 2.

Let $d$ be an arbitrary atomic formula of the form $x - 0 \leq (<)n$ or $0 - x \leq (<)n$ in $D_2'$. If $d$ is not in category 1 or 2, the shortest path in $G_{g \wedge D_1}$ between $x$ and 0 must include an edge from $G_g$. Because each edge in $G_g$ is either leaving 0 or arriving at 0, and the path is acyclic, so there is only one edge from $G_g$ at one of the two ends of the path. All the rest edges are from $G_{D_1}$ and consecutive. Because $D_1$ is canonical, we can find an edge in $G_{D_1}$ equivalent to the rest edges. So two edges, one from $G_{D_1}$ and another from $G_g$, also form a shortest path. That is, the atomic formula is equal to $d_1 \bullet d_2$ or $d_2 \bullet d_1$ for $d_1$ is in $D_1$, and $d_2$ is in $g$. So these formulas are in category 3.

For an arbitrary atomic formula $x - y \leq (<)n$ $(x, y \neq 0)$ in $D_2'$, there must be an equivalent acyclic shortest path $p$ in $G_{g \wedge D_1}$. So the path passes through the node 0 because $p$ contains at least one edge from $G_g$. We can divide $p$ into two parts: from $x$ to 0 and from 0 to $y$. These two parts are also shortest ones because $p$ is shortest. These two parts are respectively corresponding to two atomic formulas of the form $x - 0 \leq (<)n$ or $0 - x \leq (<)n$ in the category 1,2, or 3. So $x - y \leq (<)n$ is in category 4.                                                     □

### 3.3   Static Method for Irrelevant Atomic Formula Detection

The static method uses the information about clock resetting and testing to detect irrelevant atomic formulas. Before state-space exploration, we need to collect this information.

**Definition 2. Greater-test-free clock** *Let $l$ be a location. Let $e_1, e_2, \ldots, e_n$ be all the transitions leaving $l$. A clock $x$ is called greater-test-free on $l$ if the following conditions hold. (i) The time guard of each $e_i$ and the location invariant of $l$ contains no atomic formula of the form $x \geq (>)n$. (ii) For each $i$, either $e_i$ resets $x$ or $x$ is also greater-test-free on the target location of $e_i$.*

Intuitively, a clock is greater-test-free on $l$ means that for each transition path leaving $l$, $x$ will not be test by constraint $x \geq (>)n$ before it is reset. This information can be calculated before the state-space exploration. For a parallel composition of timed automata, a clock is tested and reset only by one local automaton. So calculating this information can be done with low cost.

**Lemma 1.** *Let $g$ be a time guard containing no atomic formula of the form $x_0 \geq (>)n$. Let $D_1$ and $D_2$ be two canonical time zones containing the same set of formulas which are not of the form $x_0 - y \leq (<)n$ and $(D_1 \wedge g \neq \emptyset) \wedge (D_2 \wedge g \neq \emptyset)$. Let $D_1'$ and $D_2'$ be canonical zones respectively equivalent to $D_1 \wedge g$ and $D_2 \wedge g$. The zones $D_1'$ and $D_2'$ contain the same set of atomic formulas which are not of the form $x_0 - y \leq (<)n$ $(y \in C \cup \{0\})$.*

*Proof.* Let $d$ be an atomic formula in $D_1'$ which are not of the form $x_0 - y \leq (<)n$. From theorem 2, $d$ must be in one of the following categories.

If $d$ is in category 1 or 2, $d$ is also in $D_2$ or $g$. So we have $D_2' \Rightarrow d$.

If $d$ is in category 3 and of the form $0 - y \leq (<)n$, $d$ is equal to $d_1 \bullet d_2$, where $d_1$ in $g$ and $d_2$ in $D_1$. Because there is no formula of the form $0 - x_0 \leq (<)n$ in $g$, $d_2$ is not of the form $x_0 - y \leq (<)n$. So $d_2$ is also in $D_2$. We have $D_2' \Rightarrow d$.

If $d$ is in category 3 and of the form $y - 0 \leq (<)n$ $(y \neq x_0, 0)$, $d$ can be equal to $d_1 \bullet d_2$ for some $d_1$ in $D_1$ and $d_2$ in $g$. Because $d_1$ is not of the form $x_0 - z \leq (<)$, $d_1$ is also in $D_2$. So we have $D_2' \Rightarrow d$.

If $d$ is in category 4, $d$ is equal to $d_1 \bullet d_2$ for some $d_1, d_2$ of the form $x \leq (<)n$ and $0 - y \leq (<)n$ in the first three categories. Because $d$ is not of the form $x_0 - y \leq (<)n$, $d_1$ is not of the form $x_0 - y \leq (<)n$ either. We have $D_2' \Rightarrow d_1$ because $d_1$ is in the first three categories. Also because $d_2$ is in the first three categories and of the form $0 - y \leq (<)n$, we have $D_2' \Rightarrow d_2$. So $D_2' \Rightarrow d$.

Thus we have that for each formula $x - y \leq (<)n$ in $D_1'$, $D_2' \Rightarrow x - y \leq (<)n$ if $x \neq x_0$. Similarly, we can prove that for each formula $x - y \leq (<)n$ in $D_2'$, $D_1' \Rightarrow x - y \leq (<)n$ if $x \neq x_0$. So we conclude that, $D_1'$ and $D_2'$ have the same set of formulas which are not of the form $x_0 - y \leq (<)n$.                                     □

**Theorem 3.** *Let $(l, D)$ be a symbolic state where $D$ is a canonical zone. Let $x_0$ be a clock. All the atomic formulas of the form $x_0 - y \leq (<)c$ ($y \in C \cup \{0\}$) in $D$ are irrelevant if $x_0$ is greater-test-free on $l$.*

*Proof.* In this proof, let $l$ be a location, let $x_0$ be a greater-test-free clock on $l$, let $D_1$ and $D_2$ be two canonical zones containing same set of atomic formulas of the form $x - y \leq (<)n$, where $x \neq x_0$. Let $g$ be a time guard containing no formula of the form $0 - x_0 \leq (<)n$.

Let $r$ be a set of clocks. We have that $r(D_1)$ and $r(D_2)$ are same if $x_0 \in r$.

If $g \wedge D_1$ is empty, from Theorem 1, there is an atomic formula $d_1$ in $D_1$ such that $g \wedge d_1 = \emptyset$. Formula $d_1$ can not be of the form $x_0 - y \leq (<)n$ because $g$ contains no atomic formula of the form $0 - x_0 \leq (<)n$. So $d_1$ is also in $D_2$. We have that $D_2 \wedge g$ is also empty.

If $g \wedge D_1$ is not empty, from the lemma 1, the canonical form of $g \wedge D_1$ and $g \wedge D_2$ has same set of formulas of the form $x - y \leq (<)n$, where $x \neq x_0$.

As discussed in subsection 2.2, the time zones of $sp_\delta(e)(l, D_1)$, $sp_\delta(e)(l, D_2)$ are $r(D_2^\uparrow \wedge I(l) \wedge g_e \wedge I')$ and $r(D_1^\uparrow \wedge I(l) \wedge g_e \wedge I')$ respectively. Notice that, the clock $x_0$ is greater-test-free on $l$, so $g_e$, $I(l)$ and $I'$ contains no atomic formula of the form $0 - x_0 \leq (<)n$. Based on the above conclusions and the definition of greater-test-free clocks, we conclude that for each transition path $p$ leaving $l$, the successors of $(l, D_1)$ and $(l, D_2)$ w.r.t. $p$ are either both empty, or same, or their time zones have same set of formulas of the form $x - y \leq (<)n$ for $x \neq x_0$.

Let $D'$ be the time zone derived by removing all the formulas of the form $x_0 - y \leq (<)n$ from another canonical zone $D$. Then $D$ and the canonical zone of $D'$ have the same set of formulas of the form $x - y \leq (<)$ for $x \neq x_0$. So we proved this theorem. ☐

Similarly, we can define less-test-free clocks and get another method to detect irrelevant formulas.

## 3.4   Dynamic Method for Irrelevant Atomic Formula Detection

Let $(l, D)$ be a symbolic state generated by the model-checking algorithm. Let $e_1, e_2, \ldots, e_n$ be transitions leaving $l$. Let $(l_i, D_i)$ be the symbolic states generated by the algorithm satisfying that either (1) $sp_\delta(e_i)(l, D) \subseteq (l_i, D_i)$ if $sp_\delta(e_i)(l, D) \neq \emptyset$, or (2) $D_i = \emptyset$ if $sp_\delta(e_i)(l, D)$ is empty. Let $d$ be an atomic formula in $D$. Let $D'$ be the time zone derived by removing $d$ from $D$. The atomic formula $d$ is called removable w.r.t. $e_i$ if $sp(e_i)(l, D') \subseteq (l_i', D_i)$. The formula $d$ is irrelevant if it is removable w.r.t. all the leaving transitions.

As described in subsection 2.2, the successor of $(l, D)$ w.r.t. $e_i$ is $(l', r(D^\uparrow \wedge I(l) \wedge g_i \wedge I_i'))$. Let $g_i'$ be the time guard equivalent to $I(l) \wedge g_i \wedge I_i'$.

The successor is empty if and only if $D^\uparrow \wedge g_i'$ is empty. From Theorem 1, there must be an atomic formula $d$ in $D^\uparrow$ such that $d \wedge g_i' = \emptyset$. The formula $d$ is also in $D$. All the other atomic formulas are removable w.r.t. $e_i$.

Now we will present a way to find removable atomic formulas if the successor w.r.t. $e_i$ is not empty. The time zone of $sp_\delta(e_i)(l, D)$ is $(l', r(D^\uparrow \wedge g_i'))$. So $r(D^\uparrow \wedge g_i') \subseteq D_i$.

1. Let $D'_i = r^{-1}(D_i)$, where $r$ is the set of clocks reset by the transition $e_i$. The definition of the operator $r^{-1}$ can be seen in [4]. $D'_i$ is the largest time zone satisfying that $r(D'_i) = D_i$. We have $D^\uparrow \wedge g'_i \subseteq D'_i$.
2. Calculating the canonical zone $D_1$ of $D^\uparrow \wedge g'_i$. According to Theorem 2, each formula in $D_1$ is in one of the four categories. There are 0, 1 or 2 formulas in $D^\uparrow$ associated with each formula in $D_1$. If we calculate $D_1$ based on Theorem 2, for each formula in $D_1$, we can record which formulas in $D$ are associated with it.
3. For each formula $d$ in $D'_i$, find the formula $d'$ in $D_1$ such that $d' \Rightarrow d$, then mark the formulas in $D^\uparrow$ associated with $d'$. Because $D_1$ is canonical and $D_1 \subseteq D'_i$, for each formula $d$ in $D'_i$, we can find a formula $d'$ in $D_1$ such that $d' \Rightarrow d$.
4. All unmarked constraints in $D$ are removable w.r.t. $e_i$.

Thus, we present an algorithm to detect the irrelevant formulas of a symbolic state when all of the direct successors are calculated.

## 4   Improved Algorithm

Based on the methods for irrelevant atomic formula detection as we presented in the previous sections, we get an improved algorithm depicted in Firgure 3. The algorithm explores the state space in a breadth-first way. The difference between this algorithm and the basic one is as follows.

1. Each time a symbolic state is generated, the algorithm removes the irrelevant atomic formulas detected by the static method.
2. Once the algorithm generates all the direct successors of a symbolic state $(l, D)$, it removes the irrelevant atomic formulas in $D$ detected by the dynamic method.

This algorithm can be used to check whether a location is reachable. The result of this algorithm is exact because removing irrelevant formulas will not introduce new 'reachable' location.

This algorithm can be extended to check whether a symbolic state $(l, D)$ is reachable in case $D$ is in $\mathcal{G}(C)$. We can add a new location $l_{dest}$ and a virtual transition from $l$ to $l_{dest}$ with $D$ as its time guard . Thus, the reachability problem of $(l, D)$ is now transformed into the reachbility problem of $l_{dest}$. For a parallel composition of timed automata, the virtual transition can be simulated by a set of synchronized local transitions.

In this algorithm, we only use the dynamic method to detect irrelevant formulas of a symbolic state when all of its successors are generated. We can also use this method recursively. When the algorithm detects and removes some irrelevant formulas from a symbolic state, it can recursively detect the irrelevant formulas of the predecessor of this state. To do this, we must record the predecessors of each symbolic state. Recursive detection is feasible only if the benefit of removing more irrelevant formulas surpass the cost of maintaining the predecessor-successor relations. We haven't implemented recursive detection. Later, recursive detection can be an option of our tool.

Calculate which clocks are greater-test-free or less-test-free on each location.
PASSED    := {}
WAITING := $\{(l_0, D_0)\}$
repeat begin
      get a state $(l, D)$ from WAITING and remove it from WAITING;
      if $D \not\subseteq D'$ for all $(l, D') \in PASSED$ then
        begin
            add $(l, D)$ to PASSED;
            for each $e_i$ in the set $\{e_i|\ l$ is the source loaction of $e_i\}$ do
            begin
                Let $S_i = sp_\delta(e_i)(l, D)$;
                if $S_i$ is non-empty and the location of $S_i$ is $l'$
                   return YES;
                if there exists a state $S'$ in PASSED satisfying that $S_i \subseteq S'$
                   Let $S_i := S'$;
                else begin
                   Remove irrelevant formulas detected by static method from $S_i$;
                   Add $S_i$ into WAITING;
                end
            end
            Detect and remove the irrelevant formulas of $(l, D)$ by dynamic method;
        end
      end
until WAITING = {}
return **NO**.

**Fig. 3.** The improved algorithm

The dynamic method (non-recursive) used in this algorithm need no extra memory. For the static method, we need some extra memory to record the information about greater-test-free and less-test-free clocks. The extra memory cost is small comparing to the memory needed for generated states.

## 5   Case Studies

We have incorporated the techniques presented in this paper into our experimental tool. We applied this tool to several examples using an Intel P4(1GHz) computer with 256M memory and 512M virtual memory. The tables in Figure 4 are the performance data of our tool when we use the tool to check Fischer's protocol, CSMA protocol, and FDDI protocol. The columns 'Basic', 'Inactive' and 'Irrelevant' are respectively performance data for the basic algorithm, the algorithm with 'inactive clock reduction' optimization and the algorithm with 'Irrelevant constraints removing'. The space requirement is expressed by the numbers of the generated symbolic states. We only check FDDI with 2,3, and 4 workstations because our experimental tool can only handle at most 16 clock variables. Because the number generated states are significantly reduced, the

| Systems | Basic | Inactive | Irrelevant | System | Basic | Inactive | Irrelevant |
|---------|-------|----------|------------|--------|-------|----------|------------|
| Fischer 2 | 27 | 21 | 18 | CSMA 3 | 126 | 69 | 54 |
| Fischer 3 | 157 | 103 | 65 | CSMA 4 | 913 | 387 | 199 |
| Fischer 4 | 965 | 567 | 220 | CSMA 5 | 6303 | 2226 | 644 |
| Fischer 5 | 6591 | 3631 | 727 | CSMA 6 | 43911 | 14931 | 2057 |
| Fischer 6 | 50431 | 26799 | 2378 | CSMA 7 | N/A | N/A | 6026 |
| Fischer 7 | N/A | N/A | 7737 | CSMA 8 | N/A | N/A | 16907 |
| Fischer 8 | N/A | N/A | 25080 | CSMA 9 | N/A | N/A | 45836 |
| Fischer 9 | N/A | N/A | 81035 | CSMA 10 | N/A | N/A | 120845 |
| Fischer 10 | N/A | N/A | 260998 | CSMA 11 | N/A | N/A | 311310 |
| FDDI 2 | 57 | 27 | 15 | FDDI 3 | 160 | 56 | 21 |
| FDDI 4 | 345 | 95 | 27 | | | | |

**Fig. 4.** Performance data checking Fisher's, CSMA/CD, and FDDI protocol

used CPU time is not long. The checks for these systems are performed with seconds or minutes.

The performance data shows that our technique can significantly reduce the space requirement in these cases. There are some cases, like Bang&Olufson Collision Detection Protocol[9], in which our technique results in no optimization.

Our technique works very well if there are several greater-test-free or less-test-free clocks in the system. Our experimental tool even over-performs UPPALL when checking Fischer's protocol and CSMA/CD protocol. Notice that our technique is an 'exact' one, so we can compare our performance data with those of UPPAAL without the option -A. As reported in the homepage of UPPAAL, the tool UPPAAL can check Fischer's protocol of 7 processes and CSMA protocol of 7 senders.

## 6   Conclusions

In this paper, we present a technique to reduce the space requirement of reachability analysis on timed automata. We found that some of the constraints in the generated symbolic state are irrelevant to the evolution of the timed automaton. Removing these constraints can reduce the number of symbolic states generated by the analysis algorithm.

Two methods are presented to detect irrelevant formulas during state-space exploration. The first one is called static detection method. This method finds irrelevant formulas based on the information about clock resetting and testing. This information can be calculated statically for each timed automaton before state-space exploration. This method is a generalization of 'inactive clock reduction' technique. It can detect all the irrelevant formulas associated with inactive clocks. It is more powerful because it can also detect irrelevant formulas associated with active but greater-test-free or less-test-free clocks. The performance data also shows that our technique results in better optimization.

A dynamic detection method is also proposed in this paper. If all the direct successors of a state are generated, this method detects irrelevant atomic formulas of the state according to its successors. This method can find some irrelevant formulas which can not be found by the static method. In the example depicted in Fig 2, the irrelevant formula $x - y < 3$ can not be found by the static method, but can be found by the dynamic one.

Our technique operates on the time zones individually. So we believe that this technique can be easily combined with other optimization techniques.

The optimized algorithm in this paper can check whether a location is reachable. The reachability of a symbolic state can be transformed into location reachability problem by adding virtual transitions.

# References

[1] Kim G Larsen, Paul Pettersson and Wang Yi. UPPAAL: Status & Developments. In Orna Grumberg, editor, Proceedings of the 9th International Conference on Computer-Aided Verification. Haifa, Israel, LNCS 1254, pages 456–459. Springer-Verlag, June 1997. 34, 36

[2] C.Daws, A.Olivero, S.Tripakis, and S.Yovine. The tool Kronos. In DIMACS Workshop on Verification and Control of Hybrid Systems, LNCS 1066. Springer-Verlag, October 1995. 34, 36

[3] T.A.Henzinger and P.-H. Ho. Hytech: The Cornell hybrid technology tool. In Proc. of Workshop on Tools and Algorithms for the Construction and Analysis of Systems, 1995. BRICS report series NS-95-2. 34, 36

[4] Howard Wang-Toi. Symbolic Approximations for Verifying Real-Time Systems. PhD thesis, Stanford University, 1994. 36, 42

[5] Kim G. Larsen, Fredrik Larsson, Paul Pettersson, and Wang Yi. Efficient Verification of Real-Time Systems: Compact Data Structures and State-Space Reduction. In Proc. of the 18th IEEE Real-Time Systems Symposium, IEEE Computer Society Press, December 1997. 34

[6] Johan Bengtsson, Bengt Jonsson, Johan Lilius, and Wang Yi. Partial Order Reductions for Timed Systems. In Proc. of the 9th International Conference on Concurrency Theory, September 1998. 34

[7] Jianhua ZHAO, He XU, Xuandong LI, Tao ZHENG, and Guoliang ZHENG. Partial Order Path Technique for Checking Parallel Timed Automata. In Proc. of the 7th International Symposium, FTRTFT 2002, LNCS 2469, Oldenburg, Germany, September 2003. 34

[8] Conrado Daws and Sergio Yovine. Model checking of real-time reachability properties using abstractions. In Bernard Steffen, editor, Proc. of the 4th workshop on Tools and Algorithms for the Construction and Analysis of Systems. LNCS 1384. Springer-Verlag, 1998 34

[9] Kalus Havelund, Arne Skou, Kim G. Larsen and Kristian Lund  Formal Modelling and Analysis of an Audio/Video Protocol: An Industrial Case Study Using Uppaal. In Proc. of 18th IEEE Real-Time Systems Symposium, IEEE Computer Society Press, December 1997. 44

# Adding Symmetry Reduction to UPPAAL[*]

Martijn Hendriks[1], Gerd Behrmann[2], Kim Larsen[2],
Peter Niebert[3][**], and Frits Vaandrager[1]

[1] Nijmeegs Instituut voor Informatica en Informatiekunde
University of Nijmegen
The Netherlands
{martijnh,fvaan}@cs.kun.nl
[2] Department of Computing Science
Aalborg University
Denmark
{behrmann,kgl}@cs.auc.dk
[3] Laboratoire d'Informatique Fondementale, CMI
Université de Provence
France
peter.niebert@lif.univ-mrs.fr

**Abstract.** We describe a prototype extension of the real-time model
checking tool UPPAAL with symmetry reduction. The symmetric data
type scalarset, which is also used in the MURφ model checker, was added
to UPPAAL's system description language to support the easy static detection of symmetries. Our prototype tool uses state swaps, described
and proven sound earlier by Hendriks, to reduce the space and memory
consumption of UPPAAL. Moreover, the reduction strategy is canonical,
which means that the symmetries are optimally used. For all examples
that we experimented with (both academic toy examples and industrial
cases), we obtained a drastic reduction of both computation time and
memory usage, exponential in the size of the scalar sets used.

## 1 Introduction

Model checking is a semi-automated technique for the validation and verification
of all kinds of systems [8]. The approach requires the construction of a *model* of
the system and the definition of a *specification* for the system. A model checking tool then computes whether the model satisfies its specification. Nowadays,
model checkers are available for many application areas, e.g., hardware systems
[10, 22], finite-state distributed systems [17], and timed and hybrid systems
[21, 27, 25, 16].

---

[*] Supported by the European Community Project IST-2001-35304 (AMETIST),
http://ametist.cs.utwente.nl.
[**] Peter Niebert suggested the method for efficient computation of canonical representatives at an AMETIST project meeting, and was therefore invited to join the list
of authors after acceptance of the paper.

Despite the fact that model checkers are relatively easy to use compared to manual verification techniques or theorem provers, they are not being applied on a large scale. An important reason for this is that they must cope with the *state space explosion* problem, which is the problem of the exponential growth of the state space as models become larger. This growth often renders the mechanical verification of realistic systems practically impossible: there just is not enough time or memory available. As a consequence, much research has been directed at finding techniques to fight the state space explosion. One such a technique is the exploitation of behavioral symmetries [18, 23, 20, 19, 12, 7]. The exploitation of *full* symmetries can be particularly profitable, since its gain can approach a factorial magnitude.

There are many timed systems which clearly exhibit full symmetry, e.g., Fischer's mutual exclusion protocol [1], the CSMA/CD protocol [24, 27], industrial audio/video protocols [13], and distributed algorithms, for instance [4].

Motivated by these examples, the work presented in [14] describes how UP-PAAL, a model checker for networks of timed automata [21, 3, 2], can be enhanced with symmetry reduction. The present paper puts this work to practice: a prototype of UPPAAL with symmetry reduction has been implemented. The symmetric data type *scalarset*, which was introduced in the MURφ model checker [10], was added to UPPAAL's system description language to support the easy static detection of symmetries. Furthermore, the *state swaps* described and proven sound in [14] are *optimally* used to reduce the space and time consumption of the model checking algorithm. Run-time data is reported for the examples mentioned above, showing that symmetry reduction in a timed setting can be very effective.

*Related work.* Symmetry reduction is a well-known technique to reduce the resource requirements for model checking algorithms, and it has been successfully implemented in model checkers such as MURφ [10, 19], SMV [22], and SPIN [17, 6]. As far as we know, the only model checker for timed systems that exploits symmetry is RED [25, 26]. The symmetry reduction technique used in RED, however, gives an over approximation of the reachable state space (this is called the *anomaly of image false reachability* by the authors). Therefore, RED can only be used to ensure that a state is *not* reachable when it is run with symmetry reduction, whereas symmetry enhanced UPPAAL can be used to ensure that a state is reachable, or that it is not reachable.

*Contribution.* We have added symmetry reduction as used within MURφ, a well-established technique to combat the state space explosion problem, to the real-time model checking tool UPPAAL. For researchers familiar with model checking it will come as no surprise that this combination can be made and indeed leads to a significant gain in performance. Still, the effort required to actually add symmetry reduction to UPPAAL turned out to be substantial.

The soundness of the symmetry reduction technique that we developed for UPPAAL does not follow trivially from the work of Ip and Dill [19] since the description languages of UPPAAL and MURφ, from which symmetries are extracted

```
(1)   passed := ∅
(2)   waiting := Q_0
(3)   while waiting ≠ ∅ do
(4)         get q from waiting
(5)         if q ⊨ φ then return YES
(6)         else if q ∉ passed then
(7)               add q to passed
(8)               waiting := waiting ∪ { q' ∈ Q | (q,q') ∈ Δ }
(9)         fi
(10) od
(11) return NO
```

**Fig. 1.** A general forward reachability analysis algorithm

automatically, are quite different. In fact, the proof that symmetry reduction for UPPAAL is sound takes up more than 20 pages in [14].

The main theoretical contribution of our work is an efficient algorithm for the computation of a canonical representative. This is not trivial due to UPPAAL's symbolic representation of sets of clock valuations.

Many timed systems exhibit symmetries that can be exploited by our methods. For all examples that we experimented with, we obtained a drastic reduction of both computation time and memory usage, exponential in the size of the scalar sets used.

*Outline.* Section 2 presents a very brief summary of model checking and symmetry reduction in general, while Sections 3 and 4 introduce symmetry reduction for the UPPAAL model checker in particular. In Section 5, we present run-time data of UPPAAL's performance with and without symmetry reduction, and Section 6 summarizes and draws conclusions.

A full version of the present paper including proofs of lemma 1 and of theorem 2 is available as [15].

## 2   Model Checking and Symmetry Reduction

This section briefly summarizes the theory of symmetry presented in [19], which is reused in a timed setting since (i) it has proven to be quite successful, and (ii) it is designed for reachability analysis, which is the main purpose of the UPPAAL model checker. We simplify (and in fact generalize) the presentation of [19] using the concept of bisimulations.

In general, a transition system is a tuple $(Q, Q_0, \Delta)$, where $Q$ is a set of states, $Q_0 \subseteq Q$ is a set of initial states, and $\Delta \in Q \times Q$ is a transition relation between states. Figure 1 depicts a general forward reachability algorithm which, under the assumption that $Q$ is finite, computes whether there exists a reachable state $q$ that satisfies some given property $\phi$ (denoted by $q \models \phi$).

Due to the state space explosion problem, the number of states of a transition system frequently gets too big for the above algorithm to be practical. We would like to exploit structural properties of transition systems (in particular symmetries) to improve its performance. Here the well-known notion of bisimulation comes in naturally:

**Definition 1 (Bisimulation).** *A* bisimulation *on some transition system, say* $(Q, Q_0, \Delta)$, *is a relation* $R \subseteq Q \times Q$ *such that, for all* $(q, q') \in R$,

1. $q \in Q_0$ *if and only if* $q' \in Q_0$,
2. *if* $(q, r) \in \Delta$ *then there exists an* $r'$ *such that* $(q', r') \in \Delta$ *and* $(r, r') \in R$,
3. *if* $(q', r') \in \Delta$ *then there exists an* $r$ *such that* $(q, r) \in \Delta$ *and* $(r, r') \in R$.

Suppose that, before starting the reachability analysis of a transition system, we know that a certain equivalence relation $\approx$ is a bisimulation and respects the predicate $\phi$ in the sense that either all states in an equivalence class satisfy $\phi$ or none of them does. Then, when doing reachability analysis, it suffices to store and explore only a single element of each equivalence class. To implement the state space exploration, a *representative function* $\theta$ may be used that converts a state to a representative of the equivalence class of that state:

$$\forall_{q \in Q} \, (q \approx \theta(q)) \tag{1}$$

Using $\theta$, we may improve the algorithm in Figure 1 by replacing lines 2 and 8, respectively, by:

$$(2) \; waiting := \{\, \theta(q) \mid q \in Q_0 \,\}$$

$$(8) \; waiting := waiting \cup \{\, \theta(q') \mid (q, q') \in \Delta \,\}$$

It can easily be shown that the adjusted algorithm remains correct: for all (finite) transition systems the outcomes of the original and the adjusted algorithm are equal. If the representative function is "good", which means that many equivalent states are projected onto the same representative, then the number of states to explore, and consequently the size of the passed set, may decrease dramatically. However, in order to apply the approach, the following two problems need to be solved:

– A suitable bisimulation equivalence that respects $\phi$ needs to be statically derived from the system description.
– An appropriate representative function $\theta$ needs to be constructed that satisfies formula (1). Ideally, $\theta$ satisfies $q \approx q' \Rightarrow \theta(q) = \theta(q')$, in which case it is called *canonical*.

In this paper, we use symmetries to solve these problems. As in [19], the notion of *automorphism* is used to characterize symmetry within a transition system. This is a bijection on the set of states that (viewed as a relation) is a bisimulation. Phrased alternatively:

**Definition 2 (Automorphism).** *An* automorphism *on a transition system* $(Q, Q_0, \Delta)$ *is a bijection* $h : Q \rightarrow Q$ *such that*

1. $q \in Q_0$ *if and only if* $h(q) \in Q_0$ *for all* $q \in Q$, *and*
2. $(q, q') \in \Delta$ *if and only if* $(h(q), h(q')) \in \Delta$ *for all* $q, q' \in Q$.

Let $H$ be a set of automorphisms, let **id** be the identity function on states, and let $G(H)$ be the closure of $H \cup \{$**id**$\}$ under inverse and composition. It can be shown that $G(H)$ is a group, and it induces a bisimulation equivalence relation $\approx$ on the set of states as follows:

$$q \approx q' \iff \exists_{h \in G(H)} (h(q) = q') \tag{2}$$

We introduce a symmetric data type to let the user explicitly point out the symmetries in the model. Simple static checks can ensure that the symmetry that is pointed out is not broken. Our approach to the second problem of coming up with good representative functions consists of "sorting the state" w.r.t. some ordering relation on states using the automorphisms. For instance, given a state $q$ and a set of automorphisms, find the smallest state $q'$ that can be obtained by repeatedly applying automorphisms and their inverses to $q$. It is clear that such a $\theta$ satisfies the correctness formula (1), since it is constructed from the automorphisms only.

## 3   Adding Scalarsets to UPPAAL

The tool UPPAAL is a model checker for networks of timed automata extended with discrete variables (bounded integers, arrays) and blocking, binary synchronization as well as non-blocking broadcast communication (see for instance [21]). In the remainder of this section we illustrate by an example UPPAAL's description language extended with a *scalarset* type constructor allowing symmetric data types to be syntactically indicated. Our extension is based on the notion of scalarset first introduced by Ip and Dill in the finite-state model checking tool MURϕ [10, 19]. Also our extension is based on the C-like syntax to be introduced in the forthcoming version 4.0 of UPPAAL.

To illustrate our symmetry extension of UPPAAL we consider Fischer's mutual exclusion protocol. This protocol consists of $n$ process identical up to their unique process identifiers. The purpose of the protocol is to insure mutual exclusion on the critical sections of the processes. This is accomplished by letting each process write its identifier (**pid**) in a global variable (**id**) before entering its critical section. If after some given lower time bound (say 2) **id** still contains the **pid** of the process, then it may enter its critical section.

A scalarset of size $n$ may be considered as the subrange $\{0, 1, \ldots, n - 1\}$ of the natural numbers. Thus, the $n$ process identifiers in the protocol can be modeled using a scalarset with size $n$. In addition to the global variable **id**, we use the array **active** to keep track of all active locations of the processes[1]. Global declarations are the following:

---

[1] This array is actually redundant and not present in the standard formulations of the protocol. However, it is useful for showing important aspects of our extension.

process Fischer (const proc_id pid)

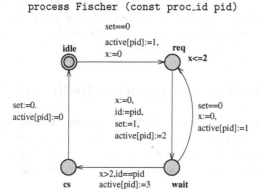

**Fig. 2.** The template for Fischer's protocol

```
typedef scalarset[3] proc_id;    // a scalarset type with size 3
proc_id id;                      // declaration of a proc_id
                                 //   variable
bool set;                        // declaration of a boolean
int active[proc_id];             // declaration of an array
                                 //   indexed by proc_id
```

The first line defines proc_id to be a scalarset type of size 3, and the second line declares id to be a variable over this type. Thus scalarset is in our extension viewed as a type constructor. In the last line we show a declaration of an array indexed by elements of the scalarset proc_id.

At this point the only thing missing is the declaration of the actual processes in the system. In the description language of UPPAAL, processes are obtained as instances of parameterized process templates. In general, templates may contain several different parameters (e.g. bounded integers, clocks, and channels). In our extension we allow in addition the use of scalarsets as parameters. In the case of Fischer's protocol the processes of the system are given as instances of the template depicted in Figure 2. The template has one local clock, x, and no local variables. Note that the header of the template defines a (constant) scalarset parameter pid of type proc_id. Access to the critical section cs is governed by suitable updates and tests of the global scalarset variable id together with upper and lower bound time constraints on when to proceed from requesting access (req) respectively proceed from waiting for access (wait). Note that all transitions update the array active to reflect the current active location of the process. The instantiation of this template and declaration of all three process in the system can be done as follows:

```
FischerProcs = forall i in proc_id : Fischer(i);
system FischerProcs;
```

The `forall` construct iterates over all elements of a declared scalarset type. In this case the iteration is over `proc_id` and a set of instances of the template `Fischer` is constructed and bound to `FischerProcs`. In the second line the final system is defined to be precisely this set.

# 4   Using Scalarsets for Symmetry Reduction

As a preliminary to this section we briefly mention the state representation of UPPAAL. A state is a tuple $(l, v, Z)$, where $l$ is the location vector, $v$ is the integer variable valuation, and $Z$ is a zone, which is a convex set of clock valuations that can efficiently be represented by a *difference bounded matrix* (DBM) [5, 9].

## 4.1   Extraction of Automorphisms

This subsection is a very brief summary of [14], to which we refer for further details. The new syntax described in the previous section enables us to derive the following information from a system description:

- A set $\Omega$ of scalarset types.
- For each $\alpha \in \Omega$: (i) a set $V_\alpha$ of variables of type $\alpha$, and (ii) a set $D_\alpha$ of pairs $(a, n)$ where $a$ is an array and $n$ is a dimension of $a$ that must be indexed by variables of type $\alpha$ to ensure soundness. We assume that arrays that are indexed by scalarsets do not contain elements of scalarsets. The reason is that this would make computation of a canonical representative as hard as testing for graph isomorphism.
- A partial mapping $\gamma : P \times \Omega \hookrightarrow \mathbb{N}$ that gives for each process $p$ and scalarset $\alpha$ the element of $\alpha$ with which $p$ is instantiated. This mapping is defined by quantification over scalarsets in the process definition section.

This information enables us to derive so-called *state swaps*. Let $Q$ be the set of states of some UPPAAL model, and let $\alpha$ be a scalarset type in the model with size $n$. A state swap $swap_{i,j}^\alpha : Q \to Q$ can be defined for all $0 \le i < j < n$, and consists of two parts:

- The *multiple process swap* swaps the contributions to the state of all pairs of processes $p$ and $p'$ if they originate from the same template and $\gamma(p, \alpha) = i$, $\gamma(p', \alpha) = j$ and $\gamma(p, \beta) = \gamma(p', \beta)$ for all $\beta \ne \alpha \in \Omega$. Swapping such a pair of symmetric processes consists of interchanging the active locations and the values of the local variables and clocks (note that this is not a problem since the processes originate from the same template).
- The *data swap* swaps array entries $i$ and $j$ of all dimensions that are indexed by scalarset $\alpha$ (these are given by the set $D_\alpha$). Moreover, it swaps the value $i$ with the value $j$ for all variables in $V_\alpha$.

Consider the instance of Fischer's mutual exclusion protocol (as described in the previous section) with three processes. There are three swap functions:

$swap_{0,1}^{\text{proc\_id}}$, $swap_{0,2}^{\text{proc\_id}}$ and $swap_{1,2}^{\text{proc\_id}}$. Now consider the following state of the model (the active location of the $i$-th process is given by $l_i$ and the local clock of this process is given by $x_i$):

$$l \qquad : l_0 = \texttt{idle},\ l_1 = \texttt{wait},\ l_2 = \texttt{cs}$$
$$v \qquad : \texttt{id} = 2,\ \texttt{set} = 1$$
$$Z \qquad : x_0 = 4,\ x_1 = 3,\ x_2 = 2.5$$
$$\texttt{active}\,;\,\texttt{active}[0] = 0,\ \texttt{active}[1] = 2,\ \texttt{active}[2] = 3$$

When we apply $swap_{0,2}^{\text{proc\_id}}$ to this state, the result is the following state:

$$l \qquad : l_0 = \texttt{cs},\ l_1 = \texttt{wait},\ l_2 = \texttt{idle}$$
$$v \qquad : \texttt{id} = 0,\ \texttt{set} = 1$$
$$Z \qquad : x_0 = 2.5,\ x_1 = 3,\ x_2 = 4$$
$$\texttt{active}\,;\,\texttt{active}[0] = 3,\ \texttt{active}[1] = 2,\ \texttt{active}[2] = 0$$

The process swap swaps $l_0$ with $l_2$, and $x_0$ with $x_2$. The data swap first changes the value of the variable id from 2 to 0, since $\texttt{id} \in V_{\text{proc\_id}}$, and then swaps the values of $\texttt{active}[0]$ and $\texttt{active}[2]$. Applying $swap_{1,2}^{\text{proc\_id}}$ to this state gives the following state:

$$l \qquad : l_0 = \texttt{cs},\ l_1 = \texttt{idle},\ l_2 = \texttt{wait}$$
$$v \qquad : \texttt{id} = 0,\ \texttt{set} = 1$$
$$Z \qquad : x_0 = 2.5,\ x_1 = 4,\ x_2 = 3$$
$$\texttt{active}\,;\,\texttt{active}[0] = 3,\ \texttt{active}[1] = 0,\ \texttt{active}[2] = 2$$

Note that this swap does not change the value of id, since the scalarset elements 1 and 2 are interchanged and id contains scalarset element 0.

A number of syntactic checks have been identified that ensure that the symmetry suggested by the scalarsets is not broken. These checks are very similar to those originally identified for the MUR$\varphi$ verification system [19]. For instance, it is not allowed to use variables of a scalarset type for arithmetical operations such as addition. The next soundness theorem has been proven in [14]:

**Theorem 1 (Soundness).** *Every state swap is an automorphism.*

As a result, the representative function $\theta$ can be implemented by minimization of the state using the state swaps. Note that every state swap resembles a transposition of the state. Hence, the equivalence classes induced by the state swaps originating from a scalarset with size $n$ consist of at most $n!$ states. The maximal theoretical gain that can be achieved using this set of automorphisms is therefore in the order of a factor $n!$.

### 4.2   Computation of Representatives

The representative of a state is defined as the minimal element of the symmetry class of that state w.r.t. a total order $\prec$ on the symmetry class. In general, the DBM representation of zones renders an efficient *canonical* minimization algorithm impossible, since minimization of a general DBM for any given total order using state swaps is at least as difficult as testing for graph isomorphism for strongly regular graphs [14]. If we assume, however, that the timed automaton that is analyzed resets its clocks to zero only, then the zones (DBMs) that are generated by the forward state space exploration satisfy the nice *diagonal property*. This property informally means that the individual clocks can always be ordered using the order in which they were reset. To formalize this, three binary relations on the set of clocks parameterized by a zone $Z$ are defined:

$$x \preccurlyeq_Z y \iff \forall_{\nu \in Z}\, \nu(x) \le \nu(y) \tag{3}$$

$$x \approx_Z y \iff \forall_{\nu \in Z}\, \nu(x) = \nu(y) \tag{4}$$

$$x \prec_Z y \iff (x \preccurlyeq_Z y \land \neg(x \approx_Z y)) \tag{5}$$

The diagonal property is then defined as follows.

**Lemma 1 (Diagonal Property).** *Consider the state space exploration algorithm described in figure 6 of [21]. Assume that the clocks are reset to the value 0 only. For all states $(l, v, Z)$ stored in the waiting and passed list and for all clocks $x$ and $y$ holds that either $x \prec_Z y$, or $x \approx_Z y$ or $y \prec_Z x$.*

Using the reset order on clocks and the diagonal property, we can define a total order, say $\prec$, on all states within a symmetry class whose minimal element can be computed efficiently. To this end we first assume a fixed indexing of the set of clocks $X$: a bijection $\rho : X \to \{1, 2, \ldots, |X|\}$. Now note that $\approx_Z$ is an equivalence relation that partitions $X$ in $P = \{X_1, X_2, \ldots, X_n\}$. We define a relation on the cells of $P$ as follows:

$$X_i \le X_j \iff (\forall_{x \in X_i, y \in X_j}\, x \preceq_Z y) \tag{6}$$

Clearly this is a total order on $P$. Let $X_i$ be a cell of $P$. The *code* of $X_i$, denoted by $C^*(X_i)$, then is the lexicographically sorted sequence of the indices of the clocks in $X_i$ (the set $\{\rho(x) \mid x \in X_i\}$). The *zone code* of the zone which induced $P$ is then defined as follows.

**Definition 3 (Zone Code).** *Let $Z$ be a zone and let $P = \{X_1, X_2, \ldots, X_n\}$ be the partitioning of the set of clocks $X$ under $\approx_Z$ such that $i \le j \Rightarrow X_i \le X_j$ (we can assume this since $\le$ is a total order on $P$). The* zone code *of $Z$, denoted by $C(Z)$, is the sequence $(C^*(X_1), C^*(X_2), \ldots, C^*(X_n))$.*

Note that every zone has exactly one zone code since the indices of equivalent clocks are sorted. Moreover, zone codes can lexicographically be ordered, since they are sequences of number sequences. This order is then used in the following way to define a total order on the states in a symmetry class (the orders on the location vectors and variable valuations are just the lexicographical order on sequences of numbers):

```
(1)  for all α ∈ Ω do
(2)          for i = 1 to |α| do
(3)                  for j = 1 to |α| − i do
(4)                          if swap^α_{j−1,j}(q) ≺ q then
(5)                                  q := swap^α_{j−1,j}(q)
(6)                  od
(7)          od
(8)  od
```

**Fig. 3.** Minimization of state $q$ using the bubble-sort algorithm. The size of scalarset type $\alpha$ is denoted by $|\alpha|$

$$(l, v, Z) \prec (l', v', Z')$$
$$\Longleftrightarrow \qquad (7)$$
$$(l < l') \vee (l = l' \wedge v < v') \vee (l = l' \wedge v = v' \wedge \mathcal{C}(Z) < \mathcal{C}(Z'))$$

We minimize the state w.r.t. the order of equation (7) using the state swaps by applying the bubble-sort algorithm to it, see Figure 3. It is clear that this representative computation satisfies the soundness equation (1), since states are transformed using the state swaps only, which are automorphisms by Theorem 1. We note that $swap^\alpha_{j−1,j}(q)$ is not computed explicitly for the comparison in the fourth line of the algorithm; using the statically derived $\gamma$, $D_\alpha$ and $V_\alpha$ (see section 4.1) we are able to tell whether swapping results in a smaller state.

The following theorem states the main technical contribution of our work. Informally, it means that the detected symmetries are optimally used.

**Theorem 2 (Canonical Representative).** *The algorithm in Figure 3 computes a canonical representative.*

Note that we assumed that arrays that are indexed by scalarsets do not contain elements of scalarsets. Otherwise, computation of a canonical representative is as hard as graph isomorphism, but this is entirely due to the discrete part of the model, and not to the clock part.

## 5   Experimental Results

This section presents and discusses experimental data that was obtained by the UPPAAL prototype on a dual Athlon 2000+ machine with 3 GB of RAM. The measurements were done using the tool `memtime`, for which a link can be found at the UPPAAL website `http://www.uppaal.com/`.

In order to demonstrate the effectiveness of symmetry reduction, the resource requirements for checking the correctness of Fischer's mutual exclusion protocol were measured as a function of the number of processes for both regular UPPAAL and the prototype, see Figure 4. A conservative extrapolation of the data shows that the verification of the protocol for 20 processes without symmetry reduction would take 115 days and 1000 GB of memory, whereas this verification can be

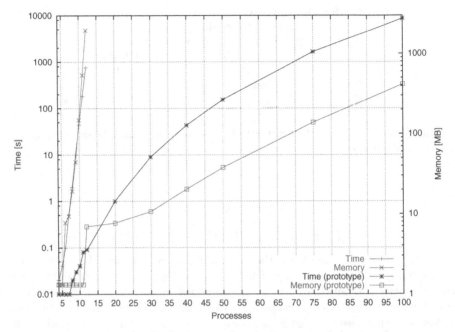

**Fig. 4.** Run-time data for Fischer's mutual exclusion protocol showing the enormous gain of symmetry reduction. The step in the graph of the memory usage is probably due to the the fact that UPPAAL allocates memory in chunks of a few megabyte at a time

done within approximately one second using less than 10 MB of memory with symmetry reduction.

Similar results have been obtained for the CSMA/CD protocol ([24, 27]) and for the timeout task of a distributed agreement algorithm[2] [4]. To be more precise, regular UPPAAL's limit for the CSMA/CD protocol is approximately 10 processes, while the prototype can easily handle 50 processes. Similarly, the prototype can easily handle 30 processes for the model of the timeout task, whereas regular UPPAAL can only handle 6.

Besides the three models discussed above, we also investigated the gain of symmetry reduction for two more complex models. First, we experimented with the previously mentioned agreement algorithm, of which we are unable to verify an interesting instance even with symmetry reduction due to the size of the state space. Nevertheless, symmetry reduction showed a very significant improvement. Second, we experimented with a model of Bang & Olufsen's audio/video proto-col [13]. The mentioned paper describes how UPPAAL is used to find a bug in the protocol, and it describes the verification of the corrected protocol for two (sym-metric) senders. Naturally, we added another sender – verification of the model

---

[2] Models of the agreement algorithm and its timeout task are available through the URL http://www.cs.kun.nl/~martijnh/

**Table 1.** Comparing the time and memory consumption of the relations for the agreement algorithm and for Bang & Olufsen's audio/video protocol with two and three senders. The exact parameters of the agreement model are the following: $n = 2$, $f = 1$, ones $= 0$, $c_1 = 1$, $c_2 = 2$ and $d$ varied (the value is written between the brackets). Furthermore, the measurements were done for the verification of the agreement invariant only. Three verification runs were measured for each model and the best one w.r.t. time is shown

| Model | Time [s] | | Memory [MB] | |
|---|---|---|---|---|
| | No reduction | Reduction | No reduction | Reduction |
| Agreement (0) | 1 | 3 | 33 | 45 |
| Agreement (1) | 21 | 16 | 294 | 180 |
| Agreement (2) | 80 | 23 | 905 | 245 |
| Agreement (3) | 231 | 32 | 2126 | 321 |
| B&O (2) | 2 | 1 | 16 | 10 |
| B&O (3) | 265 | 36 | 1109 | 181 |

for three senders was impossible at the time of the first verification attempt – and we found another bug, whose source and implications we are investigating at the time of this writing. Table 1 shows run-time data for these models.

# 6   Conclusions

The results we obtained with our prototype are clearly quite promising: with relatively limited changes/extensions of the UPPAAL code we obtain a rather drastic improvement of performance for systems with symmetry that can be expressed using scalarsets.

An obvious next step is to do experiments concerning profiling where computation time is spent, and in particular how much time is spent on computing representatives. In the tool Design/CPN [18, 20, 11] (where symmetry reduction is a main reduction mechanism) there have been interesting prototype experiments with an implementation in which the (expensive) computations of representatives were launched as tasks to be solved in parallel with the main exploration algorithm.

The scalarset approach that we follow in this paper only allows one to express total symmetries. An obvious direction for future research will be to study how other types of symmetry (for instance as we see it in a token ring) can be exploited.

# References

[1] M. Abadi and L. Lamport. An old-fashioned recipe for real time. ACM Transactions on Programming Languages and Systems, 16(5):1543–1571, September 1994. 47

[2]  R. Alur, C. Courcoubetis, and D.L. Dill. Model checking in dense real time. Information and Computation, 104:2–34, 1993. 47

[3]  R. Alur and D.L. Dill. Automata for modeling real-time systems. In 17th International Colloquium on Automata, Languages, and Programming, pages 322–335, 1990. 47

[4]  H. Attiya, C. Dwork, N. Lynch, and L. Stockmeyer. Bounds on the time to reach agreement in the presence of timing uncertainty. Journal of the ACM, 41(1):122–152, 1994. 47, 56

[5]  R. Bellman. Dynamic Programming. Princeton University Press, 1957. 52

[6]  D. Bosnacki, D. Dams, and L. Holenderski. A heuristic for symmetry reductions with scalarsets. In J.N. Oliveira and P. Zave, editors, FME 2001, number 2021 in LNCS, pages 518–533. Springer–Verlag, 2001. 47

[7]  E. M. Clarke, S. Jha, R. Enders, and T. Filkorn. Exploiting symmetry in temporal logic model checking. Formal Methods in System Design, 9(1/2):77–104, 1996. 47

[8]  E.M. Clarke, O. Grumberg, and D.A. Peled. Model Checking. The MIT Press, 2000. 46

[9]  D. Dill. Timing assumptions and verification of finite-state concurrent systems. In J. Sifakis, editor, Proc. of Automatic Verification Methods for Finite State Systems, number 407 in LNCS, pages 197–212. Springer–Verlag, 1989. 52

[10] D. L. Dill, A. J. Drexler, A. J. Hu, and C. Han Yang. Protocol verification as a hardware design aid. In IEEE International Conference on Computer Design: VLSI in Computers and Processors, pages 522–525. IEEE Computer Society, 1992. 46, 47, 50

[11] L. Elgaard. The Symmetry Method for Coloured Petri Nets - Theory, Tools, and Practical Use. PhD thesis, Department of Computing Science, University of Aarhus, Denmark, July 2002. 57

[12] E.A. Emerson and A.P. Sistla. Symmetry and model checking. In CAV '93, number 697 in LNCS. Springer–Verlag, 1993. 47

[13] K. Havelund, A. Skou, K.G. Larsen, and K. Lund. Formal modelling and analysis of an audio/video protocol: An industrial case study using UPPAAL. In 18th IEEE Real-Time Systems Symposium, pages 2–13, 1997. 47, 56

[14] M. Hendriks. Enhancing UPPAAL by exploiting symmetry. Technical Report NIII-R0208, NIII, University of Nijmegen, October 2002. 47, 48, 52, 53, 54

[15] M. Hendriks, G. Behrmann, K. G. Larsen, P. Niebert, and F. W. Vaandrager. Adding symmetry reduction to UPPAAL. Technical Report NIII-R03xx, NIII, University of Nijmegen, 2003. To appear. 48

[16] T. A. Henzinger, P. Ho, and H. Wong-Toi. HYTECH: A model checker for hybrid systems. Software Tools for Technology Transfer, 1:110–122, 1997. 46

[17] G. J. Holzmann. The SPIN model checker. IEEE Transactions on Software Engineering, 23(5):279–295, 1997. 46, 47

[18] P. Huber, A. M. Jensen, L. O. Jepsen, and K. Jensen. Reachability trees for high-level petri nets. Theoretical Computer Science, 45(3):261–292, 1986. 47, 57

[19] C.N. Ip and D.L. Dill. Better verification through symmetry. In D. Agnew, L. Claesen, and R. Camposano, editors, Computer Hardware Description Languages and their Applications, pages 87–100, Ottawa, Canada, 1993. Elsevier Science Publishers B.V., Amsterdam, The Netherlands. Journal version appeared in Formal Methods in System Design, 9(1/2):41–75, 1996. 47, 48, 49, 50, 53

[20] K. Jensen. Condensed state spaces for symmetrical Coloured Petri Nets. Formal Methods in System Design, 9(1/2):7–40, 1996. 47, 57

[21] K. G. Larsen, P. Pettersson, and W. Yi. UPPAAL in a nutshell. International Journal on Software Tools for Technology Transfer, pages 134–152, 1998. 46, 47, 50, 54

[22] K. L. McMillan. Symbolic Model Checking. PhD thesis, Carnegie Mellon University, Pittsburgh, May 1992. 46, 47

[23] P.H. Starke. Reachability analysis of petri nets using symmetries. Syst. Anal. Model. Simul./5, 8(4):293–303, 1991. 47

[24] A. S. Tanenbaum. Computer Networks. Prentice–Hall, 1996. 47, 56

[25] F. Wang. Efficient data structure for fully symbolic verification of real-time software systems. In S. Graf and M. Schwartzbach, editors, TACAS '00, number 1785 in LNCS, pages 157–171. Springer–Verlag, 2000. 46, 47

[26] F. Wang and K. Schmidt. Symmetric symbolic safety-analysis of concurrent software with pointer data structures. In D.A. Peled and M.Y. Vardi, editors, FORTE '02, number 2529 in LNCS, pages 50–64. Springer–Verlag, 2002. 47

[27] S. Yovine. KRONOS: a verification tool for real-time systems. International Journal on Software Tools for Technology Transfer, 1(2), 1997. 46, 47, 56

# TIMES: A Tool for Schedulability Analysis and Code Generation of Real-Time Systems

Tobias Amnell, Elena Fersman, Leonid Mokrushin,
Paul Pettersson, and Wang Yi[*]

Department of Information Technology
Uppsala University, P.O. Box 337, SE-751 05 Uppsala, Sweden
{tobiasa,elenaf,leom,paupet,yi}@it.uu.se

**Abstract.** TIMES is a tool suite designed mainly for symbolic schedulability analysis and synthesis of executable code with predictable behaviours for real-time systems. Given a system design model consisting of (1) a set of application tasks whose executions may be required to meet mixed timing, precedence, and resource constraints, (2) a network of timed automata describing the task arrival patterns and (3) a preemptive or non-preemptive scheduling policy, TIMES will generate a scheduler, and calculate the worst case response times for the tasks. The design model may be further validated using a model checker e.g. UPPAAL and then compiled to executable C-code using the TIMES compiler. In this paper, we present the design and main features of TIMES including a summary of theoretical results behind the tool. TIMES can be downloaded at www.timestool.com.

## 1 Introduction

In classic scheduling theory, real time tasks (processes) are usually assumed to be periodic, i.e. tasks arrive (and will be computed) with fixed rates periodically. Analysis based on such a model of computation often yields pessimistic results. To relax the stringent constraints on task arrival times, we have proposed to use automata with timing constraints to model task arrival patterns [1]. This yields a generic task model for real time systems. The model is expressive enough to describe concurrency and synchronization, and real time tasks which may be periodic, sporadic, preemptive or non-preemptive, as well as precedence and resource constraints. We believe that the model may serve as a bridge between scheduling theory and automata-theoretic approaches to system modeling and analysis. The standard notion of schedulability is naturally generalized to automata. An automaton is schedulable if there exists a scheduling strategy such that all possible sequences of events accepted by the automaton are schedulable in the sense that all associated tasks can be computed within their deadlines. It has been shown that the schedulability checking problem for such models is decidable [1]. A recent work [6] shows that for fixed priority scheduling strategy,

---

[*] Corresponding author.

K.G. Larsen and P. Niebert (Eds.): FORMATS 2003, LNCS 2791, pp. 60–72, 2004.

the problem can be efficiently solved by reachability analysis on timed automata using only 2 extra clock variables. The analysis can be done in a similar manner to response time analysis in classic Rate-Monotonic Scheduling.

The first main function of TIMES is developed based on these recent results on schedulability analysis. Its second main function is code generation. Code generation is to transform a validated design model to executable code whose execution preserves the behaviour of the model. Given a system design model in TIMES including a set of application tasks, task constraints, tasks arrival patterns and a scheduling policy adopted on the target platform, TIMES will generate a scheduler and calculate the worst-case response times for all tasks. The model may be further validated by a model-checker e.g. UPPAAL [9], and then compiled to executable C-code. We assume that the generated code will be executed on a platform on which every annotated task in the design model will not take more than the given computing time. Further assume that the platform guarantees the synchronous hypothesis in the sense that the times for handling system functions e.g. collecting external events can be ignored compared with the computing times and deadlines for the annotated tasks. Under these assumptions on the platform, code generation is essentially to resolve non-determinism in the design model. In TIMES, time non-determinism is resolved by the maximal progress assumption, that is, whenever a transition is enabled, it should be taken. External non-determinism in accepting events is resolved using priority order.

The rest of the paper is organized as follows: the next section describes the core of the input TIMES language and its informal semantics. Section 3 summarizes briefly the main theoretical work on schedulability analysis and code synthesis. Section 4 describes the main features of TIMES, the tool architecture and the main components in the implementation. Section 5 concludes the paper with a summary of ongoing work and future development.

## 2   Task Models in TIMES

The two central concepts in TIMES are *task* and *task model*. A task (or task type) is an executable program (e.g. in C) with task parameters: worst case execution time and deadline. A task may have different *task instances* that are copies of the same program with different inputs. A task model is a task arrival pattern such as periodic and sporadic tasks. In TIMES, timed automata are used to describe task arrival patterns.

### 2.1   Tasks Parameters and Constraints

Following the literature [4], we consider three types of task constraints.

**Timing Constraints.** A typical timing constraint on a task is deadline, i.e. the time point before which the task should complete its execution. We assume that the *worst case execution times* (WCET) of tasks are known (or pre-specified). We characterize a task as a pair of natural numbers denoted $(C, D)$ with $C \leq D$,

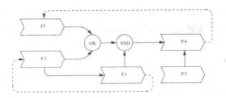

**Fig. 1.** Example of cyclic AND/OR precedence graph

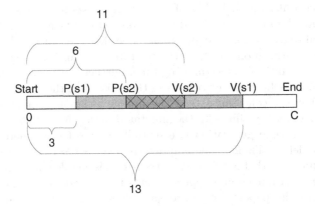

**Fig. 2.** An example semaphore access pattern

where $C$ is the WCET of $P$, $D$ is the relative deadline for $P$. In general, the execution time of a task can be an interval $[C_B, C_W]$ where $C_B$ and $C_W$ are the best and worst case execution times. The deadline $D$ is a relative deadline meaning that when task $P$ is released, it should finish within $D$ time units.

**Precedence Constraints.** The execution of a task set may have to respect some precedence relations. These relations are usually described through a precedence graph in which nodes represent tasks and edges represent precedence relation. In TIMES, we use cyclic AND/OR-precedence graphs in which we distinguish *ordinary* and *inter-iterative* edges (denoted $--\rightarrow$) [3] such that inter-iterative precedence constraints apply to all task instances except for the first one. An example of such graph is shown in Figure 1.

According to the graph, $P_4$ can start its execution only if it is preceded by $P_3$ and either $P_1$ or $P_2$. The first instance of task $P_1$ can start its execution at any time while any further instance of task $P_1$ must be preceded by task $P_4$.

**Resource Constraints.** Tasks may share resources or data variables protected by semaphores. A task must follow its given semaphore access pattern to lock and unlock semaphores, which is the resource constraint on the task. The access to semaphores will be scheduled using priority ceiling protocols e.g. the highest locker protocol [10]. A semaphore access pattern for a task is a list of timed semaphore-operations in the form: $\{S_i(P_i, V_i)\}$ where $S_i$ is the semaphore name, $P_i$ is the accumulated execution time needed for the task to reach the lock-operation on $S_i$ and $V_i$ is the accumulated execution time needed for the task to reach the unlock-operation on $S_i$. The blocking time for $S_i$ is $V_i - P_i$. An example semaphore access pattern $\{S_1(3, 13)\}; S_2(6, 11)\}$ of a task is illustrated in Figure 2. The task will try to lock $S_1$ when it has been executed for 3 time units and it will lock it for 10 time units.

## 2.2   Timed Automata as Task Arrival Patterns

The core of the TIMES input language is timed automata extended with data variables [9] and tasks [5] and [7]. As in the UPPAAL model, each edge of such an extended automaton is labeled with three labels:

1. A *guard* containing a clock constraint and/or a predicate on data variables.
2. An *action* which can be an input or output action in the form of $a!$ and $a?$.
3. A *sequence* of assignments in the form: $x := 0$ when $x$ is a clock or $v := E$ when $v$ is a data variable, where $E$ is a mathematical expression over data variables and constants.

A location of an extended automaton may be annotated with a task or a set of tasks that will be triggered when the transition leading to the location is taken. The triggered tasks will be put in a task queue (i.e. ready queue in operating system) and scheduled to run according to a given scheduling policy. The scheduler should make sure that all the task constraints are satisfied in scheduling the tasks in the task queue. To model concurrency and synchronisation between automata, networks of automata are constructed in the standard way as in e.g. UPPAAL with the annotated sets of tasks on locations unioned.

## 2.3   Shared Data Variables

Four types of shared data variables can be used for communication and resource sharing:

1. Tasks may have shared variables with each others, protected by semaphores.
2. Tasks may read and update variables owned by the automata.
3. Automata can read (but not update) variables owned by the tasks.
4. Automata may have shared variables with each other.

# 3   Analysis and Synthesis

In TIMES, a timed automaton annotated with tasks (or network of such automata) is considered as a design model. The tool offers two main functions: schedulability analysis of design models and generation of executable code from the models.

## 3.1   Schedulability Analysis

In [7], an operational semantics for timed automata extended with tasks is developed. A semantic state of such an automaton is a triple $(l, u, q)$ where $l$ is the current control location, $u$ denotes the current values of clocks and data variables, and $q$ is the current task queue keeping all the released tasks to be executed. The semantics of an automaton is defined by a transition system in which the transition rules are parameterized by a scheduling policy to schedule the task queue when new tasks are released.

Given an extended automaton and a scheduling policy, the related schedulability analysis problem is to check whether there exists a reachable state $(l, u, q)$ of the automaton where the task queue $q$ contains a task which misses its given deadline. Such states are called *non-schedulable* states. An automaton is said to be non-schedulable with the given scheduling policy if it may reach a non-schedulable state. Otherwise the automaton is schedulable. As the number of reachable states of an extended automaton is infinite, it is not obvious that the schedulability analysis problem is decidable.

The first decidability result is presented at TACAS 2002 showing that the schedulability checking problem for the optimal scheduling policy i.e. EDF can be solved by reachability analysis on timed automata extended with subtraction on clocks. Consider an automaton $A$ and a scheduling strategy Sch. To check if $A$ is schedulable with Sch, we construct timed automata $E(\mathsf{Sch})$ (the scheduler), and $E(A)$ (the task arrival pattern), and check the reachability of a predefined error state in the product automaton of the two. If the error state is reachable, automaton $A$ is not schedulable with Sch.

The maximal number of clock variables needed in constructing the scheduler automaton is $2n$ where $n$ is the total number of schedulable task instances $\sum_{i \in \mathcal{P}} \lceil D_i / C_i \rceil$ where $\mathcal{P}$ is the set of task types, and $C_i, D_i$ are the computing time and deadline for each task type $i$.

To construct $E(A)$, the automaton $A$ is annotated with distinct synchronization actions release$_i$ on all edges leading to locations labeled with the task name $P_i$ (assume that only one task is annotated). The actions will allow the scheduler to observe when a task is released by $A$ for execution. The structure of $E(\mathsf{Sch})$ is shown in Figure 3.

The main idea is to keep track of the task queue, denoted by $q$ on each step of the reachability analysis. Therefore in the encoding $E(\mathsf{Sch})$ there is a transition with the guard $nonschedulable(q)$ from every location where the queue is not empty (i.e. from all locations except Idle) to the error state. In the encoding, the task queue $q$ is represented as a vector containing pairs of clocks $(c_i, d_i)$ for every released task instance, called execution time and deadline clock respectively. The intuitive interpretation of the locations in $E(\mathsf{Sch})$ is as follows:

- Idle - the task queue is empty,
- Arrived($P_i$) - the task instance $P_i$ has arrived,
- Run($P_j$) - the task instance $P_j$ is running,
- Finished - a task instance has finished,
- Error - the task queue is non-schedulable.

Locations Arrived($P_i$) and Finished are marked as committed, which means that they are being left directly after entering.

We use the predicate nonschedulable($q$) to denote the situation when the task queue becomes non-schedulable and naturally there is a transition labeled with the predicate leading to the error-state. The predicate is encoded as follows: $\exists P_i \in q$ such that $d_i > D_i$.

We use Sch in the encoding as a name holder for a scheduling policy to sort the tasks queue. A given scheduling policy is represented by the predicate: $P_i =$ Hd(Sch($q$)). For example, Sch can be:

- Highest priority first (FPS): $P_i \in q, \forall P_k \in q$ Pri($P_i$) $\leq$ Pri($P_k$) where Pri(P) denotes the fixed priority of $P$.
- First come first served (FCFS): $P_i \in q, \forall P_k \in q\ d_i \geq d_k$
- Earliest deadline first (EDF): $P_i \in q, \forall P_k \in q\ D_i - d_i \leq D_k - d_k$
- Least laxity first (LLF): $P_i \in q, \forall P_k \in q\ c_i - d_i + D_i - C_i \leq c_k - d_k + D_k - C_k$

For more detailed description of the automaton $E$(Sch), see [7].

**Variant Execution Times.** The analysis for tasks with constant execution times can be extended to deal with interval execution times: $[C_{iB}, C_{iW}]$ for each task $P_i$ (the best case and worst case execution times). The idea is to modify the scheduler automaton as shown in Figure 4. We use $c_i$ to keep track of the lower bound of the accumulated execution time for $P_i$, and $w_i$ to denote the accumulated difference between best and worst completion time of $P_i$. Obviously $w_i$ should be set to $C_{iW} - C_{iB}$ in the beginning of task execution. Observe that each preemption will enlarge the difference for the preempted task with lower priority by the difference for the finishing task with higher priority. Accordingly, we modify the scheduler automaton as follows: The guard on edge from location Run($P_j$) to Finished should be $C_{jB} \leq c_j \leq C_{jB} + w_j$ and variable updating should be $c_k := c_k - C_{jB}, w_k := w_k + w_j$ for all $k$ such that preempted($P_k$). The rest of the scheduler automaton remains the same as before.

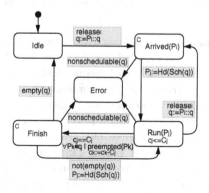

**Fig. 3.** Scheduler automaton

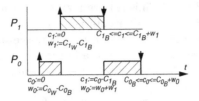

**Fig. 4.** Varying execution times

**Fixed Priority Scheduling Policy.** In a recent work [6], it is shown that the schedulability problem for Fixed Priority Scheduling Policy can be solved efficiently using ordinary timed automata with only two clock variables (in addition to the original clocks used to describe task arrivals). For models with shared data variables (e.g. data dependent control when the values of data variables of a task may influence the release time of task instances), the number of clocks needed in the analysis is $n + 1$ where $n$ is the number of tasks involved in the data sharing. More recently these results are extended to handle precedence and resource constraints [8] and implemented in TIMES.

## 3.2   Code Generation

The second main function of the tool is code generation. We consider automata extended with tasks as design models. Code generation is to transform a validated design model to executable code whose execution preserves the behaviour of the model. We assume that the generated code will be executed on a platform on which every annotated task in the design model will not take more than the given computing time. Further assume that the platform guarantees the synchronous hypothesis in the sense that the times for handling system functions e.g. collecting external events can be ignored compared with the computing times and deadlines for the annotated tasks. Under these assumptions on the platform, code generation is essentially to resolve non-determinism in the design model.

**Deterministic Semantics.** A model can exhibit two types of non-determinism: *time non-determinism*, i.e. that enabled transition can be taken at any time point within the time-zone, and *external non-determinism* i.e. that several actions may be simultaneously present from the environment. To overcome the problems introduced by this we adopt a deterministic semantics that define a subset of the behaviour. External non-determinism is resolved by defining priorities for action transitions in the controller. If several transitions are enabled in a state the one with the highest priority is taken. Time non-determinism is resolved by adopting the so-called maximal-progress assumption [11]. Maximal-progress means that the controller should take all enabled transitions until the system stabilises, i.e. no more action transitions are enabled.

**Structure of the Generated Code.** TIMES is currently able to generate code for a small generic operating system (brickOS), and code for platform independent execution. The generated code is in C and an optimising compiler is used to compile the final program. For both cases, the control structure of the timed automata is encoded into four tables and two functions. These are used by an event handling procedure which is invoked on events (such as timeouts and arrival of external events) to update the state of the controller. When an action transition has been executed the event handling procedure will continue to execute transitions until a stable state is reached, i.e. it implements the maximal progress or run-to-completion semantics.

**Code Generation for** brickOS. brickOS is a small open source operating system designed to run on the Hitachi H8 equipped RCX control brick in the LEGO®Mindstorms system. We consider brickOS to be a reasonable example of a target platform running a small operating system. On this target we let the tasks execute as separate threads which are scheduled by the underlying operating system. Due to limited support for interrupts the event handling procedure is executed every time the OS scheduler is executed (i.e. every 20 ms).

**Platform-Independent Code Generation.** The platform independent target does not rely on any specific operating system, instead it implements its own run-time system based on the scheduler automaton created for schedulability analysis. The run-time system also includes code to handle task release and execution, and an event handler that is invoked periodically to poll for new events. The current implementation of the platform independent code can only handle non-preemptive tasks.

## 4  Tool Overview

In this section, we present the main features of TIMES, the tool architecture and the main components in the implementation.

### 4.1  Features

Figure 5 illustrates a design process using TIMES. As shown in the use case, TIMES offers the following main features:

- **Editor** (see Figure 6) to graphically model a system and the abstract behaviour of its environment. A system description consists of a task set and a network of timed automata extended with the tasks.
  A task is described by the task code (in C), its (worst-case) computation time and (relative) deadline, and if applicable optional parameters for priority (for fixed priority scheduling), period (for periodic tasks), and minimal inter-arrival time (for sporadic tasks).

It is also possible to specify precedence constraints on the tasks using an editor for AND/OR precedence graphs, and resource access patterns using semaphores.

– **Simulator** (see Figure 7) to visualise the dynamic behaviour of a system model as Gantt charts and message sequence charts. The simulator can be used to randomly generate possible execution traces, or alternatively the user can control the execution by selecting the transitions to be taken. The simulator can also be used to visualise error traces produced in the analysis phase.

– **Analyser** to check that the tasks associated to a system model are guaranteed to always meet their deadline. In case schedulability analysis finds a task that may fail to meet its deadline, a trace is generated and visualised in the simulator. It is also possible to compute the worst-case response times of individual tasks. Recently, an improved schedulability analysis algorithm has been developed for tasks with fixed priorities without dependencies [6]. The schedulabilty analysis has also been extended to handle resource and precedence constraints [8]. In addition to scheduling, it is possible analyse safety and liveness properties specified as temporal logic formulae.

– **Compiler** to generate executable C code from timed automata with tasks. The compiler assumes that the target platform ensures the asynchronous hypothesis and that the task code can be executed in the specified computation time. To produce executable code, the compiler relies on a deterministic refinement of the semantics that realise a subset of the behaviour specified in the timed automata of a system model. In this way, the generated code is guaranteed to satisfy analysis results from e.g. schedulabilty analysis when executed on the target platform. The currently implemented compiler sup-

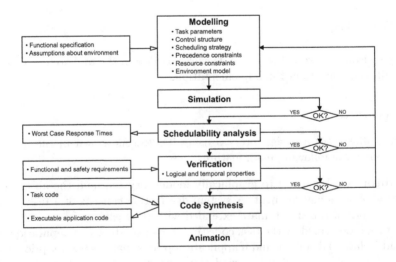

**Fig. 5.** The design process using TIMES

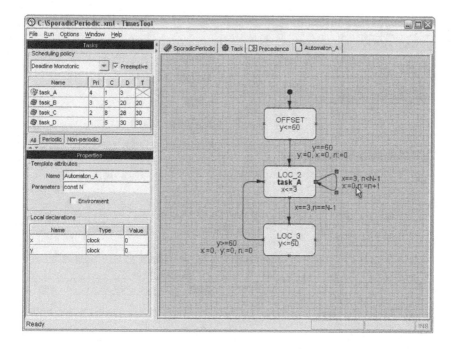

**Fig. 6.** The TIMES editor

ports code generation for: the brickOS operating system (that uses the scheduler in the brickOS runtime system), platform independent code (C code for GNU gcc, including code for a scheduling policy), and code for the Animator of TIMES.

- **Animator** to transform hybrid automata modeling the controlled environment into C code simulating the controlled objects in the environment of the embedded system. The simulated environment enables the designer to experiment with the design prior to implementation.

## 4.2 Implementation

The architecture of the TIMES tool is illustrated in Figure 8. Logically it is divided in three main parts:

- **Graphical User Interface** consisting of editors, simulator, analyser, and animator, as described above. The graphical user interface is implemented entirely in Java and uses XML to represent the system descriptions both internally and externally (on file).
- **Server** consisting of two parts: a scheduler generator implemented in Java, and a module for schedulabilty analysis based on the Uppaal engine [9] with extensions, like the rest of the Uppaal engine implemented in C++. The

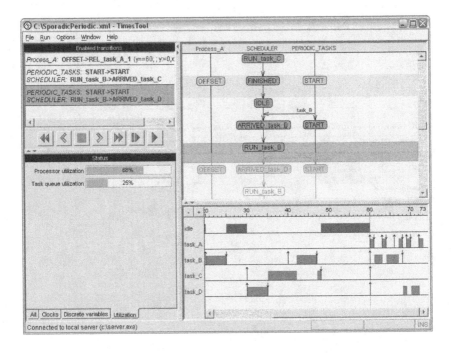

**Fig. 7.** The TIMES simulator

scheduler generator produces a scheduler automaton based on input from the editor, which is composed in parallel with an annotated version of the original system automata. The parallel composition is analysed by on-the-fly reachability techniques in the schedulabilty analysis module. Currently supported scheduling policies are: rate monotonic, deadline monotonic, fixed priority scheduling (with user defined priorities), earliest deadline first (EDF), and first come first served (FCFS). All scheduling policies support preemptive or non-preemptive task sets.

– **Compiler** that takes as input the XML system representation from the editor and the task code segments to produce executable code of the application. The generated code consists of three main parts: a set of C-functions (lookup tables) representing the automata of the system representation, a generic part storing and updating the current state according to the look-up tables, and possibly an implementation of the scheduling strategy (in case platform independent code is produced).

## 5   Applications and Current Development

**Case Studies.** Currently we are in the process of using TIMES to verify reliable message transmission with TTCAN (Timed Triggered CAN). So far, the only

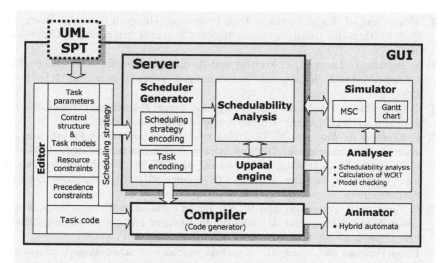

**Fig. 8.** The TIMES tool architecture

non-trivial example using TIMES is the development of the control software of
a production cell (a well-studied case in verification), consisting of an industrial
robot, a press and two transportation belts to process and move metal plates.
The robot controller is designed as a timed automaton annotated with tasks.
A complete description of the case study can be found in [2]. It is a non-trivial
application involving 12 tasks (task types), 7 automata, 17 integers, 24 booleans
and 31 clock variables (7 in the model and 24 in the scheduler). The schedulability (and a number of other requirements) of the system is verified on a machine
equipped with two 1.8 GHz AMD processors and 2 GB of main memory, running Mandrake Linux. TIMES consumes 207 MB of memory and terminates in
11 minutes. Using the option for over approximation (based on the convex-hull
approximation, the analysis requires only 13 MB and 9 seconds on the same
machine.

**UML SPT Profile.** SPT (Scheduling, Performance, and Time) specification
is a UML profile developed recently as an extension of the UML standard to
model time and time-related aspects of embedded systems. An ongoing work has
been initiated with I-Logix to develop TIMES as a plug-in tool for schedulability
analysis of UML diagrams in Rhapsody, annotated with stereotypes, constraints,
and tag definitions according to the UML SPT profile.

# References

[1] T. Amnell, E. Fersman, L. Mokrushin, P. Pettersson, and W. Yi. Times - a tool
for modelling and implementation of embedded systems. In Proc. of TACAS'02,
volume 2280 of LNCS, pages 460–464. Springer, 2002.  60

[2] Tobias Amnell, Elena Fersman, Paul Pettersson, Hongyan Sun, and Wang Yi. Code synthesis for timed automata. Nordic Journal of Computing, 9(4):269–300, 2002. 71

[3] F. Balarin, L. Lavagno, P. Murthy, and A. Sangiovanni-vincentelli. Scheduling for embedded real-time systems. IEEE Design & Test of Computers, 15(1):71–82, 1998. 62

[4] G. C. Buttazzo. Hard Real-Time Computing Systems. Predictable Scheduling Algorithms and Applications. Kulwer Academic Publishers, 1997. 61

[5] C. Ericsson, A. Wall, and W. Yi. Timed automata as task models for event-driven systems. In Proceedings of Nordic Workshop on Programming Theory, 1998. 63

[6] E. Fersman, L. Mokrushin, P. Pettersson, and W. Yi. Schedulability analysis using two clocks. In Proc. of TACAS'03, volume 2619 of LNCS, pages 224–239. Springer, 2003. 60, 66, 68

[7] E. Fersman, P. Pettersson, and W. Yi. Timed automata with asynchronous processes: Schedulability and decidability. In Proc. of TACAS'02, volume 2280 of LNCS, pages 67–82. Springer, 2002. 63, 64, 65

[8] Elena Fersman and Wang Yi. A generic approach to schedulability analysis of real-time tasks. Submitted for publication., 2003. 66, 68

[9] K. G. Larsen, P. Pettersson, and W. Yi. UPPAAL in a Nutshell. Int. Journal on Software Tools for Technology Transfer, 1(1–2):134–152, October 1997. 61, 63, 69

[10] R. Rajkumar, L. Sha, and J. P. Lehoczky. An experimental investigation of synchronisation protocols. In Proceedings 6th IEEE Workshop on Real-Time Operating Systems and Software, pages 11–17. IEEE Computer Society Press, 1998. 63

[11] Wang Yi. A Calculus of Real Time Systems. PhD thesis, Department of Computer Science, Chalmers University of Technology, 1991. 66

# Optimization of Timed Automata Models Using Mixed-Integer Programming

Sebastian Panek, Olaf Stursberg, and Sebastian Engell

Process Control Laboratory (BCI-AST)
University of Dortmund, 44221 Dortmund, Germany
{s.panek,o.stursberg,s.engell}@bci.uni-dortmund.de

**Abstract.** Research on optimization of timed systems, as e.g. for computing optimal schedules of manufacturing processes, has lead to approaches that mainly fall into the following two categories: On one side, mixed integer programming (MIP) techniques have been developed to successfully solve scheduling problems of moderate to medium size. On the other side, reachability algorithms extended by the evaluation of performance criteria have been employed to optimize the behavior of systems modeled as timed automata (TA). While some successful applications to real-world examples have been reported for both approaches, industrial scale problems clearly call for more powerful techniques and tools.

The work presented in this paper aims at combining the two types of approaches: The intention is to take advantage of the simplicity of modeling with timed automata (including modularity and synchronization), but also of the relaxation techniques and heuristics that are known from MIP. As a first step in this direction, the paper describes a translation procedure that automatically generates MIP representations of optimization problems formulated initially for TA. As a possible use of this translation, the paper suggests an iterative solution procedure, that combines a tree search for TA with the MIP solution of subproblems. The key idea is to use the relaxations in the MIP step to guide the tree search for TA in a branch-and-bound fashion.

**Keywords.** Branch-and-Bound Techniques, Discrete Optimization, Mixed-Integer Programming, Scheduling, Timed Automata.

## 1 Introduction

Optimizing the behavior of timed systems is essentially characterized by making decisions of two distinct types: one is to determine that sequence of steps (or actions) that optimizes a given performance criterion, the other is to fix the points of time at which the steps are started (and/or terminated). Often the considered performance criterion either formulates the maximization of the number of steps carried out in a given period of time, or the minimization of overall time (or more general costs) to perform a pre-specified set of steps. An example for the latter is job-shop scheduling which will be considered in this paper.

K.G. Larsen and P. Niebert (Eds.): FORMATS 2003, LNCS 2791, pp. 73–87, 2004.
© Springer-Verlag Berlin Heidelberg 2004

Several different approaches have been developed to solve optimization problems that combine logical decisions (as the sequence of the steps) with time requirements: One is to start from timed formal models, as e.g. Timed Automata (TA), and to search for the path that optimizes the performance criterion within the tree of possible evolutions of the model. The methods described in [1] and [3] are examples that follow this line. These approaches can be seen as an extension of reachability techniques for TA (see e.g. [14, 18, 9]) by a mechanism that selects preferable feasible paths according to a cost criterion.

An alternative is to formulate the sequence of steps and the time information as a system of algebraic (in-)equalities involving binary and continuous variables, and to use these equations as constraints of an optimization problem. The latter approach has been studied extensively by the optimization community in the last decades, including mathematical programming (see e.g., [10, 13, 15, 12, 11, 7]), constraint programming (e.g., [2, 11]), and evolutionary algorithms (e.g., [5]). These methods differ in their efficiency in finding feasible and optimal solutions and in the encoding of constraints. The ability to optimize some real-world examples modeled as timed systems has been demonstrated for these techniques. However, in order to solve industrial-size scheduling problems, the efficiency of available techniques must be improved.

This paper aims at going a step in this direction by combining the benefits of a TA-based approach with mathematical programming. To the authors' opinion the advantages of the earlier are (a) the simplicity of modeling (employing decomposition and synchronization) and (b) the fact that the sequential evolution of TA naturally translates into a search tree (where the depth reflects the number of transitions by which the automaton is evolved). Both points appear to be less favorable in the case that the behavior of a timed system is modelled by algebraic (in-)equalities that serve as constraints in a mixed-integer program. In addition, the formulation of transitions between different steps requires the use of binary (and continuous) auxiliary variables that usually worsen the solution performance. On the other hand, mixed-integer solvers use relaxation techniques (within a branch-and-bound procedure) that were proven to be very successful for many applications. The idea is to initially relax the integrality constraint on the binary variables, i.e., to assume that they can take any values in the interval $[0, 1]$. The solution of an optimization problem with such 'relaxed' variables is then used as a lower bound to cut off branches of the solution tree which are proved to be suboptimal. Values of such relaxed variables can also be used to determine a value assignment for the original binary variables.

To the authors' knowledge, the use of this particular heuristics has not yet been explored for the optimization of TA. Hence, the objective of this paper is to connect the modeling advantage of TA with the relaxation principle of mixed-integer approaches. As a first step in this direction, the paper describes a procedure to transform a class of optimization problems for timed automata into a corresponding mixed-integer programming formulation. In the second part, we show how this transformation can be used within an optimization algorithm that combines tree search with the idea to guide the search by relaxations.

## 2    Scheduling for Timed Automata

In this section we focus on scheduling problems as a special class of optimization problems for timed systems. However, the transformation described in Sec. 3 and the solution algorithm in Sec. 4 extend straightforwardly to other optimization tasks for timed systems. Scheduling problems typically arise in production processes, where specified quantities of different products must be available at given dates and where a limited amount of resources is available for production. The production of a certain product, called a *job*, consists of a set of *tasks*, each of which requires a set of resources for a certain period of time. Each task can consume certain amounts of intermediate products, and produces supplies to other parts of the production chain or final products. The task to be solved is to decide when a certain amount of a particular product should be processed and which resources are used.

Many special cases of this general problem are known and lead to simplified versions, for which special solution algorithms are known. In this paper, the general class of *job-shop scheduling problems* is considered, in which jobs are modeled as different sequences of tasks and are executed on different machines. Such problems are known to be NP-hard, and polynomial algorithms only exist for special cases.

In the following, we first summarize the status of mixed-integer programming approaches to scheduling problems, and then restate a variant of TA known from literature that is suitable to formulate scheduling problems.

### 2.1    Solving Scheduling Problems by Mixed-Integer Programming

In a scheduling problem, there are usually discrete as well as continuous decision variables. In the mathematical programming approach, the structure of the production process, the precedence relations, and the resource consumption of the jobs, as well as technological restrictions are modeled by (in-)equalities involving integer and real variables. Besides the decision variables, usually a large number of auxiliary variables (many of which are again integer variables) are needed. By choosing a linear cost criterion and by applying transformation techniques to nonlinear (in-)equalities [17], many problems can be described by (usually large) sets of linear constraints. The solution of these problems, termed Mixed-Integer Linear Programs (MILPs), means the computation of a set of valuations of the variables that satisfies the constraints and minimizes the cost function. For this task, efficient techniques based upon the relaxation of the integrality constraints, branch-and-bound techniques, cutting-plane methods, etc. exist. For the solution of MILPs, highly efficient commercial solvers, as e.g. CPLEX [6], are available. However, the efficiency of MILP solvers depends very much on the specific problem. While the efficient solution of some problems with several 10000 binary variables has been reported, other problems with just 100 variables can be very hard to solve [10].

## 2.2   Scheduling Problems Modeled by Timed Automata

A version of timed automata (TA) that has been used in the context of scheduling is that of *linearly priced timed automata* (LPTA) [3]. LPTA are TA extended by costs for locations and transitions. We here restate some essentials of the formal definition given in [3] since it is used as the basis of the transformation procedure described in Sec. 3.

**Definition 1.** *A LPTA is a tuple* $(L, l_0, E, I, P)$ *with a finite set $L$ of locations, the initial location $l_0$, the set $E \subset L \times \mathcal{B}(\mathbb{C}) \times Act \times \mathcal{P}(\mathbb{C}) \times L$ of transitions, where $Act$ is a set of actions, $\mathcal{B}(\mathbb{C})$ are constraints over a set $\mathbb{C}$ of clocks (given as conjunctions of atomic formulae $x \sim n$ or $x - y \sim n$ with $x, y \in \mathbb{C}$, $\sim \in \{<, \leq, =, \geq, >\}$, $n \in \mathbb{N}$), and $\mathcal{P}(\mathbb{C})$ is a set of reset assignments; $I : L \to \mathcal{B}(\mathbb{C})$ defines invariants for the locations and $P : (L \cup E) \to \mathbb{N}$ assigns prices to locations as well as transitions. A transition $(l, g, a, r, l') \in E$ between source location $l$ and target location $l'$ is denoted by $l \xrightarrow{g,a,r} l'$.*

For LPTA that are synchronized over their sets of actions, parallel composition is defined as follows:

**Definition 2.** *For two LPTA $A_i = (L_i, l_{i,0}, E_i, I_i, P_i), i = 1, 2$ with action sets $Act_1$ and $Act_2$, the parallel composition is defined as $A_1 \| A_2 = (L_1 \times L_2, (l_{1,0}, l_{2,0}), E, I, P)$ where $l = (l_1, l_2), I(l) = I_1(l_1) \wedge I_2(l_2)$, and the costs assigned to locations are combined according to $P(l) = h_L(P_1(l_1), P_2(l_2))$ with a mapping $h_L : \mathbb{Q} \times \mathbb{Q} \to \mathbb{Q}$. A transition $l \xrightarrow{g,a,r} l'$ exists for $A_1 \| A_2$ iff $g_i$, $a_i$, and $r_i$ exist for $A_i$ such that: $l_i \xrightarrow{g_i, a_i, r_i} l_i'$, $g = g_1 \wedge g_2$, $r = r_1 \cup r_2$, and $Act \subseteq Act_1 \cup \{0\} \times Act_2 \cup \{0\}$, $a := (a_1, a_2) \in Act$ (with a no-action symbol $0$). The costs assigned to transitions follow from $P((l, g, a, r, l')) = h_E(P((l_1, g_1, a_1, r_1, l_1')), P((l_2, g_2, a_2, r_2, l_2')))$ with a function $h_E : \mathbb{Q} \times \mathbb{Q} \to \mathbb{Q}$.*

With respect to the semantics, we refer to the formal definition given in [3]. Informally, an evolution of LPTA is a trace $\alpha$ consisting of a finite sequence of $n$ transitions. Along this sequence, the costs sum up according to $cost(\alpha) = \sum_{i=0}^{n} p_i$, where $p_i$ contains the cost accumulated according to $d \cdot P(l)$ while being in location $l$ for a duration $d$, and the cost contribution $P(l, g, a, r, l')$ assigned to the transition by which $l$ is left. If $(l, u)$ denotes a state of the execution trace with $u$ as a valuation of all clocks, the minimum cost of $(l, u)$ is defined as the minimal costs of all traces that lead to $(l, u)$. The optimization of an LPTA hence corresponds to the search for a trace that ends in $(l, u)$ with minimum costs.

Using this definition of LPTA, job-shop scheduling problems can be formulated easily: A job is defined as a sequence of tasks which must be processed on a limited set of resources. Each job is modeled by a separate LPTA that contains two locations per task, one that represents that the task is waiting for being processed, and one that is modeling that the task is executed on an available resource. In addition, a job automaton contains a final state denoting that the complete set of tasks is finished. A simple example of one job ($J1$) with two tasks that are executed on two resources ($M1$, $M2$) is shown in Fig. 1. While this example contains only timing constraints, the prices introduced in Def. 1 are

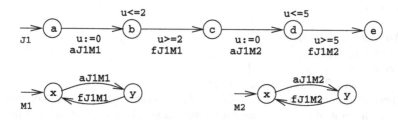

**Fig. 1.** Model for a job with two tasks that are executed on two resources $M1$ and $M2$ for 2 and 5 time units respectively. The clock $u$ is used to model timing constraints. The states of the job automaton are denoted by $a$ to $e$, and the resource states by $x$ and $y$

useful to model that processing a task on different resources leads to different costs. Each resource is modeled as a separate LPTA that contains two locations, one of which represents that the resource is allocated by a task, and one that represents that the resource is available. The transitions of a resource automaton and a task automaton synchronize each time when a task is started and finished.

If the scheduling problem is formulated in this manner, search algorithms like those published in [1, 3] can be applied. If a feasible solution exists, the result is the cost-optimal path into the desired state, which is that all specified jobs have been processed.

## 3    Transformation of TA into Mixed Integer Programs

We now present a mixed-integer linear program (MILP) formulation that is equivalent to a job-shop scheduling problem modeled by a set of task and resource LPTA. The MILP formulation retains the modularity of the automaton model, with communication realized by synchronization.

### 3.1    Model Formulation

The following formulation is structured such that it can be directly implemented in the algebraic modeling language GAMS [4]. The latter has become a standard specification language for mathematical programs and is the input format for various solvers including CPLEX. We first list the index sets, parameters and variables involved in the formulation, and then the (in-) equalities that establish the transition structure, the clock dynamics, and the synchronization of LPTA. Some of these (in-) equalities are based on the disjunctive formulations introduced in [16].

We assume here for simplicity of notation that each automaton has the same number of locations ($n_L$), clocks ($n_C$), transitions ($n_T$), clock constraints ($n_G$), and points of time ($n_K$) at which transitions occur. This assumption does not limit the generality, i.e., the extension to different sets for each automaton is straightforward.

## Index Sets

- Automata: $\mathcal{A} = \{a_1, \ldots, a_{n_A}\}$;
- Clocks: $\mathcal{C} = \{c_1, \ldots, c_{n_C}\}$;
- Locations: $\mathcal{L} = \{l_1, \ldots, l_{n_L}\}$;
- Transitions: $\mathcal{T} = \{t_1, \ldots, t_{n_T}\} \cup \{\tau\}$;
- Discrete points of time: $\mathcal{K} = \{k_1, \ldots, k_{n_K}\}$; each of these points corresponds to an instant of time at which a transition is taken, i.e., a task is started or finished. Since only jobs with a finite number of tasks are considered (and the corresponding job LPTA are acyclic) the set $\mathcal{K}$ is finite;
- Clock constraints: $\mathcal{G} = \{g_1, \ldots, g_{n_G}\}$.

## Constants and Parameters

- State invariant matrices: $A_I \in \mathbb{Q}^{n_A \times n_L \times n_G \times n_C}$ and $b_I \in \mathbb{Q}^{n_A \times n_L \times n_G}$. For a specific automaton $a$ and a location $l$, these matrices model the invariant as a polyhedron $A_I(a, l, \bullet, \bullet)\xi \leq b_I(a, l, \bullet)$, where $\xi \in \mathbb{Q}^{n_C}$ and $\bullet$ represents the dimensions of clocks and invariant conditions;
- Transition guard matrices: $A_G \in \mathbb{Q}^{n_A \times n_T \times n_G \times n_C}$ and $b_G \in \mathbb{Q}^{n_A \times n_T \times n_G}$;
- Cost rates $c_L(a, l) \in \mathbb{Q}$ assigned to locations (corresponding to the prices $P(l)$ in LPTA);
- Transition costs $c_T(a, t) \in \mathbb{Q}$ (corresponding to the prices $P((l, g, a, r, l'))$ in Def. 1);
- Reset vectors for transitions: $r(a, t, c) \in \{0, 1\}$, the components of which are zero for clocks that are reset by a transition, and which are one otherwise;
- Parameters $w(a, l, t, l') \in \{0, 1\}$ that define the automaton topology, i.e., $w(a, l, t, l') = 1$ denotes that a transition $t$ exists for automaton $a$ between location $l$ and location $l'$;
- Indicators for transition sources: $f(a, l, t) \in \{0, 1\}$, where $f(a, l, t) = 1$ denotes that location $l$ of automaton $a$ has an outgoing transition $t$;
- Synchronization indicators: $s(a, t, a', t')$, where $s(a, t, a', t') = 1$ indicates that the transition $t$ of automaton $a$ and the transition $t'$ of $a'$ are synchronized. (Obviously, these parameters are defined symmetrically: $s(a, t, a', t') = s(a', t', a, t)$ for $a \neq a'$;
- Constants $m, M \in \mathbb{Q}$, where $m$ is small and $M$ large compared to the left- and right-hand sides of the inequalities formulating the guards and invariants.

## Variables

- Variables for clock valuations at the instants when a location is reached: $x(a, c, k) \in \mathbb{Q}$;
- Variables for clock valuations at the instant when a location is left: $y(a, c, k) \in \mathbb{Q}$;
- A clock variable for each automaton: $z(a, k) \in \mathbb{Q}$. These variables are required to ensure that synchronized transitions are taken simultaneously. They are never reset to zero;

- Location indicator variables $d_L(a, l, k) \in [0, 1]$, that specify the current location of each automaton at every point of time (Note that these variables are forced to zero or one by the equations listed below.);
- Transition indicator variables $d_T(a, t, k) \in [0, 1]$ for all transitions;
- Variables that indicate the period of time during which an automaton does not change its location: $\Delta(a, k) \in \mathbb{Q}$;
- Variables that combine the information about the current locations and transitions: $d_{LT}(a, l, t, k) \in [0, 1]$.

## (In-)Equalities for Static Dependencies

- Each automaton is always only in one of its locations: $\sum_{l \in \mathcal{L}} d_L(a, l, k) = 1$ for all $a \in \mathcal{A}, k \in \mathcal{K}$.
- In every point of time in $\mathcal{K}$, each automaton takes always one of its transitions (possibly a self-loop transition): $\sum_{t \in \mathcal{T}} d_T(a, t, k) = 1$ for all $a \in \mathcal{A}, k \in \mathcal{K}$.
- Restriction to valid combinations of locations and outgoing transitions: each variable $d_{LT}(a, l, t, k)$ is set to 1 iff $d_T(a, t, k) = 1$ and $d_L(a, l, k) = 1$ for $a \in \mathcal{A}, l \in \mathcal{L}, t \in \mathcal{T}, k \in \mathcal{K}$:
$d_{LT}(a, l, t, k) \leq f(a, l, t) \cdot d_L(a, l, k)$,
$d_{LT}(a, l, t, k) \leq f(a, l, t) \cdot d_T(a, t, k)$,
$d_{LT}(a, l, t, k) \geq f(a, l, t) \cdot (d_L(a, l, k) + d_T(a, t, k) - 1)$.
- In every point of time, only one transition can occur for a source transition: $\sum_{t \in \mathcal{T}} \sum_{l \in \mathcal{L}} f(a, l, t) \cdot d_{LT}(a, l, t, k) = 1$ for all $a \in \mathcal{A}, k \in \mathcal{K}$.

## (In-)Equalities for the Linearization of Nonlinear Constraints

- Since the objective is to obtain an optimization model that exclusively contains linear constraints, products of variables have to be linearized. This is achieved by first introducing the following constraints and additional auxiliary variables:
  - $x_L^d(a, c, k, l) := x(a, c, k) \cdot d_L(a, l, k)$,
  - $y_L^d(a, c, k, l) := y(a, c, k) \cdot d_L(a, l, k)$,
  - $y_T^d(a, c, k, t) := y(a, c, k) \cdot d_T(a, t, k)$,
  - $c_L^d(a, l, k) := \Delta(a, k) \cdot d_L(a, l, k)$.

  By applying the transformations described in [16], these constraints can then be written in linear form.
- It has to be encoded that the check, whether a transition guard is satisfied, is only relevant when a transition is taken:
$\sum_{c \in \mathcal{C}} A_G(a, t, g, c) \cdot y_T^d(a, c, k, t) \leq d_T(a, t, k) \cdot b_G(a, t, g)$ for all $a \in \mathcal{A}, t \in \mathcal{T}, g \in \mathcal{G}, k \in \mathcal{K}$.
- Location invariants are checked when a location is reached:
$\sum_{c \in \mathcal{C}} A_I(a, l, g, c) \cdot x_L^d(a, c, k, l) \leq d_L(a, l, k) \cdot b_I(a, l, g)$ for all $a \in \mathcal{A}, l \in \mathcal{L}, g \in \mathcal{G}, k \in \mathcal{K}$,
- and when a location is left:
$\sum_{c \in \mathcal{C}} A_I(a, l, g, c) \cdot y_L^d(a, c, k, l) \leq d_L(a, l, k) \cdot b_I(a, l, g)$ for all $a \in \mathcal{A}, l \in \mathcal{L}, g \in \mathcal{G}, k \in \mathcal{K}$.

## Formulation of the Clock and Transition Dynamics

- Staying in locations: $y(a, c, k) = x(a, c, k) + \Delta(a, k)$ for all $a \in \mathcal{A}, c \in \mathcal{C}, k \in \mathcal{K}$.
- A location $l'$ of automaton $a$ becomes active, iff one of its predecessor locations $l$ was active and the connecting transition $t$ from $l$ to $l'$ is taken:
  $d_L(a, l', k+1) = \sum_{t \in \mathcal{T}} \sum_{l \in \mathcal{L}} w(a, l, t, l') \cdot d_{LT}(a, l, t, k)$ for all $l' \in \mathcal{L}, k \in \mathcal{K} \setminus \{k_{n_K}\}, a \in \mathcal{A}$.
- Clock resets triggered by transitions: $x(a, c, k+1) = \sum_{t \in \mathcal{T}} r(a, t, c) \cdot y_T^d(a, c, k, t)$, $a \in \mathcal{A}, c \in \mathcal{C}, k \in \mathcal{K}$.
- Assignment of the clock valuations: $z(a, k+1) = z(a, k) + \Delta(a, k)$ for all $a \in \mathcal{A}, k \in \mathcal{K}$.

## Synchronization Equations

- If required by the synchronization indicators, two transitions are synchronized:
  $s(a, t, a', t') \cdot d_T(a, t, k) \leq \sum_{k' \in \mathcal{K}} s(a, t, a', t') \cdot d_T(a', k', t')$ for all $a, a' \in \mathcal{A}$, $t, t' \in \mathcal{T}, k \in \mathcal{K}$.
- If two transitions of two automata are synchronized (as indicated by $s(a, t, a', t') = 1$), the following inequalities ensure that the clocks $z(a, k)$ and $z(a', k')$ have the same values:
  $s(a, t, a', t') \cdot (z(a, k) - z(a', k')) \leq s(a, t, a', t') \cdot M \cdot (2 - d_T(a, t, k) - d_T(a', t', k'))$
  $s(a, t, a', t') \cdot (z(a, k) - z(a', k')) \geq -s(a, t, a', t') \cdot M \cdot (2 - d_T(a, t, k) - d_T(a', t', k'))$ for all $a, a' \in \mathcal{A}, t, t' \in \mathcal{T}$, and $k, k' \in \mathcal{K}$.
  If a transition is not synchronized with any other, then all parameters $s$ referring to this transition are set to zero (i.e., no-action symbols are implicitly assumed).

In addition to the items listed here, non-negativity constraints and bounds for some variables have to be specified.

### 3.2   Objective Function

The costs of a trace of the timed automaton are defined as the sum of all transition costs plus the cost rates of locations multiplied by the durations in which the locations are active:

$$\min_{c_L^d, d_T} \Omega, \text{ with:} \tag{1}$$

$$\Omega = \sum_{a \in \mathcal{A}} \sum_{k \in \mathcal{K}} \left( \sum_{l \in \mathcal{L}} c_L^d(a, l, k) \cdot c^L(a, l) + \sum_{t \in \mathcal{T}} d_T(a, t, k) \cdot c^T(a, t) \right). \tag{2}$$

Simplified versions of this objective function are sometimes used in scheduling. One is to accumulate costs only if a job automaton reaches its final state delayed, i.e., later than a specified deadline. Another alternative is the *makespan* minimization, in which simply the time is minimized at which the last task is terminated. Then the objective function reduces to:

$$\min_{c_L^d, d_T} \Omega, \text{ where } \Omega \geq z(a, k) \ \forall \ a \in \mathcal{A}, k \in \mathcal{K}. \tag{3}$$

### 3.3  Solution Procedure

In order to determine the optimal schedule, the following steps are carried out:

1. Each automaton used for modeling the scheduling problem is transformed into a corresponding MILP representation according to the scheme presented above.
2. The MILP model has to be initialized by assigning the value one to the location indicator variables that correspond to the initial states. Similarly, the final states (in which all job automata have reached their terminal states) are specified for $k = n_k$. Furthermore, all clock variables are initialized to zero.
3. The MILP model together with a chosen objective function can then be solved by a MILP solver.
4. Finally the optimal schedule is extracted from the optimization result. The solution contains valid values for all $x$ and $y$ variables representing the starting and finishing times of tasks. These values lead straightforwardly to a valid schedule that meets all restrictions. In addition, from the variables $d_L$ and $d_T$ that have the value one, the traces of each single LPTA can be easily constructed.

### 3.4  Illustrative Example

In order to illustrate the transformation step, the simple scheduling problem introduced already in Fig. 1 is considered again. The two job automata interact with the two resource automata through synchronization labels. In automaton $J1$, the first transition labeled by aJ1M1 is used to allocate the resource $M1$. The second transition is used to free the resource through the synchronization fJ1M1. Similar synchronization is used for the second task in which $M2$ is allocated and freed. The automaton $J2$ for Job 2 is identical except of the labels and the task durations (5 time units for task 1, and 2 time units for task 2). Of course, both resource automata include corresponding synchronized transitions for the second job. This model was implemented in the GAMS language with 5 discrete points of time for each automaton. The resulting MIP model has 2450 equations, 903 single variables and 60 binary variables. The solution on a 2.4 GHz machine took 0.6 seconds and a schedule with a makespan of 9 has been found to be optimal in the 60th branch-and-bound node. It is shown as schedule $S1$ in Fig. 4.

It should be remarked that variations of the MILP encoding of LPTA are conceivable; but we found the scheme presented here to be an efficient one with respect to the solution performance. It has to be considered however, that the transformation scheme maintains the modularity and the synchronization of the

LPTA used to model the jobs and resources. If one alternatively first composes the automata and then applies the transformation scheme to the product, the components required to model the synchronization variables vanish from the MILP program.

## 4    Combining TA Optimization with MIP Relaxations

With an increasing complexity of the LPTA model, the number of binary and continuous variables in the MILP representation grows quickly, and makes an efficient solution difficult. In order to reduce this effect, we now sketch an idea that combines the tree search for LPTA with the relaxation principle used in MILP techniques.

The known procedure to determine a cost-optimal path for LPTA is to build a search tree starting from the initial state and to extend the branches according to reachability criteria. This means that a node of the tree is extended by successor nodes that represent locations which are reachable by single transitions (the guard of which is enabled). Hence, a branch of the tree corresponds to the sequence of locations that are encountered during a possible evolution of the automaton. The costs accumulated along the path are used to assess the path, and the search is directed by this assessment. In [8], the branch-and-bound principle is used in this context, that means branches for which the accumulated costs are higher as for the best solution found so far, are cut (i.e., are not further explored). However, it is obvious that this type of search can only operate with costs accumulated up to the current node, but does not consider a cost-to-go.

Existing MILP techniques also build a search tree, but a node here represents a state in which a certain subset of discrete variables is fixed to integer values, while the remainder is not. A key difference to the search for TA is that in MILP complete paths from the initial location to the target location ('all jobs are processed') are considered in each step. In each node a linear program (LP) is solved in which the discrete variables, that are not yet fixed, are treated as continuous variables. The solution of the LP defines a lower bound for the cost of the original problem. This bound can be used to select which discrete variables are fixed next to particular values. Since this criterion is applied quite successfully in MILP, the objective is to embed it into the tree search for LPTA.

The following procedure is suggested. The tree search including the branch-and-bound principle is carried out as sketched above. However, in each encountered node the following is done: The LPTA model is transformed into the corresponding MILP representation using the scheme described in Sec. 3. The degrees of freedom that are fixed already by the previous evolution of the LPTA model result in fixed variables of the MILP model; specifically, the corresponding variables $d_T(a, l, k)$ and $\Delta(a, k)$ are treated as parameters with constant values. The remaining discrete (respectively binary) variables are relaxed, i.e., are treated temporarily as continuous variables. The optimization is then solved as a linear program which returns a lower bound for the overall cost. The difference between this lower bound and the accumulated cost for the current node of the tree search can be interpreted as an estimation of the cost-to-go. The values of the relaxed variables that correspond to the lower bound are taken as hints which further evolution of the automaton should be investigated first. In the simplest case that can be done by rounding the solutions for the relaxed variables to the nearest integer values and translating these values back into a path of the automaton. Even more importantly, the lower bounds are used to cut branches of the search tree, if the lower bound obtained for a particular node has a already a higher value as the costs of the best solution found so far.

The procedure is repeated by alternating between exploring the search tree and evaluating a relaxed MILP model. Note that by fixing variables in each step, the complexity of the LP problems decreases along a branch of the tree.

Figure 2 formulates this procedure as a high-level algorithm: Assume that $A$ denotes the parallel composition of all job and resource automata. Let $L_f \subset L$ denote the subset of final locations of $A$, in which all job automata have reached their final location. Furthermore, we use the notion of *zones* which are essentially polyhedra in the clock space that specify the reachable subset of the location invariants. See [3] for more details on zones. Note that we here again consider the case of makespan minimization, i.e., considering clock valuations (in terms of zones) is a sufficient to evaluate the cost criterion.

Within the algorithm, the MILP model is denoted by $M$, and we use three lists *Passed*, *Waiting*, and *Succ*. Elements of these lists are triples $(l, Z, b)$ consisting of a location $l$, a zone $Z$, and a lower bound $b$ for the costs $\Omega$. The algorithm in Fig. 2 then realizes the procedure of iteratively computing the successor location of the LPTA, computing a lower bound by solving the relaxed MILP program, and cutting branches in the search tree by comparing accumulated costs and lower bounds with the minimal cost value obtained so far.

In order to illustrate the algorithm, we reconsider the example introduced in Fig. 1 and used in Sec. 3.4. A search tree is formed for the composition of all job and resource automata starting with the initial state $(a, a, x, x)$. The final state is the one in which all jobs are completed and all resources are idle: $(e, e, x, x)$. For each node, the accumulated time is recorded, the current state is fixed in the corresponding LP model, and the latter is solved. Its solution is assigned as a lower bound to the current node. The search tree is visualized in Fig. 3. For the sake of clarity, only the minimal accumulated time (to reach the state) and the lower bound but not the complete zones are shown in the tree. Since the search strategy used in this example is best-lower-bound search, the procedure quickly finds the path leading to the final state within 9 time units – this path represents the optimal schedule with the minimal makespan. Other paths are not completely explored because lower bounds encountered at an intermediate state are greater than the best found solution. This cutting rule is the same as the one commonly used in branch-and-bound techniques; it allows here to cut off large parts of the search tree for LPTA: In our example, only 14 of 47 nodes of the tree have been explored. The dashed nodes and transitions show two suboptimal

$\Omega := \infty$
$Passed := \emptyset$
$M = \text{TRANSFORM}(A)$      // transform the LPTA into the MILP model
$M_r = \text{RELAX}(A)$      // relax the binary variables of $M$
$b_0 = \text{SOLVE\_LP}(M_r)$      // solve $min(\Omega)$ by linear programming
$Waiting := \{(l_0, Z_0, b_0)\}$
WHILE $Waiting \neq \emptyset$
    $(l, Z, b) = \text{SELECTREMOVE}(Waiting)$ such that $b$ is minimal
                               // realizes a best-lower-bound-first search strategy
   IF $l \in L_f$ THEN
      IF $min(Z) < \Omega$ THEN
         $\Omega := min(Z)$
      END
   ELSE
      IF $b \leq \Omega$ AND $min(Z) < \Omega$ THEN
         $Succ = \text{COMPUTESUCCESSORS}(A, (l, Z, b))$
                                // list of successors $(l', Z', -)$
         $Succ' := Succ \setminus (Passed \cap Succ)$
         FOR ALL $(l', Z', -) \in Succ'$ DO
            $M_r' = \text{UPDATEMr}(M_r, (l', z'))$
                             // fix variables for transitions into $(l', z')$
            $b' = \text{SOLVE\_LP}(M_r')$
            $Waiting := Waiting \cup \{(l', Z', b')\}$
         END
      END
   END
   $Passed := Passed \cup \{(l, Z, b)\}$
END

**Fig. 2.** Optimization algorithm for LPTA using relaxations and brand-and-bound principles

schedules, for which it is clear from Fig. 4 that the corresponding schedules $S2$ and $S3$ in Fig. 4 are not preferable over $S1$ due to the larger makespan.

## 5   Conclusions

The contribution of this paper is two-fold. First it introduces a procedure to transform a minimization problem formulated for LPTA into a corresponding MILP. The transformation scheme can either retain the modular structure and the synchronization of the separate automata, or one can first determine the automata product and then apply the transformation to the result. The transformation scheme can straightforwardly be written algorithmically, and can thus be fully automated. Hence, the transformation procedure as such allows specifying the optimization problem as LPTA (what appears to be easier than starting with an algebraic model directly) and to proceed then with a set of established MILP techniques.

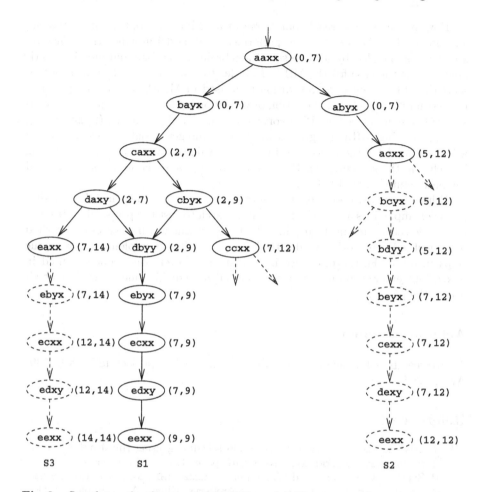

**Fig. 3.** Search tree for the composed LPTA model. The states and transitions drawn with solid lines represent the parts encountered within a best-lower-bound-first search. Elements drawn with dashed lines correspond to suboptimal schedules. Each node is decorated with a pair consisting of accumulated time and lower bound (the latter obtained from solving a linear program)

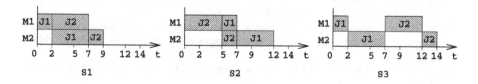

**Fig. 4.** Three out of eight possible schedules computed for the example

However, using the reachability tree of an LPTA for optimization has the advantage, that the search is not performed on a model in which the transition structure is encoded by a large set of algebraic constraints and auxiliary variables – as is the case for the MILP. In order to combine the advantages of tree search for LPTA with the relaxation idea used in MILP, we have proposed in the second part a scheme consisting of the following steps: (a) sequentially constructing a search tree for LPTA considering reachability criteria, (b) generating a corresponding MILP program with relaxed integer variables, (c) computing a lower bound for the cost-to-go by solving a linear program, and (d) cutting branches in the search tree if the accumulated costs or the lower bound exceed the best solution found before.

While this paper describes the algorithm and sketches the principle for a simplistic example, it is a matter of current research to test the proposal for a set of real-world scheduling problems, in order to determine in which cases the method has benefits. This work includes to develop an efficient implementation of the steps involved. Furthermore, we investigate how alternative concepts used in mixed-integer programming (as, e.g., specialized cutting rules) can be embedded in the procedure.

## Acknowledgement

This research is financially supported within the EU project IST-2001-35304 AMETIST.

## References

[1] Y. Abdeddaim and O. Maler. Job-shop scheduling using timed automata. In Computer-Aided Verification, LNCS 2102, pages 478–492. Springer, 2001. 74, 77
[2] P. Baptiste, C. Le Pape, and W. Nuijten. Constraint-Based Scheduling: Applying Constraint Programming to Scheduling Problems. Kluwer Acad. Publisher, 2001. 74
[3] G. Behrmann, A. Fehnker, T.S. Hune, P. Petterson, K. Larsen, and J. Romijn. Efficient guiding towards cost-optimality in UPPAAL. In Tools and Algor. for the Construction and Analysis of Syst., LNCS 2031, pages 174–188. Springer, 2001. 74, 76, 77, 83
[4] A. Brooke, D. Kendrick, A. Meeraus, and R. Raman. GAMS - A User's Guide. GAMS Development Corp., Washington, 1998. 77
[5] R. Bruns. Scheduling. In T. Baeck, D.B. Fogel, and Z. Michalewicz, editors, Handbook of Evolutionary Computation, pages F1.5:1–9. Inst. of Physics Publishing, 1997. 74
[6] CPLEX. Using the CPLEX Callable Library. ILOG Inc., Mountain View, CA, 2002. 75
[7] S. Engell, A. Maerkert, G. Sand, and R. Schultz. Production planning in a multiproduct batch plant under uncertainty. In Progress in Industrial Mathematics, pages 526–531. Springer, 2002. 74
[8] A. Fehnker. Citius, Vilius, Melius - Guiding and Cost-Optimality in Model Checking of Timed and Hybrid Systems. Dissertation, KU Nijmegen, 2002. 82

[9] S. Graf, M. Bozga, and L. Mounier. If-2.0: A validation environment for component-based real-time systems. In Computer-Aided Verification, volume 2404 of LNCS, pages 343–348. Springer, 2002. 74

[10] M. Groetschel, S. O. Krumke, and J. Rambau, editors. Online Optimization of Large Scale Systems. Springer, 2001. 74, 75

[11] I. Harjunkoski and I. E. Grossmann. Decomposition techniques for multistage scheduling problems using mixed-integer and constraint programming methods. Computers and Chemical Enginnering, 26(11):1501–1647, 2002. 74

[12] M. Ierapetritou and C. Floudas. Effective continuous-time formulation for short-term scheduling: Multipurpose batch processes. Industrial and Engineering Chemistry Research, 37:4341–4359, 1998. 74

[13] E. Kondili, C. C. Pantelides, and R. W. H. Sargent. A general algorithm for short-term scheduling of batch operations: MILP formulation. Computers and Chemical Engineering, 17:211–227, 1993. 74

[14] K. G. Larsen, P. Petterson, and W. Yi. UPPAAL in a nutshell. Int. Journal on Software Tools for Technology Transfer, 1(1):134–152, 1997. 74

[15] N. Shah, C. C. Pantelides, and R. W. H. Sargent. A general algorithm for short-term scheduling of batch operations: Computational issues. Computers and Chemical Engineering, 17(2):229–244, 1993. 74

[16] O. Stursberg and S. Panek. Control of switched hybrid systems based on disjunctive formulations. In Hybrid Systems: Computation and Control, LNCS 2289, pages 421–435. Springer, 2002. 77, 79

[17] H. P. Williams. Model Building in Mathematical Programming. Wiley, 1999. 75

[18] S. Yovine. Kronos: A verification tool for real-time systems. Int. Journal of Software Tools for Technology Transfer, 1(1/2):123–133, October 1997. 74

# Discrete-Time Rewards Model-Checked

Suzana Andova[1], Holger Hermanns[1,2], and Joost-Pieter Katoen[1]

[1] Formal Methods and Tools Group, Department of Computer Science
University of Twente, P.O. Box 217, 7500 AE Enschede, The Netherlands
[2] Department of Computer Science
Saarland University, D-66123 Saarbrücken, Germany

**Abstract.** This paper presents a model-checking approach for analyzing discrete-time Markov reward models. For this purpose, the temporal logic probabilistic CTL is extended with reward constraints. This allows to formulate complex measures – involving expected as well as accumulated rewards – in a precise and succinct way. Algorithms to efficiently analyze such formulae are introduced. The approach is illustrated by model-checking a probabilistic cost model of the IPv4 zeroconf protocol for distributed address assignment in ad-hoc networks.

## 1 Introduction

Modelling techniques such as queueing networks and probabilistic variants of Petri nets, automata networks and process algebra are convenient means to describe performance and dependability models. Based on a high-level specification of the system under investigation, the underlying model – albeit a continuous time or a discrete time Markov chain (CTMC or DTMC) – is automatically obtained and can be analyzed with well-studied means to obtain transient and stationary measures. Most of these techniques have been extended to CTMCs (or DTMCs) augmented with costs, or dually bonuses (rewards); approaches using stochastic reward nets [7], reward-based variants of process algebra [4] extensions of automata [15], logic-based approaches [8] and so on, have been proposed. These formalisms provide adequate means to specify performance and dependability *models*.

It is fair to say that the specification of performance or dependability *measures* in a high-level manner has received far less attention. In recent works, we have proposed to use appropriate extensions of temporal logic – as typically used to reason about the functional correctness of systems – for specifying constraints over such measures [2, 3]. This technique allows to specify standard (e.g., transient and stationary) and complex measures in a precise, unambiguous and lucid manner. Even more importantly, this specification technique is complemented by powerful means to automatically check constraints on measures over finite Markovian models using a light-weight extension of model checking [9]. This hides specialized algorithms from the performance engineer, supports automated measure-driven model adaptation, and allows for the checking of quantitative as

K.G. Larsen and P. Niebert (Eds.): FORMATS 2003, LNCS 2791, pp. 88–104, 2004.

well as functional properties (such as absence of deadlocks) in a single integrated framework.

The model-checking algorithms for CTMCs rely on well-developed standard numerical algorithms. Therefore even the more intricated measures – beyond standard stationary and transient measures – can be checked rather efficiently. Further work in this area has focussed on CTMCs decorated with rewards. We have introduced a logic to specify measures over such so-called continuous-time Markov reward models (CMRMs) [2, 12]. The logic allows one to express a rich spectrum of measures. For instance, when rewards are interpreted as costs, this logic can express a constraint on the probability that, given a start state, a certain goal can be reached within $t$ time units while deliberately avoiding to visit certain intermediate states, and with a total cost (i.e., accumulated reward) below a given threshold. Such path-based measures are, however, computationally expensive as they are based on determining transient reward distributions, a measure that has not been widely addressed in the literature and for which the rarely available algorithms are highly time- and/or space-consuming [12].

In this paper, we aim to avoid this inefficiency by considering discrete-time Markov chains instead, decorated with (possibly multiple) state rewards. This paper introduces a logic and model-checking algorithms for discrete time Markov reward models (DMRMs). In particular, we extend probabilistic CTL [11] with operators to reason about long-run average, and more importantly, by operators that allow to specify constraints on (i) the expected reward rate at a time instant, (ii) the long-run expected reward rate per time unit, (iii) the cumulated reward rate at a time instant – all for a specified set of states – and (iv) the cumulated reward over a time interval. The proposed logic allows to specify non-trivial, though interesting, constraints such as *"the probability to reach one of the goal states (via indicated allowed states) within n steps while having earned an accumulated reward that does not exceed r is larger than 0.92"*. We present model-checking algorithms that verify such properties in an efficient manner, and show how these can be extended to multiple rewards in a straightforward way. The approach is illustrated by checking some properties of the IPv4 zeroconf protocol for distributed address assignment in ad-hoc networks.

## 2   Discrete-Time Markov Reward Models

This section presents the basic concepts of discrete-time Markov reward models that are needed for the rest of the paper. For more details we refer to [14].

**DMRMs.** In order to enable the logical specification of measures-of-interest over performability models we consider a slight extension of traditional Markov models where states are equipped with elementary properties, the so-called atomic propositions. Let $AP$ be a fixed, finite set of atomic propositions.

**Definition 1.** *A (labelled) DTMC $\mathcal{D}$ is a tuple $(S, \mathbf{P}, L)$ where $S$ is a finite set of states, $\mathbf{P} : S \times S \to [0, 1]$ is a probability matrix such that $\sum_{s' \in S} \mathbf{P}(s, s') \in$*

$\{0,1\}$ *for all* $s \in S$, *and* $L : S \rightarrow 2^{AP}$ *is a* labelling *function that assigns to each state* $s \in S$ *the set* $L(s)$ *of atomic propositions that are valid in* $s$.

**Definition 2.** *A discrete-time reward model (DMRM)* $\mathcal{M}$ *is a pair* $(D, \rho)$ *with DTMC* $\mathcal{D} = (S, \mathbf{P}, L)$ *and* $\rho : S \rightarrow \mathbb{R}_{\geqslant 0}$ *a reward assignment function that associates a real reward (or: cost) to any state in* $S$. *Real number* $\rho(s)$ *denotes the reward earned on leaving state* $s$.[1]

**Paths.** Let $\mathcal{M}$ be a DMRM with underlying DTMC $\mathcal{D} = (S, \mathbf{P}, L)$ and reward function $\rho$. An infinite path $\sigma$ is a sequence $s_0 \rightarrow s_1 \rightarrow s_2 \rightarrow \dots$ where $s_i \in S$ and $\mathbf{P}(s_i, s_{i+1}) > 0$ for $i \geqslant 0$. For $i \in \mathbb{N}$ let $\sigma[i] = s_i$, the state occupied after $i$ transitions. A finite path $\sigma$ with length $n$ is a sequence $s_0 \rightarrow \dots \rightarrow s_n$ with $s_i \in S$ and $\mathbf{P}(s_i, s_{i+1}) > 0$ for $0 \leqslant i < n$. $Path(s)$ denotes the set of (finite and infinite) paths starting in $s$. Let $\Pr_s$ denote the unique probability measure on sets of paths that start in state $s$ [3]. The *cumulative reward* along finite path $\sigma$ with length $n$ is defined as $\rho(\sigma) = \sum_{i \geqslant 0}^{n-1} \rho(s_i)$. Note that rewards are considered on leaving a state, i.e., $\rho(s_n)$ is not considered in the cumulative reward of $\sigma$.

*Example 1.* Consider the DMRM $\mathcal{M}$ depicted below with $S = \{s_1, s_2, s_3, s_4\}$, $L(s_1) = L(s_2) = \{a\}$, $L(s_3) = \{b\}$ and $L(s_4) = \{a, c\}$, reward structure $\rho$ defined by $\rho(s_1) = 2, \rho(s_2) = 3, \rho(s_3) = 0, \rho(s_4) = 2$, and the probability matrix defined by:

$$\mathbf{P} = \begin{bmatrix} 0.2 & 0.5 & 0.3 & 0 \\ 0 & 0 & 0.1 & 0.9 \\ 0.4 & 0.3 & 0.3 & 0 \\ 0 & 0.6 & 0 & 0.4 \end{bmatrix}$$

An example finite path is $\sigma = s_1\, s_2\, s_3\, s_2\, s_4$; we have $\sigma[0] = s_1$ and $\sigma[3] = s_2$, and the cumulative reward $\rho(\sigma)$ equals 8. The probability of $\sigma$, $\Pr_{s_1}(\sigma) = \frac{1}{2} \cdot \frac{1}{10} \cdot \frac{3}{10} \cdot \frac{9}{10}$.

**Transient and Limiting Behavior.** Transient analysis studies the system at a certain time instant $n$. Let $\pi(s, s', n)$ denote the probability that the system is in state $s'$ after $n$ steps given that the system started in state $s$. These transition probabilities can be calculated using the Chapman-Kolmogorov equations:

$$\pi(s, s', n) = \sum_{t \in S} \pi(s, t, i) \cdot \pi(t, s', n-i) \text{ for } 0 \leqslant i \leqslant n, \tag{1}$$

where $\pi(s, s', 0) = 0$ if $s' \neq s$ and $\pi(s, s, 0) = 1$. When $n$ tends to infinity, one considers the limiting (i.e., long-run) behaviour of DTMCs. The limiting behaviour of a DTMC strongly depends on the structure of the considered chain, more

---

[1] Here we consider rewards to be constant, but there do exist variants in which rewards are random variables.

specifically, on the capacity of states reaching each other within finitely many steps. It is well known that in case of an irreducible finite aperiodic DTMC the limit $\lim_{n \to \infty} \pi(s, s', n)$ exists and precisely characterises the limiting probabilities $\pi(s, s')$, also called steady-state probabilities, of the DTMC [13]. If the considered DTMC is irreducible and periodic then this limit does not exist. In that case, one considers the long-run fraction of time that the system spends in state $s'$ when starting in state $s$:

$$\pi(s, s') = \lim_{n \to \infty} \frac{1}{n+1} \sum_{i=0}^{n} \pi(s, s', i) \tag{2}$$

The probabilities $\pi(s, s')$ can (in both cases) be characterised as the unique solution of the following system of linear equations:

$$\pi(s) = \sum_{s' \in S} \pi(s') \cdot \mathbf{P}(s, s') \text{ such that } \sum_{s \in S} \pi(s) = 1. \tag{3}$$

For irreducible aperiodic DTMCs, $\pi(s, s')$ coincides with $\lim_{n \to \infty} \pi(s, s', n)$, see e.g., [14]. Although the initial state does not have any influence on the value of $\pi(s, s')$, we keep this notation because in the case of reducible DTMCs the initial state has influence on the limiting behaviour. Let $\pi(s, S')$ denote $\sum_{s' \in S'} \pi(s, s')$ for $S' \subseteq S$.

**Reward Measures.** For DMRMs the following reward measures are considered, see also [18]. Assume that the system starts in state $s$.

- The expected reward rate per time-unit up to time instant $n$, denoted $g(s, n)$, and its limiting counterpart, the long-run expected reward rate per time-unit, denoted $g(s)$. They are defined as follows:

$$g(s, n) = \frac{1}{n+1} \sum_{i=1}^{n} E(\rho(\sigma_s[i])) \text{ and } g(s) = \lim_{n \to \infty} g(s, n)$$

  where (the random variable) $\sigma_s$ ranges over $Path(s)$.
- The instantaneous reward at time instant $n$: $\rho(s, s', n) = \pi(s, s', n) \cdot \rho(s')$. For $S' \subseteq S$ let $\rho(s, S', n) = \sum_{s' \in S'} \rho(s, s', n)$, i.e., $\rho(s, S', n)$ is the instantaneous reward at time instant $n$ in the set $S'$.
- The expected accumulated reward until the $n$-th transition is defined as follows: $y(s, n) = \sum_{i=0}^{n-1} \rho(s, S, i)$. According to the definition of path reward, the sum goes up to $n-1$, i.e., the reward of the last state of the path is ignored. An alternative characterisation of this reward measure is: $y(s, n) = \sum_{\sigma \in Path(s, s', n)} \rho(\sigma) \cdot \Pr_s(\sigma)$, where $Path(s, s', n)$ denotes the set of finite paths of length $n$ that start in $s$ and end in $s'$.

**Multiple Rewards.** If various measures-of-interest are to be determined for a Markov model, typically several different reward structures are imposed, see e.g., [18, Section II]. For $k > 0$ reward structures, a DMRM is a $(k+1)$-tuple

$(\mathcal{D}, \rho_1, \ldots, \rho_k)$ with $\mathcal{D}$ a DTMC and $\rho_j$ a reward assignment function, for $0 < j \leqslant k$. The reward measures defined above can now be all considered for each of the different reward assignments $\rho_j$ in a fairly straightforward manner. Let $\rho_j(\sigma)$ be the accumulated $j$-th reward along finite path $\sigma$, i.e., for $\sigma = s_0 \to \ldots \to s_n$ we have $\rho_j(\sigma) = \sum_{i \geqslant 0}^{n-1} \rho_j(s_i)$. The instantaneous $j$-th reward at time instant $n$ is defined by: $\rho_j(s, s', n) = \pi(s, s', n) \cdot \rho_j(s')$. The other reward measures are generalised in a similar manner.

## 3   Probabilistic Reward CTL

This section introduces the logic Probabilistic Reward CTL (PRCTL) that is aimed at the specification of performability measures over discrete-time Markov reward models. To simplify the presentation we first recall Probabilistic CTL (PCTL) by Hansson and Jonsson [11], and extend it by a long-run average operator.

**PCTL with Long-Run Average.** Let $a \in AP$, $p \in [0, 1]$, $n$ be a natural (or $\infty$) and binary comparison operator $\trianglelefteq \; \in \{\leqslant, <, \geqslant, >\}$. The syntax of PCTL is:

$$\Phi ::= \text{tt} \; \Big| \; a \; \Big| \; \Phi \wedge \Phi \; \Big| \; \neg \Phi \; \Big| \; \mathcal{L}_{\trianglelefteq p}(\Phi) \; \Big| \; \mathcal{P}_{\trianglelefteq p}(\Phi \mathcal{U}^{\leqslant n} \Phi)$$

The other boolean connectives are derived in the usual way. For the sake of simplicity, we do not consider the next state operator in this paper. The standard (i.e. unbounded) until formula is obtained by taking $n$ equal to $\infty$, i.e., $\Phi \mathcal{U} \Psi = \Phi \mathcal{U}^{\leqslant \infty} \Psi$. Temporal operators like $\Diamond$, $\Box$ and their timed variants $\Diamond^{\leqslant n}$ or $\Box^{\leqslant n}$ can be derived, e.g., $\mathcal{P}_{\trianglelefteq p}(\Diamond^{\leqslant n} \Phi) = \mathcal{P}_{\trianglelefteq p}(\text{tt} \, \mathcal{U}^{\leqslant n} \Phi)$ and $\mathcal{P}_{\geqslant p}(\Box \Phi) = \mathcal{P}_{\leqslant 1-p}(\Diamond \neg \Phi)$. Let $Sat(\Phi) = \{ s \mid s \models \Phi \}$ be the set of states that satisfy $\Phi$. The semantics of PCTL is defined by [11]:

| | | |
|---|---|---|
| $s \models \text{tt}$ for all $s \in S$ | $s \models \Phi \wedge \Psi$ | iff $s \models \Phi \wedge s \models \Psi$ |
| $s \models a$ iff $a \in L(s)$ | $s \models \mathcal{L}_{\trianglelefteq p}(\Phi)$ | iff $\pi(s, Sat(\Phi)) \trianglelefteq p$ |
| $s \models \neg \Phi$ iff $s \not\models \Phi$ | $s \models \mathcal{P}_{\trianglelefteq p}(\Phi \mathcal{U}^{\leqslant k} \Psi)$ | iff $Prob(s, \Phi \mathcal{U}^{\leqslant k} \Psi) \trianglelefteq p$ |

$\mathcal{P}_{\trianglelefteq p}(\Phi \mathcal{U}^{\leqslant k} \Psi)$ asserts that the probability measure of the paths that start in $s$ and that satisfy $\Phi \mathcal{U}^{\leqslant k} \Psi$ meets the bound $\trianglelefteq p$. The state formula $\mathcal{L}_{\trianglelefteq p}(\Phi)$ asserts that the long-run average fraction of time for the set of $\Phi$-states meets the bound $\trianglelefteq p$. Here, $Prob(s, \Phi \mathcal{U}^{\leqslant k} \Psi) = \Pr_s \{ \sigma \in Path(s) \mid \sigma \models \Phi \mathcal{U}^{\leqslant k} \Psi \}$. Formula $\Phi \mathcal{U}^{\leqslant n} \Psi$ asserts that $\Psi$ will be satisfied within $n$ steps and that all preceding states satisfy $\Phi$, i.e.:

$$\sigma \models \Phi \mathcal{U}^{\leqslant n} \Psi \quad \text{iff} \quad \exists j \leqslant n. \, (\sigma[j] \models \Psi \wedge \forall i < j. \, \sigma[i] \models \Phi)$$

Some example properties that can be expressed in PCTL for our running example are $\mathcal{P}_{\geqslant 0.3}(\Diamond b)$ (a $b$-state can be reached with probability at least 0.3), $\mathcal{P}_{\geqslant 0.3}(a \, \mathcal{U}^{\leqslant 3} b)$ (a $b$-state can be reached with probability at least 0.3 by at most 3 hops along $a$-states), and $\mathcal{L}_{\leqslant 0.5}(a)$ (the long-run average fraction of time spent in $a$-states is at most 0.5).

**Syntax of PRCTL.** We now extend the logic PCTL with ample means to specify properties that do not only address probabilistic aspects but in addition allow to specify constraints over reward measures. Some of the new operators are inspired by Baier *et al.* [3] who introduced a performability logic for continuous-time Markov reward models (with state rewards).

Let $J \subseteq \mathbb{R}_{\geq 0}$ be an interval on the real line, $n$ a natural number, $p \in [0,1]$ and $N \subseteq \mathbb{N} \cup \{\infty\}$ an interval of natural numbers (or infinity). The syntax of PRCTL is defined by the following syntax clauses:

$$\Phi ::= \text{tt} \ \Big| \ a \ \Big| \ \Phi \wedge \Phi \ \Big| \ \neg \Phi \ \Big| \ \mathcal{L}_{\trianglelefteq p}(\Phi) \ \Big| \ \mathcal{P}_{\trianglelefteq p}(\Phi \, \mathcal{U}_J^N \, \Phi) \ \Big|$$
$$\mathcal{E}_J^n(\Phi) \ \Big| \ \mathcal{E}_J(\Phi) \ \Big| \ \mathcal{C}_J^n(\Phi) \ \Big| \ \mathcal{Y}_J^n(\Phi)$$

The intuitive interpretation of these operators is as follows. Formula $\mathcal{E}_J^n(\Phi)$ asserts that the expected reward rate in $\Phi$-states up to $n$ transitions – reached at the $n$-th epoch – lies within the interval $J$. Formula $\mathcal{E}_J(\Phi)$ expresses that the long-run expected reward rate per time-unit for $\Phi$-states meets the bounds of $J$. The formula $\mathcal{C}_J^n(\Phi)$ asserts that the instantaneous reward in $\Phi$-states at the $n$-th epoch meets the bounds of $J$. Formula $\mathcal{Y}_J^n(\Phi)$ asserts that the expected accumulated reward in $\Phi$-states until the $n$-th transition meets the bounds of $J$.

**Semantics of PRCTL.** The semantics of the state-formulas of PRCTL that are common with PCTL is identical to the semantics for PCTL as presented above. The semantics of the new operators is defined by:

$$s \models \mathcal{E}_J^n(\Phi) \ \text{iff} \ g(s, Sat(\Phi), n) \in J \qquad s \models \mathcal{C}_J^n(\Phi) \ \text{iff} \ \rho(s, Sat(\Phi), n) \in J$$
$$s \models \mathcal{E}_J(\Phi) \ \text{iff} \ g(s, Sat(\Phi)) \in J \qquad s \models \mathcal{Y}_J^n(\Phi) \ \text{iff} \ y(s, Sat(\Phi), n) \in J$$

where we have that for $S' \subseteq S$:

$$g(s, S', n) = \frac{1}{n+1} \sum_{i=0, \sigma_s[i] \in S'}^{n} E(\rho(\sigma_s[i])) \ \text{and} \ g(s, S') = \lim_{n \to \infty} g(s, S', n).$$

Note that $g(s, n)$ as defined earlier coincides with $g(s, S, n)$. Stated in words, $g(s, S', n)$ denotes the expected reward rate up to the $n$-th epoch given that we are only interested in states belonging to the set $S'$. The expected accumulated reward for states in $S'$ until the $n$-th transition is defined by: $y(s, S', n) = \sum_{i=0}^{n-1} \rho(s, S', i)$. Note that $y(s, n)$ as defined earlier coincides with $y(s, S, n)$.

Formula $\Phi \, \mathcal{U}_J^N \, \Psi$ asserts that $\Psi$ will be satisfied within $j \in N$ steps, that all preceding states satisfy $\Phi$, and that the accumulated reward until reaching the $\Psi$-state lies in the interval $J$. Formally:

$$\sigma \models \Phi \, \mathcal{U}_J^N \, \Psi \ \text{iff} \ \exists j \in N. \ \left( \sigma[j] \models \Psi \wedge \forall i < j. \sigma[i] \models \Phi \wedge \sum_{i=0}^{j-1} \rho(\sigma[i]) \in J \right)$$

*Example 2.* Some example properties that can be expressed in PRCTL for our running example are, $\mathcal{P}_{\geq 0.3}(a \, \mathcal{U}_{(23,\infty)}^{\leq 3} \, b)$ (a $b$-state can be reached with probability at least 0.3 by at most 3 hops along $a$-states accumulating costs of more than 23), and $\mathcal{Y}_{[3,5]}^3(a)$ (the accumulated costs expected within 3 hops is at least 3 and at most 5).

**Multiple Rewards.** The logic PRCTL can easily be enhanced such that properties over models equipped with multiple reward structures can be treated. Suppose $\mathcal{M} = (\mathcal{D}, \rho_1, \ldots, \rho_k)$ is a DMRM with $k > 0$ reward structures, and let $0 < j \leqslant k$. The reward operators of PRCTL can be generalised in a straightforward manner such that constraints on all $k$ reward structures can be expressed in a single formula. For instance, the formula $\mathcal{E}_{J_1,\ldots,J_k}(\Phi)$ expresses that the long-run expected reward rate per time-unit for $\Phi$-states meets the bounds of $J_1$ for reward structure $\rho_1$, $\ldots$, the bounds of $J_k$ for reward structure $\rho_k$. Its semantics is defined by: $s \models \mathcal{E}_{J_1,\ldots,J_k}(\Phi)$ if and only if $g_j(s, Sat(\Phi)) \in J_j$ for all $0 < j \leqslant k$. The other operators can be generalised in a similar manner.

If extending to multiple rewards, it is actually possible to encode the time constraint $N$ (in $\mathcal{U}_J^N$) into a reward constraint over a simple auxiliary reward structure.

## 4   Model-Checking Algorithms

Given a state $s$ of DMRM $\mathcal{M}$ and a PRCTL-formula $\Phi$, the question to be addressed is how to check whether or not $\Phi$ holds for state $s$, i.e., whether $s \models \Phi$ or $s \not\models \Phi$. The basic procedure is the same as for model-checking CTL [9]: the set $Sat(\Phi)$ of all states satisfying $\Phi$ is computed recursively and we have that $s \models \Phi$ if and only if $s \in Sat(\Phi)$. The recursive computation basically boils down to a bottom-up traversal of the parse tree of the formula $\Phi$. For the propositional fragment of PRCTL this goes along the lines of CTL. For determining $Sat(\mathcal{L}_{\trianglelefteq p}(\Phi))$ we use the method of [3]. Model-checking time-bounded until-formulae is based on the path graph generation.

**Path Graph Generation.** The basic concept of the algorithm is to compute the "unfolding" of the DMRM under consideration while keeping track of the accumulated reward so far. Nodes in the tree that have the same accumulated rewards are grouped together in a single vertex. The resulting directed acyclic graph $(V, E, v_0)$ with finite (non-empty) set of vertices $V$, set of edges $E$ and a distinguished initial vertex $v_0$, is called *path graph*. We have $V \subseteq \mathbb{N} \times \mathcal{P}(S \times (0,1])$. The initial vertex $v_0$ equals $(0, \{(s_0, 1)\})$ where $s_0$ is the state of the DMRM to be investigated, 0 is the accumulated reward so far, and 1 denotes that the probability to be in state $s_0$ equals one (when starting in $s_0$). In general, vertex $v = (k, S_k)$ with $S_k = \{(s_1, p_1), \ldots, (s_m, p_m)\}$ denotes that starting from state $s_0$ each state $s_i$ ($0 < i \leqslant m$) can be reached with probability $p_i > 0$ (possibly via more than one path) while having earned a cumulative reward $k$. A path graph is basically an unfolding of the DMRM – while keeping track of the accumulated reward – and thus may be infinite. Since we are interested in paths of a certain length, viz. $n$, we "cut off" the unfolding at depth $n$. Formally, we consider the sets $V_0, \ldots, V_n$, where $V_i \subseteq V$ is the set of all vertices (of the above form) that can be reached in exactly $i$ steps. Thus, for $v = (k, S_k) \in V_n$ with $S_k = \{(s_1, p_1), \ldots, (s_m, p_m)\}$ we have that $\sum_i p_i$ equals the probability to gain $k$ reward in $n$ transitions when starting in state $s_0$. Figure 1 presents the

```
1. V₀ := { 0 };                      // only nodes with cumulative reward zero
2. S₀ := { (s₀, 1) };          // state s₀ can be reached with probability one
3. for (i := 0; i < n; i++)                  // i is current level number
4.    foreach m ∈ Vᵢ                         // check all rewards at level i
5.       foreach (s, p) ∈ Sₘ
6.          m' := m + ρ(s);                  // new cumulative reward
7.          foreach s' with P(s, s') > 0     // all direct successors of s
8.             if m' ∉ Vᵢ₊₁                   // encountering fresh reward m'
9.             then Vᵢ₊₁ := Vᵢ₊₁ ∪ { m' };       // add new vertex
10.                Sₘ' := { (s', p·P(s, s')) };
11.             elseif (s', p') ∈ Sₘ'         // shared node encountered?
12.                then p' := p' + p·P(s, s');
13.                else Sₘ' := Sₘ' ∪ { (s', p·P(s, s')) };
14.             endif;
15.          endforeach;
16.       endforeach;
17.    endforeach;
18. endfor;
```

**Fig. 1.** Path graph generation algorithm

pseudo-code of a variation of the path graph generation algorithm [16]. For the sake of simplicity, we let $V_i$ be a set of naturals such that if $m \in V_i$ then there is a vertex $v = (m, S_m) \in V_i$.

*Example 3.* The path graph for our running example DMRM for three steps ($n = 3$), starting from state $s_1$ is:

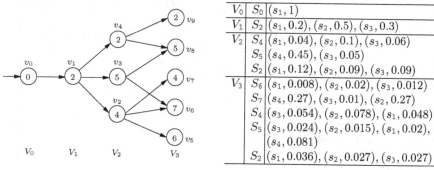

| $V_0$ | $S_0$ | $(s_1, 1)$ |
|---|---|---|
| $V_1$ | $S_2$ | $(s_1, 0.2), (s_2, 0.5), (s_3, 0.3)$ |
| $V_2$ | $S_4$ | $(s_1, 0.04), (s_2, 0.1), (s_3, 0.06)$ |
| | $S_5$ | $(s_4, 0.45), (s_3, 0.05)$ |
| | $S_2$ | $(s_1, 0.12), (s_2, 0.09), (s_3, 0.09)$ |
| $V_3$ | $S_6$ | $(s_1, 0.008), (s_2, 0.02), (s_3, 0.012)$ |
| | $S_7$ | $(s_4, 0.27), (s_3, 0.01), (s_2, 0.27)$ |
| | $S_4$ | $(s_3, 0.054), (s_2, 0.078), (s_1, 0.048)$ |
| | $S_5$ | $(s_3, 0.024), (s_2, 0.015), (s_1, 0.02),$ |
| | | $(s_4, 0.081)$ |
| | $S_2$ | $(s_1, 0.036), (s_2, 0.027), (s_3, 0.027)$ |

**Time-Bounded Transient Rewards: Fix-Point Characterization.** Verification algorithms for until-formulae (in CTL and PCTL) are inspired by a fix-point characterization [9, 11]. Checking the bounded until-operator in PRCTL amounts to computing the least solution of the following set of equations: $Prob(s, \Phi \, \mathcal{U}_J^N \, \Psi)$ equals 1 if $s \in Sat(\Psi)$ and $0 \in N$ and $0 \in J$,

$$Prob(s, \Phi \, \mathcal{U}_J^N \, \Psi) = \sum_{s' \in S} P(s, s') \cdot Prob(s', \Phi \, \mathcal{U}_{J \ominus \rho(s)}^{N \ominus 1} \, \Psi) \qquad (4)$$

if $s \in Sat(\Phi)$, $\sup N > 0$, and $\rho(s) \geqslant \sup J > 0$, and equals 0 otherwise. Here, $N \ominus n = \{ m-n \mid m \in N, m \geqslant n \}$ and $J \ominus r = \{ j-r \mid j \in J, j \geqslant r \}$ for some $r \in \mathbb{R}$. Stated in words, the probability to reach a $\Psi$-state from $s$ in $n \in N$ steps by earning a reward $r \in J$ equals the probability to move to a direct successor $s'$ of $s$ multiplied by the probability to reach a $\Psi$-state from $s'$ in $N \ominus 1$ transitions by earning $J \ominus \rho(s)$ reward. Model-checking the until-operator in PRCTL thus amounts to determining the least solution of this set of linear equations.

**Time-Bounded Transient Rewards: Algorithm.** The algorithm to check time- and (possibly) reward-bounded until-formulas is based on the path graph generation algorithm presented above. We discuss our algorithm for the case that $N$ is a singleton set, say $N = \{n\}$, and later discuss how this can be adapted to arbitrary sets. Suppose we have to check whether $s_0 \models \mathcal{P}_{\unlhd q}(\Phi \mathcal{U}_J^n \Psi)$ assuming $s_0 \models \Phi$. In order to do so, the following adaptations are made to the algorithm of Figure 1:

```
1.  V₀ := { 0 };                      // only nodes with cumulative reward zero
2.  S₀ := { (s₀, 1) };                // state s₀ can be reached with probability one
3.  for (i := 0; i < n; i++)
4.      foreach m ∈ Vᵢ                          // check all rewards at level i
5.          foreach (s, p) ∈ Sₘ
6.              if (m + ρ(s) ≤ sup J)           // reward bound not exceeded?
7.              then m' := m + ρ(s);                    // new cumulative reward
8.                  foreach s' with P(s, s') > 0    // all direct successors of s
9.                      if (i < n−1 ∧ s' ∈ Sat(Φ)) ∨ (i = n−1 ∧ s' ∈ Sat(Ψ))
10.                         if m' ∉ Vᵢ₊₁          // encountering fresh reward m'
11.                         then Vᵢ₊₁ := Vᵢ₊₁ ∪ { m' };      // add new vertex
12.                             Sₘ' := { (s', p·P(s, s')) };
13.                         elseif (s', p') ∈ Sₘ'      // shared node encountered?
14.                             then p' := p' + p·P(s, s');
15.                             else Sₘ' := Sₘ' ∪ { (s', p·P(s, s')) };
16.                         endif;
17.                     endif;
18.                 endforeach;
19.             endif;
20.         endforeach;
21.     endforeach;
22. endfor;
23. pr := 0;
24. foreach m ∈ (Vₙ ∩ J)
25.     foreach (s, p) ∈ Sₘ pr := pr + p; endforeach;
26. endforeach;
27. return pr ⊴ q.
```

**Fig. 2.** Checking whether $s_0 \models \mathcal{P}_{\unlhd q}(\Phi \mathcal{U}_J^n \Psi)$

- in selecting the successor states of $s$ (line 8 in Figure 1) we only consider $\Phi$-states if $i < n-1$ and only $\Psi$-states if $i = n-1$ (i.e., in the last step); this guarantees that all paths considered satisfy $\Phi \mathcal{U}^n \Psi$ and all other paths are ignored;
- it is checked whether the reward bound $\sup J$ is exceeded (line 6);
- in order to decide whether $s_0 \models \mathcal{P}_{\trianglelefteq q}(\Phi \mathcal{U}_J^n \Psi)$ we check (after finishing the outermost iteration) whether the total probability to end up in states with an accumulated reward in $J$ meets the bound $\trianglelefteq q$, i.e., whether

$$\sum_{(s,p)\in S_m \wedge m\in(V_n \cap J)} p \trianglelefteq q,$$

This requires an iteration over all vertices in $V_n \cap J$.

The resulting algorithm is presented in Figure 2. Note that the original path graph generation algorithm by Qureshi and Sanders [16] is obtained by checking the PRCTL formula $tt\,\mathcal{U}_k^n\,tt$.

The following optimization can be made. For checking properties with lower-bounds on the required probabilities (i.e., $\trianglelefteq \in \{\geqslant >\}$), the computation can be terminated as soon as the total probability of all vertices at level $i$ (with $i < n$) is less than $q$ (or at most $q$, respectively). In that case, it is certain that the PRCTL-formula is refuted – as the total probability will not further increase by going from level $i$ to $i+1$.

*Example 4.* Consider the formula $\mathcal{P}_{\geqslant 0.3}(a\,\mathcal{U}_{[6,10]}^4\,c)$ for state $s_1$. Stated in words, we want to check whether the probability to reach a $c$-state via an $a$-path (a path only consisting of $a$-states) in exactly 4 steps while earning a reward between 6 and 10 exceeds 0.3. Note that the path graph for $n = 4$ is the extension of the path graph of the previous example with the level $V_4$. The pruned path graph that is obtained after running our adapted path graph generation algorithm is:

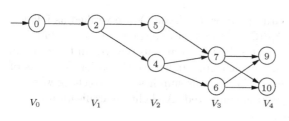

| | | |
|---|---|---|
| $V_0$ | $S_0$ | $(s_1, 1)$ |
| $V_1$ | $S_2$ | $(s_1, 0.2), (s_2, 0.5)$ |
| $V_2$ | $S_4$ | $(s_1, 0.04), (s_2, 0.1)$ |
| | $S_5$ | $(s_4, 0.45)$ |
| $V_3$ | $S_6$ | $(s_1, 0.008), (s_2, 0.02)$ |
| | $S_7$ | $(s_4, 0.36), (s_2, 0.18)$ |
| $V_4$ | $S_9$ | $(s_4, 0.234)$ |
| | $S_{10}$ | $(s_4, 0.162)$ |

As the sum of the probabilities in the vertices of $V_n$ (0.234+0.162) exceeds 0.3 we conclude that $s_1 \models \mathcal{P}_{\geqslant 0.3}(a\,\mathcal{U}_{[6,10]}^4\,c)$.

**Time Intervals.** The next question is how we can adapt the algorithm if the number of transitions is not fixed (like $n$ above) but an interval, i.e., $N = [n, n']$ with $0 \leqslant n \leqslant n'$. Obviously, in the worst case we have to generate all levels of the path graph from 0 to $n'$. Since the above algorithm does not keep track of probabilities achieved in earlier levels (but only in the last two levels), a new variable is introduced to accumulate these probabilities from level $n$ to $n'$. Furthermore, we cut-off all transitions emanating from a node for which the formula under consideration becomes true. Thus, if during model-checking $\mathcal{P}_{\trianglelefteq p}(\Phi \mathcal{U}_J^N \Psi)$ we encounter that $\Phi \mathcal{U}_J^n \Psi$ is valid for a generated path from $s_0$ to $s$, no further investigation of the sub-tree starting in $s$ is done as all such paths have the path $s_0, \dots, s$ as prefix.

**Unbounded Time Case.** For model checking an until-formula with an unbounded time (or reward) interval, the algorithm in Fig. 2 is not always terminating. To solve this problem we do the following. If the DMRM contains a strongly connected component (SCC) $B$ with only $\Phi$-states having reward 0, we transform (as a preprocessing step) the DMRM into an equivalent DMRM that does not have such SCCs. This basically amounts to compute $Prob(s, \Diamond \neg B)$ for each possible entrance state $s$ of $B$. This amounts to solving a system of linear equations [14].

**Multiple Rewards.** The previous algorithm can easily be extended in order to deal with DMRMs with more than one reward structure. Suppose we have $k$ reward structures and we are about to check whether $s_0 \models \mathcal{P}_{\trianglelefteq p}(\Phi \mathcal{U}_{J_1,\dots,J_k}^N \Psi)$. In this case, vertices in the path graph are tuples $(l_1, \dots, l_k, S_l)$ with $S_l = \{(s_1, p_1), \dots (s_m, p_m)\}$ as before that are to be interpreted as follows: starting from state $s_0$ each state $s_i$ $(0 < i \leqslant m)$ can be reached with probability $p_i$ while having earned reward $l_j$ according to reward structure $\rho_j$ $(0 < j \leqslant k)$. The algorithm is now obtained by interpreting $m$ and $m'$ as $k$-dimensional vectors and interpreting the statement in which these variables occur accordingly. For instance, $m + \rho(s) \leqslant \sup J$ should now be read as $\forall 0 < j \leqslant k . m_j + \rho_j(s) \leqslant \sup J_j$. Note that the time complexity of the algorithm is increased by a factor $k$.

**Reward Measures.** The reward-operators $\mathcal{C}$, $\mathcal{E}$, and $\mathcal{Y}$ are verified as follows. In order to decide whether $s \in Sat(\mathcal{C}_J^n(\Phi))$ we first determine the set $Sat(\Phi)$, i.e., the states that fulfill $\Phi$, and then sum the instantaneous rewards in these states at epoch $n$ (when starting in $s$) – which boils down to a transient analysis of the underlying DTMC – and check whether this sum lies in $J$. To check whether $s \in Sat(\mathcal{E}_J(\Phi))$, recursively $Sat(\Phi)$ is computed. A slight generalisation of [14, Theorem 3.23] now yields that

$$s \in Sat(\mathcal{E}_J(\Phi)) \text{ if and only if } \sum_{s' \in Sat(\Phi)} \pi(s, s') \cdot \rho(s') \in J$$

This thus boils down to solving a system of linear equations. For $\mathcal{E}_J^n(\Phi))$ we have:

$$s \in Sat(\mathcal{E}_J^n(\Phi)) \text{ if and only if } \sum_{s' \in Sat(\Phi)} \pi^*(s, s', n) \cdot \rho(s') \in J$$

where $\pi^*(s, s', n) = \frac{1}{n+1} \sum_{i=0}^n \pi(s, s', i)$. Finally, checking the $\mathcal{Y}$-operator amounts to solving a system of linear equations (again). The quantity

$y(s, Sat(\Phi), n)$ is characterized as the smallest solution of the following system of linear equations:

$$E(s,n) = \begin{cases} 0 & \text{if } n = 0 \\ \rho(s) + \sum_{s' \in S} \mathbf{P}(s,s') \cdot E(s', n-1) & \text{if } s \in Sat(\Phi) \ \wedge \ n > 0 \\ \sum_{s' \in S} \mathbf{P}(s,s') \cdot E(s', n-1) & \text{if } s \notin Sat(\Phi) \ \wedge \ n > 0 \end{cases}$$

**Complexity Analysis.** If real numbers are permitted as rewards, the time complexity of the algorithm (cf. Fig. 2) is exponential in $|S|$ due to the fact that all paths (of some length) may have distinct accumulated rewards. If, however, we restrict to naturals or rationals – which mostly suffices – as rewards, checking a time- and reward-bounded until-formula has a time complexity in $\mathcal{O}(n \cdot \sup J \cdot |S|^3)$.

## 5   Case Study: The IPv4 Zeroconf Protocol

As a case study, we consider the IPv4 zeroconf protocol, a simple protocol proposed by the IETF [6], aimed at the self-configuration of IP network interfaces in ad-hoc networks. The probababilistic behaviour of this protocol modeled as an DMRM is adopted from [5].

**The IPv4 Zeroconf Protocol.** The IPv4 zeroconf protocol is designed for a home local network of appliances (microwave oven, laptop, VCR, DVD-player etc.) each of which supplied with a network interface to enable mutual communication. Such ad-hoc networks must be hot-pluggable and self-configuring. Among others, this means that when a new appliance (interface) is connected to the network, it must be configured with a *unique* IP address automatically. The zeroconf protocol solves this task in the following way. A host that needs to be configured randomly selects an IP address, $U$ say, out of the 65024 available addresses and broadcasts a message (called *probe*) saying "Who is using the address $U$?". If the probe is received by a host that is already using the address $U$, it replies by a message indicating that $U$ is in use. After receiving this message the host to be configured will re-start: it randomly selects a new address, broadcasts a probe, etc.

Due to message loss or a busy host, a probe or a reply message may not arrive at some (or all) other hosts. In order to increase the reliability of the protocol, a host is required to send $n$ probes, each followed by a listening period of $r$ time units. Therefore, the host can start using the selected IP address only after $n$ probes have been sent and no reply has been received during $n \cdot r$ time units. Note that after running the protocol a host may still end up using an IP address already in use by another host, e.g., because all probes were lost. This situation, called *address collision*, is highly undesirable since it may force a host to kill active TCP/IP connections.

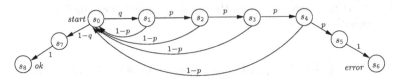

**Fig. 3.** DMRM model of the IPv4 zeroconf protocol (for $n$=4 probes)

**Modeling the Protocol.** The protocol behaviour of a single host is modeled by a DTMC that is adapted from [5]. The DTMC consists of $n+5$ states (cf. Figure 3 for $n = 4$) where $n$ is the maximal number of probes needed (as above). The initial state is $s_0$ (labeled *start*). In state $s_{n+4}$ (labeled *ok*) the host finally ends up with an unused IP address; in state $s_{n+2}$ (labeled *error*) it ends up with an address that is already in use, i.e., an address collision. State $s_i$ ($0 < i \leqslant n$) is reached after issuing the $i$-th probe. In state $s_0$ the host randomly chooses an IP address. With probability $q = m/65024$, where $m$ is the number of hosts in the network when connecting the host to the network, this address is already in use. With probability $1-q$ the host chooses an unused address and ends up in state $s_{n+3}$. Then it issues $n-1$ probes and waits $n \cdot r$ time units before using this address. If the chosen IP address is already in use, state $s_1$ is reached. Now two situations are possible. With probability $p$, no reply is received during $r$ time units (as either the probe or its reply has been lost), and a next probe is sent, resulting in state $s_2$. If, however, a reply has arrived in time, the host returns to the initial state and re-starts the protocol. The behaviour in state $s_i$ ($2 \leqslant i < n$) is similar. If in state $n$, however, no reply has received within $r$ time units after sending the $n$-th probe, an address collision occurs.

We consider three reward structures for this model:

- The first reward assignment (denoted $\rho_1$) represents waiting times and is defined by: $\rho_1(s_i) = r$ for $0 < i \leqslant n$, $\rho_1(s_0) = 0$ assuming that the host randomly selects an address promptly, $\rho_1(s_{n+3}) = n \cdot r$, $\rho_1(s_{n+2}) = \rho_1(s_{n+4}) = 0$, and $\rho_1(s_{n+1}) = E$, where $E$ denotes some large number that represents the highly undesirable situation of an address collision.
- The second reward assignment (denoted $\rho_2$) is used to keep track of the number of probes that are sent in total. It is defined by: $\rho_2(s_i) = 1$ for $0 < i \leqslant n$, $\rho_2(s_{n+3}) = n$ and 0 otherwise.
- Finally, the third reward assignment (denoted $\rho_3$) is used to keep track of the number of failed attempts to acquire an unused address. It is defined by: $\rho_3(s_0) = 1$ and 0 otherwise.

**Properties of Interest.** The reward-based operators are subscripted with three reward intervals that refer to the three reward structures defined above. Intervals equal to $[0, \infty)$ are represented by a small line; if all reward intervals equal $[0, \infty)$ then the subscript is omitted. For instance, $\Diamond_{-[4,10]-}$error asserts that the protocol eventually ends up with an address collision (state *error*) where between 4 and 10 probes have been sent; no constraints are given on the number of collisions and the required time. We consider the following properties that are of interest for the IPv4 zeroconf protocol and provide their formal specification in PRCTL.

(i) "The probability to end up with an unused address is at least $p'''$": $\mathcal{P}_{\geqslant p'}(\Diamond ok)$. As state $ok$ is one of the BSCCs of the DTMC an alternative formulation would be $\mathcal{L}_{\geqslant p'}(ok)$. (Note that, in general, the formulae $\mathcal{P}(\Diamond \Phi)$ and $\mathcal{L}(\Phi)$ are not equivalent.)

(ii) "The probability to end up with an unused address within time $t$ exceeds $p'''$":

$$\mathcal{P}_{> p'}(\Diamond_{[0,t]--} ok)$$

(iii) "The probability to end up with an unused address after at most $k$ probes exceeds $p'''$":

$$\mathcal{P}_{> p'}(\Diamond_{-[0,k]-} ok)$$

(iv) "The probability to end up with an unused address within time $t$ while having sent at most $k$ probes exceeds $p'''$":

$$\mathcal{P}_{> p'}(\Diamond_{[0,t][0,k]-} ok)$$

(v) "The probability to select more than $k$ times a used address during the execution of the protocol is at most $p'''$":

$$\mathcal{P}_{\leqslant p'}(\Diamond_{--[k+1,\infty)} start)$$

Here we use the fact that on selecting a used address the host returns to the $start$ state. As the host also starts the protocol in that state, the lowerbound of the reward bound equals $k+1$ (rather than $k$).

Note that the first property does not refer to any reward and is in fact PCTL-formula that can be verified using any model checker for this logic.

**Verification Results.** We use the following settings for the parameters: $n = 7$, $m = 10^4$, $p = 10^{-3}$, and only present results concerning the two formulas (ii) and (iv), both requiring the application of the main algorithm (Figure 2). On the top of Figure 4 different plots are shown for varying values of the bound $t$ in formula (ii), $\mathcal{P}_{> p'}(\Diamond_{[0,t]--} ok)$. We display the border probability $p'$ where the truth value of the formula turns from false to true. These boundary probabilities are very close to 1. Therefore we use a semi-logarithmic scale, and plot $1 - p'$ instead of $p'$. The value $1 - p'$ corresponds to the probability of not obtaining an unused address in time. As expected, increasing the waiting time $r$ decreases the likelihood to obtain an unused address in time; but small changes to $r$ may not induce a change of the likelihood. This phenomenon has to do with the fact that the state prior to the $ok$-state ($s_7$ in Figure 3) has a reward $n \cdot r$. For a fixed upper bound on the reward (here time $t$), a small increase of $r$ does not necessarily induce less opportunities (i.e. paths) to reach the $ok$-state,

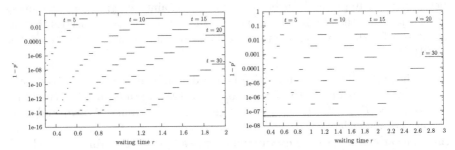

**Fig. 4.** Boundary probabilities for formula (ii) (left) and (iv) (right)

explaining the displayed discontinuities. We further note that increasing the time
bound $t$ also decreases the probability for a successful address assignment. On
the bottom of Figure 4 we depict the corresponding border probabilities for
formula (iv), $\mathcal{P}_{>p'}(\Diamond_{[0,t][0,k]}-ok)$, with the same variations on $t$, while $k = 15$
remains constant. We observe that the shape of the displayed plots is similar to
the corresponding ones for formula (ii), but the computed probabilities $p'$ are
lower by several orders of magnitude. This is a result of the constraining effect of
the second reward interval $[0, k]$, which induces that far less paths to the $ok$-state
satisfy all constraints.

## 6    Related Work

A temporal logic with accompanying efficient model-checking algorithms for
CTMCs has been introduced by Baier *et al.* [3] and was later extended to reward
models [2, 12]. Another notable approach to continuous time reward analysis is
based on path automata [15]. Basic analysis algorithms for continuous reward
models are discussed in [16] which served also as a basis for our discrete time
model checking algorithms.

In the discrete time context, we are aware of the work of Voeten [20], who
describes a rich assembly language for discrete reward measures. Instead of
state-space based analysis algorithms, discrete event simulation is proposed to
compute these measures. De Alfaro [10] also describes analysis algoritms for
DMRM-like models, focussing on long-run average behaviours rather than on
finite-horizon properties such as bounded until while allowing for nondeterminism. DMRM models have recently become somewhat *en vogue* as models for
power-aware systems. Sokolsky *et al.* [19] have proposed a process algebra to
specify power-aware systems as discrete time Markov models with nondeterminism, and have proposed model checking a $\mu$-calculus-like logic for their analysis.

## 7    Conclusion

This paper developed logics and algorithms for model checking discrete time
Markov reward models, providing means to formulate and efficiently check com-

plex measures constraints – involving expected as well as accumulated rewards. We illustrated the approach by model-checking a probabilistic cost model of the IPv4 zeroconf protocol developed for ad-hoc network address assignement.

Based on the work presented here, we are further investigating efficient algorithms for continuous time reward model checking.

## Acknowledgements

This work has taken place in the context of the SPACE project (SPecification-bAsed Performability ChEcking) that is supported by the Netherlands Organization for Scientific Research (NWO). We thank William H. Sanders for fruitful discussions at an early stage of this work. Henrik Bohnenkamp is thanked for discussions concerning the zeroconf protocol. Part of this work has been achieved during the Dagstuhl Seminar 03201 "Probabilistic Methods in Verification and Planning".

## References

[1] A. Aziz, K. Sanwal, V. Singhal and R. Brayton. Verifying continuous time Markov chains. In Computer-Aided Verification, LNCS 1102, pp. 269–276, 1996.

[2] C. Baier, B. R. Haverkort, H. Hermanns and J.-P. Katoen. On the logical characterisation of performability properties. In Automata, Languages, and Programming, LNCS 1853, pp. 780–792, 2000. 88, 89, 102

[3] C. Baier, B. R. Haverkort, H. Hermanns and J.-P. Katoen. Model-checking algorithms for continuous-time Markov chains. IEEE Transactions on Software Engineering, **29**(6):524–545, 2003. 88, 90, 93, 94, 102

[4] M. Bernardo. An algebra-based method to associate rewards with EMPA terms. In Automata, Languages and Programming (ICALP), LNCS 1256, pp. 358–368, 1997. 88

[5] H. Bohnenkamp, P. van der Stok, H. Hermanns, and F. W. Vaandrager. Cost optimisation of the IPv4 zeroconf protocol. In Intl. Conf. on Dependable Systems and Networks. IEEE CS Press. 2003. (to appear). 99, 100

[6] S. Cheshire, B. Adoba and E. Guttman. Dynamic configuration of IPv4 link-local addresses. 2002 (draft). (available at www.ietf.org/internet-drafts). 99

[7] G. Ciardo, J. Muppala and K. S. Trivedi. SPNP: Stochastic Petri Net Package. In Proc. 3rd Int'l Workshop on Petri Nets and Performance Models: 142–151, 1989. 88

[8] G. Clark, S. Gilmore, and J. Hillston. Specifying performance measures for PEPA. In Formal Methods for Real-Time and Probabilistic Systems, LNCS 1601, pp. 211–227, 1999. 88

[9] E. Clarke, O. Grumberg and D. Peled. Model Checking. MIT Press, 1999. 88, 94, 95

[10] L. de Alfaro. How to specify and verify the long-run average behavior of probabilistic systems. In IEEE Symp. on Logic in Computer Science, pp. 454-465, 1998. 102

[11] H. Hansson and B. Jonsson. A logic for reasoning about time and reliability. Form. Asp. of Comp. **6**: 512–535, 1994. 89, 92, 95

[12] B. Haverkort, L. Cloth, H. Hermanns, J.-P. Katoen and C. Baier. Model checking performability properties. In Intl. Conf. on Dependable System s and N etworks, pp. 103–113, 2002. 89, 102

[13] J. G. Kemeny and J. L. Snell. Finite M arkov Chains. Van Nostrand, 1960. 91

[14] V. G. Kulkarni. M odeling and A nalysis of Stochastic System s. Chapman & Hall, 1995. 89, 91, 98

[15] W. D. Obal and W. H. Sanders. State-space support for path-based reward variables. Perform ance E valuation, 35(3-4): 233–251, 1999. 88, 102

[16] M. A. Qureshi and W. H. Sanders. Reward model solution method with impulse and rate rewards: An algorithm and numerical results. Perform ance E valuation, 20: 413–436, 1994. 95, 97, 102

[17] A. Reibman and K. Trivedi. Transient analysis of cumulative measures of Markov model behavior. Comm un. Statist.Stochastic M odels, 5(4):683–710, 1989.

[18] R. M. Smith, K. S. Trivedi and A. V. Ramesh. Performability analysis: measures, an algorithm and a case study. IEEE Tr. on Computers, 37(4):406–417, 1988. 91

[19] O. Sokolsky, A. Philippou I. Lee, and K. Christou. Modeling and analysis of power-aware systems. In Tools and A lgorithm s for the C onstruction and A nalysis of System s, LNCS 2619, pp. 409–425, 2003. 102

[20] J. P. M. Voeten. Performance evaluation with temporal rewards. Perform ance E valuation, 50:189–218, 2002. 102

# Performance Analysis of Probabilistic Timed Automata Using Digital Clocks*

Marta Kwiatkowska[1], Gethin Norman[1], David Parker[1], and Jeremy Sproston[2]

[1] School of Computer Science
University of Birmingham, Birmingham B15 2TT, United Kingdom
{m.z.kwiatkowska,g.norman,d.a.parker}@cs.bham.ac.uk
[2] Dipartimento di Informatica
Università di Torino, 10149 Torino, Italy
sproston@di.unito.it

**Abstract.** Probabilistic timed automata, a variant of timed automata extended with discrete probability distributions, is a specification formalism suitable for describing both nondeterministic and probabilistic aspects of real-time systems, and is amenable to model checking against probabilistic timed temporal logic properties. In the case of classical (non-probabilistic) timed automata, it has been shown that for a large class of real-time verification problems correctness can be established using an integer-time model, inducing a notion of digital clocks, as opposed to the standard dense model of time. Based on these results, we address the question of under what conditions digital clocks are sufficient for the performance analysis of probabilistic timed automata. We extend previous results concerning the integer-time semantics of an important subclass of probabilistic timed automata to consider the computation of expected costs or rewards. We illustrate this approach through the analysis of the dynamic configuration protocol for IPv4 link-local addresses.

## 1 Introduction

Network protocols increasingly often rely on the use of randomness and timing delays, for example exponential back-off in Ethernet and IEEE 802.11, and IEEE 1394 FireWire root contention. Since these protocols execute in a distributed environment, it is important to also consider nondeterminism when modelling their behaviour. For example, we may wish to model a system for which the likelihood of a certain event occurring changes with respect to the amount of time elapsed. A natural model for systems that exhibit nondeterminism, probability and real-time, called *probabilistic timed automata* – a probabilistic extension of timed automata [1] – has been proposed in [19]. In the probabilistic timed automata model real-valued clocks measure the passage of time and transitions can be probabilistic, that is, be expressed as a discrete probability distribution on the set of target states. In [19] model checking algorithms for verifying the

---

* Supported in part by the EPSRC grants GR/N22960 and GR/S11107.

K.G. Larsen and P. Niebert (Eds.): FORMATS 2003, LNCS 2791, pp. 105–120, 2004.
© Springer-Verlag Berlin Heidelberg 2004

likelihood of certain temporal properties being satisfied by such system models are introduced. However, these model checking algorithms are either based on *region equivalence* [1], and hence suffer from the state-space explosion problem, or on *forwards reachability*, which leads to approximate results [19, 11]. An alternative approach, based on *backwards reachability*, is given in [20].

When modelling real-time systems, there is often a trade-off between the expressiveness of the model and the complexity of the associated solution algorithms. A *dense*-time model is more expressive than an *integer*-time model; however, it is often the case that an integer-time model is easier to verify, since it can lead to a finite-state system and allows one to apply efficient symbolic methods developed for untimed systems. We refer to the clocks of an integer-time model as *digital clocks*. Henzinger et al. [15] study the question of which real-time properties can be verified by considering system behaviours featuring only integer durations. These results are applied to timed automata in [9, 24], and it is shown that an approach using digital clocks is applicable to the verification of *closed, diagonal-free* timed automata; intuitively, these are automata whose constraints do not compare clocks or use strict comparison operators.

We have previously shown that probabilistic reachability properties, such as 'with probability 0.05 or less, the system aborts', of closed, diagonal-free probabilistic timed automata can be analysed faithfully using digital clocks [22]. The main contribution of this paper is to extend this research by showing that digital clocks are also sufficient for verifying *expected reachability* properties such as 'the expected time until a data packet is delivered is at most 0.05 seconds', or 'the expected cost of a host choosing an IP address is at most 40'.

In [12], de Alfaro presents a model-checking algorithm for verifying probabilistic and expected reachability properties of finite-state models. We implemented the algorithms of de Alfaro in the probabilistic model checking tool PRISM [17, 25], allowing us to automatically verify expected-cost properties of interest for integer-time models.

The paper proceeds by revisiting the definition of probabilistic timed automata in the next section. Expected reachability properties for probabilistic timed automata are presented in Section 3, and the correctness of the digital clock interpretation of probabilistic timed automata with respect to these properties is stated. In Section 4, we present a case study, in which PRISM is used to analyse the performance of the dynamic configuration protocol for IPv4 link-local addresses. Finally, in Section 5, we conclude the paper.

## 2    Probabilistic Timed Automata

**Time, Clocks, Zones and Distributions.** Let $\mathbb{T} \in \{\mathbb{R}, \mathbb{N}\}$ be the *time domain* of either the non-negative reals or naturals. Let $\mathcal{X}$ be a finite set of variables called *clocks* which take values from the time domain $\mathbb{T}$. A point $v \in \mathbb{T}^{|\mathcal{X}|}$ is referred to as a *clock valuation*. Let $\mathbf{0} \in \mathbb{T}^{|\mathcal{X}|}$ be the clock valuation which assigns 0 to all clocks in $\mathcal{X}$. For any $v \in \mathbb{T}^{|\mathcal{X}|}$ and $t \in \mathbb{T}$, the clock valuation $v \oplus t$

denotes the *time increment* of values in $v$ by $t$. We use $v[X := 0]$ to denote the clock valuation obtained from $v$ by resetting all of the clocks in $X \subseteq \mathcal{X}$ to 0.

Let $Zones(\mathcal{X})$ be the set of *zones* over $\mathcal{X}$, which are conjunctions of atomic constraints of the form $x \sim c$ for $x \in \mathcal{X}$, $\sim \in \{\leq, =, \geq\}$, and $c \in \mathbb{N}$. The clock valuation $v$ *satisfies* the zone $\zeta$, written $v \models \zeta$, if and only if $\zeta$ resolves to true after substituting each clock $x \in \mathcal{X}$ with the corresponding clock value from $v$. Readers familiar with timed automata will note that we consider the syntax of *closed, diagonal-free* zones, which *do not* feature atomic constraints of the form $x > c$ or $x < c$ (closed) or $x - y \sim c$ (diagonal free) for $x, y \in \mathcal{X}$, $c \in \mathbb{N}$.

A discrete probability *distribution* over a countable set $Q$ is a function $\mu : Q \to [0, 1]$ such that $\sum_{q \in Q} \mu(q) = 1$. For a possibly uncountable set $Q'$, let $\mathsf{Dist}(Q')$ be the set of distributions over countable subsets of $Q'$. For $q \in Q$, let $\mu_q \in \mathsf{Dist}(Q)$ be the distribution which assigns probability 1 to $q$.

## Syntax of Probabilistic Timed Automata.
We review the definition of probabilistic timed automata [19].

**Definition 1.** *A* probabilistic timed automaton *is a tuple* $(L, \bar{l}, \mathcal{X}, \Sigma, I, prob)$ *where: $L$ is a finite set of* locations *including the* initial location $\bar{l}$*; $\mathcal{X}$ is a set of* clocks*; $\Sigma$ is a finite set of* events*; the function* $I : L \to Zones(\mathcal{X})$ *is the* invariant condition*; and the finite set* $prob \subseteq L \times Zones(\mathcal{X}) \times \Sigma \times \mathsf{Dist}(2^{\mathcal{X}} \times L)$ *is the* probabilistic edge relation.

A *state* of a probabilistic timed automaton is a pair $(l, v)$ where $l \in L$ and $v \in \mathbb{T}^{|\mathcal{X}|}$ are such that $v \models I(l)$. Informally, the behaviour of a probabilistic timed automaton can be understood as follows. The model starts in the state $(\bar{l}, \mathbf{0})$; that is, in the initial location $\bar{l}$ with all clocks set to 0. In any state $(l, v)$, there is a nondeterministic choice of either (1) making a *discrete transition* or (2) letting *time pass*. In case (1), a discrete transition can be made according to any $(l, g, \sigma, p) \in prob$ which is *enabled*; that is, the zone $g$ is satisfied by the current clock valuation $v$. Then the probability of moving to the location $l'$ and resetting all of the clocks in $X$ to 0 is given by $p(X, l')$. In case (2), the option of letting time pass is available only if the invariant condition $I(l)$ is satisfied while time elapses. Note that we often refer to the model presented above as *closed, diagonal-free probabilistic timed automata*, in order to distinguish the zones used with those in previous work [19].

## Semantics of Probabilistic Timed Automata.
The semantics of probabilistic timed automata is defined in terms of *timed probabilistic systems*, which exhibit timed, nondeterministic and probabilistic behaviour. They are a variant of Markov decision processes [14] and Segala's probabilistic timed automata [26].

**Definition 2.** *A* timed probabilistic system $\mathsf{PS} = (S, \bar{s}, Act, \mathbb{T}, Steps)$ *consists of a set $S$ of states, an initial state $\bar{s} \in S$, a set $Act$ of actions, a time domain $\mathbb{T}$, and a probabilistic transition relation* $Steps \subseteq S \times (Act \cup \mathbb{T}) \times \mathsf{Dist}(S)$.

A *probabilistic transition* $s \xrightarrow{a,\mu} s'$ is made from a state $s \in S$ by first nondeterministically selecting an action-distribution or duration-distribution pair $(a, \mu)$ such that $(s, a, \mu) \in Steps$, and second by making a probabilistic choice of target state $s'$ according to the distribution $\mu$, such that $\mu(s') > 0$.

We consider two ways in which a timed probabilistic system's computation may be represented. A *path* represents a particular resolution of both nondeterminism *and* probability. Formally, a path of a timed probabilistic system is a finite or infinite sequence of probabilistic transitions $\omega = s_0 \xrightarrow{a_0, \mu_0} s_1 \xrightarrow{a_1, \mu_1} \cdots$. A path $\omega$ is *initialised in $s$* if $s_0 = s$. We denote by $\omega(i)$ the $(i+1)$th state of $\omega$, $last(\omega)$ the last state of $\omega$ if $\omega$ is finite, and $step(\omega, i)$ the action associated with the $i$-th step. If $\omega$ is infinite, the duration up to the $(n+1)$th state of $\omega$ is defined by $\mathcal{D}_\omega(n+1) \stackrel{\text{def}}{=} \sum\{a_i \mid 0 \le i \le n \wedge a_i \in \mathbb{T}\}$. Let $Path_{ful}(s)$ be the set of infinite paths initialised in $s$.

The second notion of a timed probabilistic system's computations is that of an *adversary*, which represents a particular resolution of nondeterminism *only*. Formally, an adversary is a function $A$ mapping every finite path $\omega$ to a pair $(a, \mu)$ such that $(last(\omega), a, \mu) \in Steps$ [28]. For any adversary $A$, let $Path_{ful}^A(s)$ denote the set of infinite paths initialised in $s$ associated with $A$. Then, we define the probability measure $Prob_s^A$ over $Path_{ful}^A(s)$ by classical techniques [16].

We restrict our attention to *time-divergent adversaries*; a common restriction imposed in real-time systems so that unrealisable behaviour (corresponding to time not advancing beyond a bound) is disregarded during analysis. We say that a path $\omega$ is *divergent* if for any $t \in \mathbb{R}$, there exists $j \in \mathbb{N}$ such that $\mathcal{D}_\omega(j) > t$.

**Definition 3.** *An adversary $A$ of a timed probabilistic system* PS *is divergent if and only if for each state $s$ the probability $Prob_s^A$ of the divergent paths of $Path_{ful}^A(s)$ is 1. Furthermore, let $Adv_{PS}$ be the set of divergent adversaries of* PS.

We now define the semantics of probabilistic timed automata defined in terms of timed probabilistic systems. Observe that the definition is parameterized both by a time domain $\mathbb{T}$ and time increment $\oplus$, and that the summation in the definition of discrete transitions is required for the cases in which multiple clock resets result in the same target location.

**Definition 4.** *Let* PTA $= (L, \bar{l}, \mathcal{X}, \Sigma, I, prob)$ *be a probabilistic timed automaton. The semantics of* PTA *with respect to the time domain $\mathbb{T}$ and the time increment $\oplus$ is the timed probabilistic system* $[\![PTA]\!]_\mathbb{T}^\oplus = (S, \bar{s}, \Sigma, \mathbb{T}, Steps)$ *where:* $S \subseteq L \times \mathbb{T}^{|\mathcal{X}|}$ *and* $(l, v) \in S$ *if and only if* $v \models I(l)$; $\bar{s} = (\bar{l}, \mathbf{0})$; *and* $((l, v), a, \mu) \in$ Steps *if and only if one of the following conditions holds:*

**Time transitions.** $a \in \mathbb{T}$, $\mu = \mu_{(l, v \oplus a)}$ *and* $v \oplus t \models I(l)$ *for all* $0 \le t \le a$;
**Discrete transitions.** $a \in \Sigma$ *and there exists* $(l, g, \sigma, p) \in prob$ *such that* $v \models g$ *and for any* $(l', v') \in S$, *we have* $\mu(l', v') = \sum_{X \subseteq \mathcal{X} \,\&\, v' = v[X := 0]} p(X, l')$.

Traditionally, the semantics of probabilistic timed automata assumes that the reals form the underlying model of time, paired with a time increment which is standard addition. The continuous semantics of a probabilistic timed automaton is a timed probabilistic system with generally uncountably many states.

**Definition 5.** *The* continuous semantics *of a probabilistic timed automaton* PTA *is defined as* $[\![PTA]\!]^{+}_{\mathbb{R}}$; *that is,* $\mathbb{T} = \mathbb{R}$ *and* $\oplus = +$.

**Higher-Level Modelling.** To aid modelling, probabilistic timed automata can be *composed in parallel* [22], and can feature *integer variables*, *urgent locations and events*, and *committed locations* (as in UPPAAL timed automata [3]). The techniques of [27] can be adapted to represent, syntactically, integer variables and committed locations within our definition of probabilistic timed automata; urgent events require a minor adjustment to the semantics of probabilistic timed automata [21].

# 3  Performance Measures

In this section, we consider two performance measures for probabilistic timed automata. The first is *probabilistic reachability*, namely the maximal and minimal probability of reaching, from the initial state, a certain set of goal or target states. For a timed probabilistic system $PS = (S, \bar{s}, Act, \mathbb{T}, Steps)$, set $F \subseteq S$ of target states, and adversary $A \in Adv_{PS}$, let:

$$p^{A}_{\bar{s}}(F) \stackrel{\text{def}}{=} Prob^{A}_{\bar{s}}\{\omega \in Path^{A}_{ful}(\bar{s}) \mid \exists i \in \mathbb{N}\,.\,\omega(i) \in F\}\,.$$

**Definition 6.** *The* maximal *and* minimal reachability probabilities *of reaching the set of states* $F$ *of the timed probabilistic system* PS *are defined as follows:*

$$p^{\max}_{PS}(F) \stackrel{\text{def}}{=} \sup_{A \in Adv_{PS}} p^{A}_{\bar{s}}(F) \quad and \quad p^{\min}_{PS}(F) \stackrel{\text{def}}{=} \inf_{A \in Adv_{PS}} p^{A}_{\bar{s}}(F)\,.$$

This performance measure has been studied in the context of probabilistic timed automata by Kwiatkowska et al. [19, 22].

The second measure we consider is *expected reachability*, which allows us to compute the expected cost (or reward) accumulated before reaching a certain set of states. Expected reachability is defined with respect to a cost function mapping actions and durations to real values, as well as a set $F \subseteq S$ of target states, and corresponds to the expected cost (with respect to the given cost function) of reaching a state in $F$. More formally, for a timed probabilistic system $PS = (S, \bar{s}, Act, \mathbb{T}, Steps)$, cost function $c : Act \cup \mathbb{T} \to \mathbb{R}$, set $F \subseteq S$ of target states, and adversary $A \in Adv_{PS}$, let $e^{A}_{\bar{s}}(cost(c, F))$ denote the usual expectation with respect to the measure $Prob^{A}_{\bar{s}}$ over $Path^{A}_{ful}(\bar{s})$, where for any $\omega \in Path^{A}_{ful}(\bar{s})$:

$$cost(c, F)(\omega) = \begin{cases} \displaystyle\sum_{i=1}^{\min\{j \mid \omega(j) \in F\}} c(step(\omega, i-11)) & \text{if } \exists j \in \mathbb{N}.\ \omega(j) \in F \\ \infty & \text{otherwise.} \end{cases}$$

The value of $cost(c, F)(\omega)$ equals the total cost, with respect to the cost function $c$, accumulated until a state in $F$ is reached along the path $\omega$. Note that we define the cost of a path which does not reach $F$ to be $\infty$, even though the total cost of the path may not be infinite. Hence, the expected cost of reaching $F$ from $s$ is finite if and only if a state in $F$ is reached from $s$ with probability 1. *Expected time reachability* (the expected time with which a given set of states can be reached) is a special case of expected reachability, corresponding to the case when $c(a) = 0$ for all $a \in Act$ and $c(t) = t$ for all $t \in \mathbb{T}$.

**Definition 7.** *The* maximal and minimal expected costs *of reaching a set of states $F$ under the cost function $c$ in the timed probabilistic system* PS *are defined as follows:*

$$e_{\mathsf{PS}}^{\max}(c, F) = \sup_{A \in Adv_{\mathsf{PS}}} e_{\bar{s}}^{A}(cost(c, F)) \quad and \quad e_{\mathsf{PS}}^{\min}(c, F) = \inf_{A \in Adv_{\mathsf{PS}}} e_{\bar{s}}^{A}(cost(c, F)).$$

We note that calculating expected reachability is equivalent to the *stochastic shortest path problem* for Markov decision processes; see for example [6].

At the level of probabilistic timed automata, one can define a cost function using a pair $(r, c_{\Sigma})$, where $r \in \mathbb{R}$ gives the rate at which cost is accumulated as time passes, and $c_{\Sigma} : \Sigma \to \mathbb{R}$ is a function assigning the cost of executing each event in $\Sigma$. The associated cost function $c_{r,c_{\Sigma}}$ is defined by $c_{r,c_{\Sigma}}(t) = t \cdot r$ for all $t \in \mathbb{T}$, and $c_{r,c_{\Sigma}}(\sigma) = c_{\Sigma}(\sigma)$ for all $\sigma \in \Sigma$. A probabilistic timed automaton equipped with a pair $(r, c_{\Sigma})$ is a probabilistic generalisation of uniformly priced timed automata [4].

For both probabilistic and expected reachability, we can consider reaching a state satisfying a formula which is a conjunction of propositions identifying locations and clock constraints of the form $x \sim c$ for $x \in \mathcal{X}$, $\sim \in \{\leq, =, \geq\}$ and $c \in \mathbb{N}$. Instead of considering these cases separately, we just note that such reachability problems can be reduced to those referring to locations only by modifying syntactically the probabilistic timed automaton of interest (see [19]).

For examples of the types of properties of probabilistic timed automata which can be expressed using expected reachability, consider the following: 'the expected time until a host can use an IP address is at most 0.05 seconds', 'the expected number of packets sent before failure is at least 300' and 'the expected number of lost messages within the first 200 seconds is at most 10'. In the case of the third example, we would first need to modify the probabilistic timed automaton under study by adding a distinct clock (to represent global time) and a location such that, from all locations, once the global clock has reached 200 seconds, the only transition is to this new location. The set of target states would then be the set containing only the new location and the cost function would equal 0 on all time transitions and events except those events corresponding to a message being lost; the costs for those actions would be set to 1.

**Performance Measures and Digital Clocks.** We now show, under the restriction that the probabilistic timed automaton under study is diagonal-free and closed, that it suffices just to consider the integer-time semantics when verifying expected reachability properties.

**Definition 8.** *For any $x \in \mathcal{X}$, let $\mathbf{k}_x$ denote the greatest constant that the clock $x$ is compared to in the zones of* PTA. *Define $\oplus_{\mathbb{N}}$ such that, for any clock* *valuation* $v \in \mathbb{N}^{|\mathcal{X}|}$ *and time duration* $t \in \mathbb{N}$, *the clock valuation* $v \oplus_{\mathbb{N}} t$ *assigns the value* $\min\{v_x + t, \mathbf{k}_x + 1\}$ *to all clocks* $x \in \mathcal{X}$. *The* integer-time semantics *of* PTA *is then defined as* $[\![\text{PTA}]\!]_{\mathbb{N}}^{\oplus_{\mathbb{N}}}$; *that is,* $\mathbb{T} = \mathbb{N}$ *and* $\oplus = \oplus_{\mathbb{N}}$.

Let PTA $= (L, \bar{l}, \mathcal{X}, \Sigma, I, prob)$ be a (closed, diagonal-free) probabilistic timed automaton. For any set of locations $L' \subseteq L$, we denote by $F_{\mathbb{T}}^{L'}$ the set of all states of $[\![\text{PTA}]\!]_{\mathbb{T}}^{\oplus}$ which correspond to these locations; that is $F_{\mathbb{T}}^{L'} = \{(l, v) \mid l \in L',\ v \in \mathbb{T}^{|\mathcal{X}|} \wedge v \models I(l)\}$.

**Theorem 1.** *For any (closed, diagonal-free) probabilistic timed automaton* PTA, *set of locations* $L' \subseteq L$ *and cost function* $c : \Sigma \cup \mathbb{R} \to \mathbb{R}$ *which satisfies* $c(t+t') = c(t) + c(t')$ *for all* $t, t' \in \mathbb{R}$:

$$e_{[\![\text{PTA}]\!]_{\mathbb{R}}^{+}}^{\max}(c, F_{\mathbb{R}}^{L'}) = e_{[\![\text{PTA}]\!]_{\mathbb{N}}^{\oplus_{\mathbb{N}}}}^{\max}(c, F_{\mathbb{N}}^{L'}) \quad and \quad e_{[\![\text{PTA}]\!]_{\mathbb{R}}^{+}}^{\min}(c, F_{\mathbb{R}}^{L'}) = e_{[\![\text{PTA}]\!]_{\mathbb{N}}^{\oplus_{\mathbb{N}}}}^{\min}(c, F_{\mathbb{N}}^{L'}).$$

The proof of the correctness of Theorem 1 can be found in [18]. Note that any cost functions defined by a pair $(r, c_\Sigma)$, where $r \in \mathbb{R}$ and $c_\Sigma : \Sigma \to \mathbb{R}$, will satisfy the condition $c(t + t') = c(t) + c(t')$ for all $t, t' \in \mathbb{R}$. The analogous result for probabilistic reachability is proved in [22] and states:

$$p_{[\![\text{PTA}]\!]_{\mathbb{R}}^{+}}^{\max}(F_{\mathbb{R}}^{L'}) = p_{[\![\text{PTA}]\!]_{\mathbb{N}}^{\oplus_{\mathbb{N}}}}^{\max}(F_{\mathbb{N}}^{L'}) \quad and \quad p_{[\![\text{PTA}]\!]_{\mathbb{R}}^{+}}^{\min}(F_{\mathbb{R}}^{L'}) = p_{[\![\text{PTA}]\!]_{\mathbb{N}}^{\oplus_{\mathbb{N}}}}^{\min}(F_{\mathbb{N}}^{L'}).$$

# 4   Case Study: Dynamic Configuration of Link-Local Addresses in IPv4

In this section, we illustrate the utility of the integer-time semantics of probabilistic timed automata with an analysis of the dynamic configuration protocol for IPv4 link-local addresses [10].

The dynamic configuration protocol for IPv4 addresses offers a distributed 'plug-and-play' solution in which IP address configuration is managed by individual devices connected to a local network. Upon connecting to the network, a device, henceforth called a *host*, first randomly chooses an IP address from a pool of 65024 available (the Internet Assigned Number Authority has allocated the addresses from 169.254.1.0 to 169.254.254.255 for the purpose of such link-local networks). The host waits a random time of between 0 and 2 seconds before sending four *Address Resolution Protocol* (ARP) packets, called *probes*, to all of the other hosts of the network. Probes contain the IP address selected by the host, operate as requests to use the address, and are sent at 2 second intervals. A host which is already using the address will respond with an ARP reply packet, asserting its claim to the address, and the original host will restart the protocol by reconfiguring its chosen address and sending new probes. If the host sends four probes without receiving an ARP reply packet, then it commences to use the chosen IP address. The host then sends confirmations of this fact to the other hosts of the network by means of two *gratuitous* ARPs, also at 2 second intervals. The protocol has an inherent degree of redundancy, for example with regard to the number of repeated ARP packets sent, in order to cope with message loss. Indeed, message loss makes possible the undesirable situation in which two or more hosts use the same IP address simultaneously.

A host which has commenced using an IP address must reply to ARP packets containing the same IP addresses that it receives from other hosts. It continues using the address unless it receives any ARP packet other than a probe (for example, a gratuitous ARP) containing the IP address that it is using currently, In such a case, the host can either *defend* its IP address, or *defer* to the host which sent the conflicting ARP packet. The host may only defend its address if it has not received a previous conflicting ARP packet within the previous ten seconds; otherwise it is forced to defer. A defending host replies by the sending an ARP packet, thereby indicating that it is using the IP address. A deferring host does not send a reply; instead, it ceases using its current IP address, and reconfigures its IP address by restarting the protocol.

As in [29], we assume a 'broadcast'-based communication medium with no routers (for example, a single wire), in which messages arrive in the order in which they are sent. In contrast to the analytic analysis of the protocol of Bohnenkamp et al. [8], we model the possibility that a device could surrender an IP address that it is using to another host; and in contrast to timed-automata-based analysis of Zhang and Vaandrager [29], we model some important probabilistic characteristics of the protocol, and consider parameters more faithful to the standard (such as the maximum number of times a device can witness an ARP packet with the same IP address as that which it wishes to use before 'backing off' and remaining idle for at least one minute).

In the standard [10], there is no mention of what a host should do with messages corresponding to its current IP address (i.e. the probes and gratuitous ARP packets specified in the standard) which are in its output buffer (i.e. those that have yet to be sent), when it reconfigures (choses a new IP address). However, when the host does reconfigure, unless it picks the same IP address, which happens with the very small probability 1/65024, these messages are not relevant. In fact, such messages will slow down the network and may even make hosts reconfigure when they do not need to. We therefore considered two different versions of the protocol: one where the host does not do anything about these messages (no_reset) and another where the host clears its buffer (removes the messages) when it is about to choose a new IP address (reset).

### 4.1    Modelling the Dynamic Configuration Protocol

**Preliminaries.** We consider in detail one *concrete host*, which is attempting to configure an IP address for a network in which, as in [8], there are 1000 *abstract hosts* (they are called abstract because we do not study their behaviour in depth) which have already configured IP addresses. Therefore, when the concrete host picks an address, the probability of this address being *fresh* (not in use by an abstract host) is 64024/65024. We also assume that the concrete host never picks the same IP address twice, as this happens only with a very small probability.

Following the above assumptions, we require only three abstract IP addresses:

**Table 1.** Integer variables used in the probabilistic timed automata

| variable | description | range |
|---|---|---|
| coll | the number of address collisions detected by the concrete host | $0 \dots 10$ |
| iph | the current address of the concrete host | $1 \dots 2$ |
| defend | equals 1 when the host is defending its address (0 otherwise) | $0 \dots 1$ |
| probes | the number of probes/ARPs sent by the concrete host | $0 \dots N$ |
| ip | the address of the ARP packet currently being sent | $0 \dots 2$ |
| $n$ | the number of packets in the concrete host's output buffer | $0 \dots 8$ |
| $b[i]$ | the address of packet $i$ in the concrete host's output buffer | $0 \dots 1$ |
| $m_0$ | the number of packets containing an IP address of type 0 in all of the buffers of the abstract hosts | $0 \dots 20$ |
| $m_1$ | the number of packets containing an IP address of type 1 in all of the buffers of the abstract hosts | $0 \dots 8$ |

0 – an address of an abstract host which the concrete host previously chose;
1 – an address of an abstract host which is the concrete host's current choice;
2 – a fresh address which is the concrete host's current choice.

As in the standard [10], we suppose that it takes between 0 and 1 second to send a packet between hosts (where the choice of the exact time delay is non-deterministic). Since the abstract hosts have already picked their IP address, by supposing that they always defend their addresses, the concrete host will never receive probes. It then follows that we do not need to record the type of message being sent, but instead only the IP address in the message, and whether it is sent from the concrete host to the abstract hosts or vice versa.

As in [29], we consider the case in which hosts use output buffers to store the packets they want to send. We have chosen the size of the buffers such that the probability of any buffer becoming full is negligible. We suppose that the concrete host can send a packet to all the abstract hosts at the same time and only one of the abstract hosts can send a packet to the concrete host at a time.

The set of variables of our probabilistic timed automata includes both clocks ($x$, $y$ and $z$) and *integer variables* which are described in Table 1. Note that the range of the integer variable *probes* is changed for different verification instances, and since the abstract IP address 2 corresponds to a fresh address chosen by the concrete host we need only two buffers for the abstract hosts (corresponding to addresses of type 0 and 1).

**Probabilistic Timed Automata for the Protocol.** In the following, we describe the modelling of the **reset** version of the protocol only. We use two probabilistic timed automata, one to model the concrete host and one to model the environment (the abstract hosts and the output buffers of *all* hosts).

The model for the concrete host is shown in Figure 1. The host commences in the location RECONF (the double border indicates it is the initial location); this

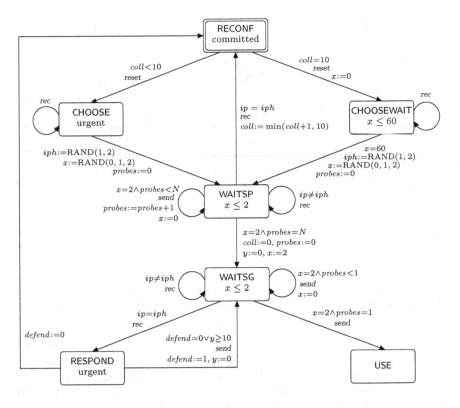

**Fig. 1.** Probabilistic timed automaton for the concrete host

is a committed location, and therefore must be left immediately. In RECONF, the host chooses a new IP address by moving to the location CHOOSE if it has experienced less than ten address collisions, and to CHOOSE_WAIT otherwise. These transitions are labelled with the event *reset* to inform the environment that the host's buffer is to be reset (all messages in its buffer are to be removed).

In both CHOOSE and CHOOSEWAIT, the address selection is represented by the assignment $iph:=\text{RAND}(1,2)$, which corresponds to the host randomly selecting an IP address (using the probabilities given at the start of this section). The assignment to the clock $x$ (a uniform choice between $\{0,1,2\}$) approximates the random delay of between 0 and 2 made by the host before sending the first probe. Note that, in CHOOSEWAIT, since the host has already experienced at least ten address collisions, it waits 60 seconds before choosing a new address.

In the location WAITSP the host sends $N$ probes at 2 second intervals (the self-loop labelled with *send*). The host may also receive packets by means of the event *rec*. If it receives a packet which has a different IP address ($ip\neq iph$), then the host ignores the packet (and remains in WAITSP); however, if it has the same address, the host immediately reconfigures (moves to RECONF). When

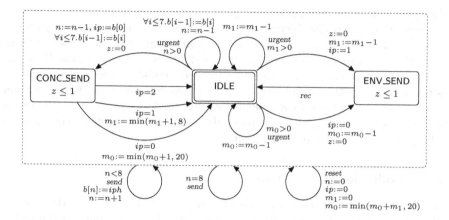

**Fig. 2.** Probabilistic timed automaton for the environment

sending the $N$th probe, the host proceeds to location WAITSG, waits 2 seconds and then sends two gratuitous ARPs (re-using the variable *probes* to count these ARPs). After these ARPs have been sent, the host moves to USE. However, if while in WAITSG the host receives a packet with the same IP address, it moves to RESPOND. In this location, the host can decide to reconfigure (return to RECONF), or defend its IP address (by sending an ARP packet) if it has either not yet defended the address (*defend*=0) or 10 seconds have passed since it previously defended the address ($y \geq 10$). This defence takes the form of sending of a defending packet, as denoted by the send labelled transition from RESPOND to WAITSG.

The model for the environment is shown in Figure 2. The dotted box labelled with three transitions which surrounds the model denotes that these transitions are available in *all* of the locations of the model. More precisely, in all locations, the environment may: receive a *send* event from the concrete host and, if the host's buffer is not full ($n < 8$), the corresponding packet is added to the buffer (otherwise it is lost); receive a *reset* event and clear the buffer of the concrete host ($n$:=0) and, since we assume that the concrete host will never choose the same IP address twice, sets the IP address in any packet being sent or to be sent to type 0 (i.e. $ip$:=0, $m_1$:=0 and $m_0$:= $\min(m_0+m_1, 20)$).

The behaviour of the environment commences in the location IDLE. The transition which probabilistically moves to either IDLE or CONC_SEND corresponds to the environment sending a packet from the concrete host's buffer. The *urgent* labelling denotes that the transition should be taken as soon as it is enabled, i.e. it should be taken as soon as there is something to send. Similarly, the transitions which move probabilistically to either IDLE or ENV_SEND correspond to an abstract host sending a packet, and are again urgent. There are two such transitions, since the address in the packet can either be of type 0 ($m_0>0$) or 1 ($m_1>0$). For each of these transitions, the loop (remaining in

IDLE) corresponds to the packet being lost by the medium, while the other edge corresponds to the packet being sent correctly (therefore the required buffers are updated when one of these transitions is taken). Note that, since each of these transitions corresponds to a message from a different host, when more than one of these transitions is enabled, there is a nondeterministic choice as to which one is taken. We vary the probability of message loss depending on the verification instance. Once in either CONC_SEND or ENV_SEND, after a delay of between 0 and 1 seconds, the model returns to IDLE; this corresponds to the message taking between 0 and 1 seconds to send.

## 4.2   Verification Using PRISM

In this section, we outline our results of using PRISM [17] to verify the integer-time model of the probabilistic timed automata of the dynamic configuration protocol given in Section 4.1. In the experiments, we fixed the number of hosts at 1000 and varied both the number of probes a host sends ($N$), and the probability of message loss. Further details, including analysis for a network of 20 hosts, can be found at the PRISM web page [25]. The algorithms used by PRISM for both probabilistic and expected reachability are taken from the literature; for probabilistic reachability see [7], and for expected reachability see [12, 13].

To apply model-checking methods we must ensure that the model under study has only finitely-many states and is finitely branching. From the construction given in Section 3, the integer-time model will have finitely many states. To ensure finite branching, we restrict the delays from $\mathbb{N}$ to some finite set. More precisely, we allow delays of duration 1 only. Then, since any transition of duration $t \in \mathbb{N}$ can be modelled by a sequence of transitions of duration 1 and we restrict our attention to divergent adversaries, nothing is lost by omitting delays greater than 1 or equal to 0.

Note that, because we have abstracted certain aspects of the network (for example, the time taken to send a message), the presented results will give upper and lower bounds on the performance of the protocol, for example the actual reachability probability will lie in between the minimum and maximum reachability probabilities computed for the model under study.

**Probabilistic Reachability.** The probabilistic reachability property we consider is the (minimum and maximum) probability of the host using an IP address which is already in use by another host. The results obtained in the case of maximum probabilistic reachability are given in Table 2. For results concerning minimum reachability probabilities see the PRISM web page [25].

The results obtained show the expected result: increasing the number of probes sent decreases the probability of the host using an IP address which is already in use (recall that the number specified by the standard is four). When the probability of message loss is 0, Table 2 shows that the maximum probability is 0 for the the model reset (the model where the host clears its buffer) provided the host sends more than one probe. On the other hand, for the model no_reset (when the host does not clear its buffer), even if the host sends more than one

**Table 2.** Maximum probabilistic reachability results for the IPv4 protocol

| number of probes sent | probability of message loss | | | | | | | |
|---|---|---|---|---|---|---|---|---|
| | 0 | | 0.1 | | 0.01 | | 0.001 | |
| | no_reset | reset | no_reset | reset | no_reset | reset | no_reset | reset |
| 1 | 0.01538 | 0.01538 | 0.01538 | 0.01538 | 0.01538 | 0.01538 | 0 .01538 | 0.01538 |
| 2 | 8.0e-5 | 0 | 0.00298 | 0.00296 | 3.8e-4 | 3.1e-4 | 1.1e-4 | 3.1e-5 |
| 3 | 1.2e-6 | 0 | 5.6e-4 | 5.6e-4 | 7.2e-6 | 6.2e-6 | 1.3e-6 | 6.2e-8 |
| 4 | 4.2e-7 | 0 | 1.1e-4 | 1.1e-4 | 5.0e-7 | 1.2e-7 | 4.1e-7 | 1.2e-10 |
| 5 | 8.5e-9 | 0 | 2.0e-5 | 2.0e-5 | 9.8e-9 | 2.4e-9 | 8.4e-9 | 2.5e-13 |
| 6 | 2.2e-9 | 0 | 3.9e-6 | 3.9e-6 | 1.9e-9 | 4.9e-11 | 2.2e-9 | 4.9e-16 |

probe, this maximum reachability probability is greater than 0. To understand this result, consider the fact that, if a host does not clear its buffer, then there is a chance that the probes corresponding to its new IP address will get delayed, and hence the host will not receive a reply to these probes until after it starts using the address (as the probability is 0, the host will eventually get a reply).

In the cases when message loss is greater than 0, the results again demonstrate that, by allowing the host to clear its buffer, the performance of the protocol improves; that is, the maximum reachability probability decreases (our experiments also show that the minimum probability increases, see [25]).

**Expected Reachability.** We consider the expect cost of a host choosing an IP address and using it. As in [8], the cost is defined as the time to start using an IP address plus an additional cost ($10^6$) associated with the host using an address which is already in use. Note that the choice of the value of this additional cost will depend on how damaging it is for two hosts to use the same IP address, which in turn depends on the network and the nature of its devices.

The results for the model reset are presented in Figure 3. Note that, the model no_reset produced similar results, although the minimum costs are smaller and the maximum costs are larger (see [25] for further details). This is to be expected, since the results for probabilistic reachability show that, when the host does not clear its buffer, there is a greater chance of it using an IP address which is already in use, and hence of incurring a greater cost.

These results are similar to those of [8]: as the message loss probability increases, one must increase the number of probes sent in order to reduce the expected cost; however, by sending too many probes the expected cost may then start to increase. The rationale for this is that, although increasing the number of probes sent decreases the probability of the host using an IP address which is already in use (that is, decreases the chance of incurring the additional cost), it increases the expected time to choose an IP address (sending more probes takes more time).

118    Marta Kwiatkowska et al.

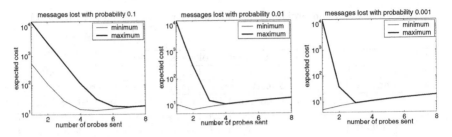

**Fig. 3.** Expected cost results for the IPv4 protocol

## 5 Conclusions

We have presented results demonstrating that digital clocks are sufficient for analysing a large class of probabilistic timed automata and performance properties. Since many of today's protocols include both timing and probabilistic behaviour, this approach is widely applicable, a fact which we illustrate by analysing the performance of the IPv4 dynamic configuration protocol.

Future work could consider extending the cost functions in order to vary the rate of cost accumulation in different locations, as in priced or weighted timed automata [5, 2]. There are still limitations as to the size of the models that can be considered using digital clocks. In the case of probabilistic reachability, a generally more efficient approach is to consider *zones*, and in particular the backwards reachability approach introduced in [20]. The application of zones to the verification of priced timed automata [23] may be instructive to this line of research.

## References

[1] R. Alur and D. L. Dill. A theory of timed automata. Theoretical Computer Science, 126(2):183–235, 1994. 105, 106
[2] R. Alur, S. Torre, and G. Pappas. Optimal paths in weighted timed automata. In Proc. HSCC '01, volume 2034 of LNCS, pages 49–62. Springer, 2001. 118
[3] G. Behrmann, A. David, K. Larsen, O. Möller, P. Pettersson, and W. Yi. UPPAAL - present and future. In Proc.CDC '01. IEEE Computer Society Press, 2001. 109
[4] G. Behrmann, A. Fehnker, T. Hune, K. Larsen, P. Pettersson, and J. Romijn. Efficient guiding towards cost-optimality in UPPAAL. In Proc. TACAS '01, volume 2031 of LNCS, pages 174–188. Springer, 2001. 110
[5] G. Behrmann, A. Fehnker, T. Hune, K. Larsen, P. Pettersson, J. Romijn, and F. Vaandrager. Minimum-cost reachability for priced timed automata. In Proc. HSCC '01, volume 2034 of LNCS, pages 147–161. Springer, 2001. 118
[6] D. Bertsekas and J. Tsitsiklis. An analysis of stochastic shortest path problems. Mathematics of Operations Research, 16(3):580–595, 1991. 110
[7] A. Bianco and L. de Alfaro. Model checking of probabilistic and nondeterministic systems. In Proc. FSTTCS '95, volume 1026 of LNCS, pages 499–513. Springer, 1995. 116

[8] H. Bohnenkamp, P. v. d. Stok, H. Hermanns, and F. Vaandrager. Cost-optimisation of the IPv4 zeroconf protocol. In Proc. IPDS 2003. IEEE CS Press, 2003. 112, 117

[9] D. Bosnacki. Digitization of timed automata. In Proc. FM ICS '99, pages 283–302, 1999. 106

[10] S. Cheshire, B. Adoba, and E. Guttman. Dynamic configuration of IPv4 link-local addresses. Draft, August 2002. Available from: www.ietf.org/internet-drafts/ draft-ietf-zeroconf-ipv4-linklocal-07.txt. 111, 112, 113

[11] C. Daws, M. Kwiatkowska, and G. Norman. Automatic verification of the IEEE 1394 root contention protocol with KRONOS and PRISM. In Proc. FM ICS '02, volume 66(2) of ENTCS. Elsevier Science, 2002. 106

[12] L. de Alfaro. Formal Verification of Probabilistic Systems. PhD thesis, Stanford University, 1997. 106, 116

[13] L. de Alfaro. Computing minimum and maximum reachability times in probabilistic systems. In Proc. CONCUR '99, volume 1664 of LNCS, pages 66–81. Springer, 1999. 116

[14] C. Derman. Finite-State Markovian Decision Processes. Academic Press, 1970. 107

[15] T.A. Henzinger, Z. Manna, and A. Pnueli. What good are digital clocks? In Proc. ICALP '92, volume 623 of LNCS, pages 545–558. Springer, 1992. 106

[16] J. Kemeny, J. Snell, and A. Knapp. Denumerable Markov Chains. Graduate Texts in Mathematics. Springer, 2nd edition, 1976. 108

[17] M. Kwiatkowska, G. Norman, and D. Parker. PRISM: Probabilistic symbolic model checker. In Proc. TOOLS '02, volume 2324 of LNCS, pages 200–204, 2002. 106, 116

[18] M. Kwiatkowska, G. Norman, D. Parker, and J. Sproston. Performance analysis of probabilistic timed automata using digital clocks. Technical Report CSR-03-6, School of Computer Science, University of Birmingham, 2003. 111

[19] M. Kwiatkowska, G. Norman, R. Segala, and J. Sproston. Automatic verification of real-time systems with discrete probability distributions. Theoretical Computer Science, 282:101–150, 2002. 105, 106, 107, 109, 110

[20] M. Kwiatkowska, G. Norman, and J. Sproston. Symbolic computation of maximal probabilistic reachability. In Proc. CONCUR '01, LNCS, pages 169–183. Springer, 2001. 106, 118

[21] M. Kwiatkowska, G. Norman, and J. Sproston. Probabilistic model checking of the IEEE 802.11 wireless local area network protocol. In Proc. PAPM /PROBMIV '02, volume 2399 of LNCS, pages 169–187. Springer, 2002. 109

[22] M. Kwiatkowska, G. Norman, and J. Sproston. Probabilistic model checking of deadline properties in the IEEE 1394 FireWire root contention protocol. Formal Aspects of Computing, 14(3):295–318, 2003. 106, 109, 111

[23] K. Larsen, G. Behrmann, E. Brinksma, A. Fehnker, T. Hune, P. Pettersson, and J. Romijn. As cheap as possible: Efficient cost-optimal reachability for priced timed automata. In Proc. CAV '01, volume 2102 of LNCS, pages 493–505. Springer, 2001. 118

[24] J. Ouaknine and J. Worrell. Universality and language inclusion for open and closed timed automata. In Proc. HSCC '03, volume 2623 of LNCS, pages 375–388. Springer, 2003. 106

[25] PRISM web page. http://www.cs.bham.ac.uk/~{}dxp/prism/. 106, 116, 117

[26] R. Segala. Modeling and Verification of Random ized Distributed Real-Time Systems. PhD thesis, Massachusetts Institute of Technology, 1995. 107

[27] S. Tripakis. The formal analysis of timed systems in practice. PhD thesis, Université Joseph Fourier, 1998. 109

[28] M. Vardi. Automatic verification of probabilistic concurrent finite state programs. In Proc. FOCS '85, pages 327–338. IEEE Computer Society Press, 1985. 108

[29] M. Zhang and F. Vaandrager. Analysis of a protocol for dynamic configuration of IPv4 link local addresses using UPPAAL. Technical Report, NIII, University of Nijmegen, 2003. 112, 113

# An Interval-Based Algebra
# for Restricted Event Detection

Jan Carlson and Björn Lisper

Department of Computer Science and Engineering
Mälardalen University, Sweden
{jan.carlson,bjorn.lisper}@mdh.se

**Abstract.** In this article, we propose an interval based algebra for detection of complex events. The algebra includes a strong restriction policy
in order to comply with the resource requirements of embedded or real-
time applications. We prove a set of algebraic properties to justify the
novel restriction policy and to establish the relation between the unrestricted algebra and the restricted version. Finally, we present an efficient
algorithm that implements the proposed algebra.

## 1   Introduction

A wide range of applications, including active databases, traffic monitoring systems and rule based embedded systems, are based on the detection of events
that trigger an appropriate response from the system. Events can be simple,
e.g., sampled directly from the environment or occuring within the system, but
it is often necessary to react to more sophisticated situations involving a number
of simpler events that occur in accordance with some pattern.

A standard way in which to allow systems to react to sophisticated situations
is to introduce complex events by means of an event algebra. These complex
events can then be used to trigger actions just like simple events. A benefit of
this method is that the mechanisms handling event detection are separated from
the rest of the system logic.

Since our primary interest concerns embedded applications and systems with
strict timeliness requirements, it is essential that the event detection can be
implemented with limited resources. As a result, the algebra must be restricted
so as to only detect a subset of all possible occurrences of complex events. This
can be achieved by applying a suitable restriction policy, as will be described in
the next section.

A great many event algebras have been proposed for different applications.
Most of them include operators such as disjunction, sequence, conjunction and
some form of negation, but the semantics of these operators vary. Further, many
systems add to these some operators of their own. Restriction policies are typically informally defined and little effort spent determining the algebraic properties of the algebra.

K.G. Larsen and P. Niebert (Eds.): FORMATS 2003, LNCS 2791, pp. 121–133, 2004.
© Springer-Verlag Berlin Heidelberg 2004

We propose an interval based event algebra with well-defined formal semantics, and with a restriction policy strong enough to make it effectively implementable. We also state a number of algebraic properties, including a clear description of the relation between the unrestricted algebra and the restricted version. Finally, we present an efficient algorithm that implements the proposed algebra.

The rest of this paper is organised as follows: Section 2 introduces techniques commonly used in event algebras and presents related work. The algebra is defined in Section 3, followed by a presentation of the algebraic properties in Section 4. Section 5 presents the algorithm, and Section 6 concludes the paper.

## 2    Event Algebras

The following operations, or variants of them, are found in most event algebras. *Disjunction* of $A$ and $B$ means that either of $A$ and $B$ occurs, here denoted $A \lor B$. *Conjunction* means that both events have occurred, possibly not simultaneously, and is denoted $A + B$. The *negation*, denoted $A - B$, occurs when there is an occurrence of $A$ during which there is no occurrence of $B$. Finally, a *sequence* of $A$ and $B$ is an occurrence of $A$ followed by an occurrence of $B$, and is denoted $A; B$.

Examples of how event algebras are used in the area of active datebases include SAMOS [5], Snoop [3] and Ode [6]. These three systems differ primarily in the choice of detection mechanism. SAMOS is based on Petri nets, while Snoop uses event graphs. In Ode, event definitions are equivalent to regular expressions and can be detected by state automata.

A formalized schema for this type of event detection, including a definition of the operations and restriction policies of Snoop using this schema, has been defined by Mellin and Andler [10]. Liu et al. uses Real Time Logic to define the semantics of an event detection system. As a result, the conditions for event occurrences can be transformed into timing constraints and handled by general timing constraint monitoring techniques [9].

The event algebra developed by Hinze and Voisard is designed to suite event notification service systems in general [7]. Their algebra contains time restricted sequence and conjunction, which permits events like *A occurs less than t time units before B* to be expressed.

In the area of knowledge representation, similar techniques are used to reason about event occurrences. Interval Calculus introduce formalised concepts for properties, actions and events, where events are expressed in terms of conditions for their occurrence [2]. Event Calculus [8] also deals with the occurrences of events, but, as in the Interval Calculus, the motivation is slightly different from ours. Rather than detecting complex events as they occur, the focus of Event Calculus is the inferences that can be made from the fact that certain events have occurred.

## 2.1   Restricted Detection

A very straightforward definition of the sequence operator is that the sequence $A; B$ should occur whenever $A$ occurs and then $B$ occurs. Using this definition, three occurrences of $A$ followed by two occurrences of $B$ would generate six occurrences of the sequence. While this may be acceptable, or even desirable, in some applications, the memory requirements (each occurrence of $A$ must be remembered forever) and the increasing number of simultaneous events means that it is unsuitable in many cases. Also, it is argued that many applications are interested only in a subset of the instances that are generated by this definition.

One way to deal with this is to define the event algebra in two steps. The operations are defined in an unrestricted, straightforward way like in our example above. Then a restriction policy is defined. This acts like a filter, so that only a subset of the occurrences allowed by the unrestricted definition are detected. For example, the restriction policy could state that only the latest occurrence of $A$ are allowed to create an occurrence of $A; B$ when $B$ occurs.

This type of restriction based methods are for example used by Snoop [3] and in the algebra proposed by Hinze and Voisard [7]. Zimmer and Unland present a formal restriction framework in which the event algebras of Snoop, SAMOS and Ode are compared [11].

## 2.2   Interval-Based Event Detection

Single point detection means that every complex event, including those that occur during a time interval, is associated with a single time point (the time of detection, i.e, the end of the occurrence interval). Galton and Augusto [4] showed that this results in unintended semantics for some operation compositions.

For example, using single point detection an instance of the event $A; (B; C)$ is detected if $B$ occurs first, and then $A$ followed by $C$. The reason is that these occurrences cause a detection of $B; C$ which is associated with the occurrence time of $C$. Since $A$ occurs before this time point, an occurrence of $A; (B; C)$ is detected.

This problem can be solved by associating the occurrence of a complex event with the occurrence interval, rather than the time of detection. In this setting, the sequence $A; B$ can be defined to occur only if the intervals of $A$ and $B$ are non-overlapping. In our example, no occurrence of $A; (B; C)$ would be detected, since $A$ occurs within the interval associated with the occurrence of $B; C$.

Most event algebras, especially in the area of active databases, use single point detection. An interval based version of Snoop has been developed by Adaikkalavan and Chakravarthy [1], and the work by Mellin and Andler is also based on intervals [10].

# 3   The Event Algebra

The system is assumed to have a pre-defined set of primitive events that it should be able to react to. These events can be external (sampled from the environment or originating from another system) or internal (such as the violation of a condition over the system state, or a timeout), but the detection mechanism does not distinguish between these categories.

We assume occurrences of primitive events to be instantaneous and atomic, and allow occurrences to carry values. This value could for example identify at which external device the event occurred, or be some measured value from the environment. The values are not manipulated in any way by the detection mechanism, but simply forwarded to the part of the system that reacts to the detected events. An occurrence of a primitive event is represented by the tuple $\langle v, \tau \rangle$, where $v$ is the value ($v$ belongs to some arbitrary domain of values), and $\tau$ is the time of the occurrence. We assume a discrete time modelled by the natural numbers.

## 3.1   Basic Concepts

From the simple events, represented by a set $\mathcal{I}$ of identifiers, expressions representing complex events can be constructed as follows.

**Definition 1.** *Given a set $\mathcal{I}$ of identifiers we define:*

- *If $A \in \mathcal{I}$, then $A$ is an event expression.*
- *If $A$ and $B$ are event expressions, so are $A \vee B$, $A + B$, $A - B$ and $A; B$.*

The complex event expressions in the definition represent disjunction, conjunction, negation and sequence, respectively.

**Definition 2.** *An* event instance *is a set of value-time tuples. A primitive event instance is a singleton set. For an event instance $a$, we define:*

$$start(a) = Min_{\langle v, \tau \rangle \in a} (\tau)$$
$$end(a) \;\; = Max_{\langle v, \tau \rangle \in a} (\tau)$$

From the definition follows that for any primitive event instance $a$, $start(a) = end(a)$. Non-primitive event instances are considered to occur throughout an interval from the earliest of the included primitive event instances, to the latest one.

All instances of a particular event form an event stream. The semantics of the algebra, presented below, associates with each event expression a corresponding event stream.

**Definition 3.** *An* event stream *is a set of event instances. An event stream $A$ is said to be* non-simultaneous *if all instances have different end times. A primitive event stream is a non-simultaneous event stream containing only primitive event instances.*

## 3.2   Unrestricted Semantics

**Definition 4.** *For an event stream $S$ and an event instance $a$, define $empty(S, a)$ to hold iff there is no $s \in S$ such that $start(a) \leq start(s)$ and $end(s) \leq end(a)$.*

The following four functions over event streams form the core of the algebra semantics, as they define the basic characteristics of the four operations.

**Definition 5.** *For event streams $S$ and $T$, define:*

$$dissem(S, T) = S \cup T$$
$$consem(S, T) = \{s \cup t \mid s \in S \wedge t \in T\}$$
$$negsem(S, T) = \{s \mid s \in S \wedge empty(T, s)\}$$
$$seqsem(S, T) = \{s \cup t \mid s \in S \wedge t \in T \wedge end(s) < start(t)\}$$

**Definition 6.** *An* interpretation *is a function that maps each identifier in $\mathcal{I}$ to a primitive event stream.*

**Definition 7.** *The unrestricted meaning of an event expression for a given interpretation $S$ is defined as follows:*

$$[A]^{S} \quad = S(A) \text{ if } A \in \mathcal{I}$$
$$[A \vee B]^{S} = dissem([A]^{S}, [B]^{S})$$
$$[A + B]^{S} = consem([A]^{S}, [B]^{S})$$
$$[A - B]^{S} = negsem([A]^{S}, [B]^{S})$$
$$[A; B]^{S} \quad = seqsem([A]^{S}, [B]^{S})$$

To simplify the presentation, we will use the notation $[A]$ instead of $[A]^{S}$ whenever the choice of $S$ is obvious or arbitrary.

## 3.3   Restricted Semantics

As discussed in the introduction, due to efficiency considerations we have to restrict the detection to a subset of the instances defined by the unrestricted semantics. As a first step, we remove simultaneous instances of an event stream (i.e., instances $a$ and $a'$ of the same event stream with $end(a) = end(a')$). In order not to lose the desired algebraic properties, this filtering must be done carefully.

**Definition 8.** *Let remsim be any function over event streams such that the following holds. For an event stream $S$, $remsim(S)$ is a minimal subset of $S$ such that for any element $s \in S$ there is an element $s' \in remsim(S)$ with $start(s) \leq start(s')$ and $end(s) = end(s')$.*

Informally, from a number of instances with the same end time, we keep only one with maximal start time. Using discrete time ensures that such a function exists.

For all operations except sequence, this restriction is enough to allow an efficient implementation (negation does not need any restriction at all). For sequence, however, we also have to deal with the problem that in the unrestricted version, each occurrence of the first argument is used over and over again in combination with all subsequent instances of the second argument. This means that every instance of the first argument must be stored throughout the system lifetime, thus precluding an implementation with limited resources.

**Definition 9.** *Let restrict be any function over event streams such that the following holds. For an event stream $S$, restrict$(S)$ is a minimal subset of $S$ such that for any element $s \in S$ there is an element $s' \in$ restrict$(S)$ with start$(s) \leq$ start$(s')$ and end$(s') \leq$ end$(s)$.*

Informally, when detecting a sequence $A; B$, an instance of $A$ can only be combined with the earliest possible instance of $B$. Similarly, an instance of $B$ can only be combined with the latest possible instance of $A$. This is similar, but not equivalent, to the recent context of Snoop.

**Definition 10.** *The restricted meaning of an event expression for a given interpretation $S$ is:*

$$
\begin{aligned}
\llbracket A \rrbracket^S &= S(A) \text{ if } A \in \mathcal{I} \\
\llbracket A \vee B \rrbracket^S &= remsim(dissem(\llbracket A \rrbracket^S, \llbracket B \rrbracket^S)) \\
\llbracket A + B \rrbracket^S &= remsim(consem(\llbracket A \rrbracket^S, \llbracket B \rrbracket^S)) \\
\llbracket A - B \rrbracket^S &= negsem(\llbracket A \rrbracket^S, \llbracket B \rrbracket^S) \\
\llbracket A; B \rrbracket^S &= restrict(seqsem(\llbracket A \rrbracket^S, \llbracket B \rrbracket^S))
\end{aligned}
$$

As in the unrestricted version, we will use the notation $\llbracket A \rrbracket$ instead of $\llbracket A \rrbracket^S$ whenever the choice of $S$ is obvious or arbitrary.

*Example 1.* To illustrate the difference between the unrestricted and the restricted semantics, these tables show the event instances of $A$ and $B$ (which we assume to be primitive, so $[A] = \llbracket A \rrbracket$ and $[B] = \llbracket B \rrbracket$), together with the corresponding instances of the complex events $A+B$ and $A; B$, using both unrestricted and restricted semantics.

| Expression | Instances |
|---|---|
| $[A]$ | |
| $[B]$ | |
| $[A + B]$ | |
| $\llbracket A + B \rrbracket$ | |

| Expression | Instances |
|---|---|
| $[A]$ | |
| $[B]$ | |
| $[A; B]$ | |
| $\llbracket A; B \rrbracket$ | |

# 4    Algebraic Properties

A main concern regarding the restriction policy has been to ensure that the restricted algebra should comply with the algebraic laws that intuitively should hold for an event algebra. Disjunction and sequence should be associative, conjunction should be distributive over disjunction, etc. This is not the only requirement, however, since it would be trivially satisfied by a restriction policy that simply filters away all instances. The restriction policy should remove as few instances as possible, while still ensuring the desired algebraic properties and allowing an implementation with bounded resources. More specifically, we want a theoretical description of the relation between the unrestricted semantics and the restricted version.

The following theorem justifies the proposed restriction policy. The subset result is not trivial, since with a different restriction policy $[\![B]\!] \subset [B]$ could easily mean that $[\![A - B]\!] \supset [A - B]$. The second statement ensures that our restriction policy does not remove too much. Every removed instance leaves some trace in the restricted version, as the interval between the start and end time of the removed instance must be non-empty.

**Theorem 1.** *For any event expression A, the following holds:*

*i)* $[\![A]\!] \subseteq [A]$
*ii)* $a \in [A] \;\Rightarrow\; \exists_{a' \in [\![A]\!]} \, (start(a) \leq start(a') \wedge end(a') \leq end(a))$

*Proof.* We prove the theorem by structural induction over expressions. As a base case, both statements hold trivially for any primitive event expression since $[\![A]\!] = [A]$ when $A \in \mathcal{I}$. For the inductive case, assume that both statements hold for event expressions $A_1$ and $A_2$. From Definition 5, and the fact that $restrict(P) \subseteq P$ and $remsim(S) \subseteq S$, it follows that statement *i)* holds for $A_1 \vee A_2$, $A_1 + A_2$ and $A_1; A_2$.

In order to show that statement *i)* holds for negation, take an arbitrary $a \in [\![A_1 - A_2]\!]$. Then $a \in [\![A_1]\!]$ and $empty([\![A_2]\!], a)$. By assumption *i)*, this means that $a \in [A_1]$ and assumption *ii)* implies $empty([A_2], a)$. Thus, $a \in negsem([A_1], [A_2])$, so $a \in [A_1 - A_2]$ which means that we have $[\![A_1 - A_2]\!] \subseteq [A_1 - A_2]$.

Continuing the inductive case with statement *ii)*, we consider first the case of sequence. We take an arbitrary $a \in [A_1; A_2]$ which implies $a = a_1 \cup a_2$ where $a_1 \in [A_1]$ and $a_2 \in [A_2]$ with $end(a_1) < start(a_2)$. By assumption *ii)*, there are instances $a_1' \in [\![A_1]\!]$ and $a_2' \in [\![A_2]\!]$ such that $start(a_i) \leq start(a_i')$ and $end(a_i') \leq end(a_i)$ for $i \in \{1, 2\}$ and thus $a_1' \cup a_2' \in seqsem([\![A_1]\!], [\![A_2]\!])$. Then, by the definition of *restrict*, there must be some element $a' \in restrict(seqsem([\![A_1]\!], [\![A_2]\!]))$ with $start(a_1') \leq start(a')$ and $end(a') \leq end(a_2')$. So, we have found an instance $a' \in [\![A_1; A_2]\!]$ for which $start(a) = start(a_1) \leq start(a_1') \leq start(a')$ and $end(a') \leq end(a_2') \leq end(a_2) = end(a)$.

For negation, we take an arbitrary $a \in [A_1 - A_2]$. This implies $a \in [A_1]$ and $empty([A_2], a)$, which by assumption *i)* means that $empty([\![A_2]\!], a)$. By assumption *ii)*, there is an instance $a' \in [\![A_1]\!]$ with $start(a) \leq start(a')$ and $end(a') \leq end(a)$. We have $empty([\![A_2]\!], a')$, and thus $a' \in negsem([\![A_1]\!], [\![A_2]\!])$.

So, we have found an instance $a' \in [\![A_1 - A_2]\!]$ for which $start(a) \leq start(a')$ and $end(a') \leq end(a)$.

The proofs for disjunction and conjunction are similar to the cases above, and have been left out due to space limitations. Together, this proves by induction that both statements hold for any event expression $A$.     □

In order to reason about algebraic properties like associativity, etc. we must define a relaxed concept of equivalence. As a result of the restriction policy, the two sets $[\![A;(B;C)]\!]$ and $[\![(A;B);C]\!]$ are not necessarily equal. However, we can show that for every instance of $[\![A;(B;C)]\!]$ there is an instance of $[\![(A;B);C]\!]$ with the same start- and end time, and vice versa. This means, for example, that in systems where events are used to trigger response actions, the two expressions would trigger actions at the same time (although possibly with different values). This time based notion of equality is formalised as follows.

**Definition 11.** *For event instances a and b, event streams S and T, and event expressions A and B, define:*

$$a \cong b \quad iff \quad start(a) = start(b) \ and \ end(a) = end(b)$$
$$S \cong T \quad iff \quad \{\langle start(a), end(a)\rangle \mid a \in S\} = \{\langle start(b), end(b)\rangle \mid b \in T\}$$
$$A \cong B \quad iff \quad [\![A]\!] \cong [\![B]\!]$$

Trivially, $\cong$ in an equivalence relation. Moreover, we will show that it satisfies the substitutive condition, and hence defines structural congruence over event expressions. For the proof, we need the following lemma.

**Lemma 1.** *For event streams such that $S \cong S'$ and $T \cong T'$, we have:*

$$dissem(S,T) \cong dissem(S',T') \qquad negsem(S,T) \cong negsem(S',T')$$
$$consem(S,T) \cong consem(S',T') \qquad remsim(S) \quad \cong remsim(S')$$
$$seqsem(S,T) \cong seqsem(S',T') \qquad restrict(S) \quad \cong restrict(S')$$

*Proof.* The four equivalences regarding *dissem*, *consem*, etc. follow trivially from the fact that Definition 5 only considers start and end times. For the *remsim* equivalence, take an arbitrary $a \in remsim(S)$. Then $a \in S$ so there is an $a' \in S'$ with $a \cong a'$. The definition of *remsim* implies that there is some $b \in remsim(S')$ such that $start(a') \leq start(b)$ and $end(b) = end(a')$. In the same way, there is a corresponding element $b' \in S$ such that $b \cong b'$ so there is some element $c \in remsim(S)$ with $start(b') \leq start(c)$ and $end(c) = end(b')$.

We have two elements $a$ and $c$ in $remsim(S)$ with $start(a) \leq start(c)$ and $end(a) = end(c)$. Assuming $a \neq c$, the set $remsim(S) - \{a\}$ meets the requirement in the definition of *remsim*, contradicting the minimality. Hence, we must have $a = c$, which implies $start(a) = start(b)$. So, for an arbitrary $a \in remsim(S)$ we have found a $b \in remsim(S')$ with $a \cong b$, and hence $remsim(S) \cong remsim(S')$.

The proof of the *restrict* equivalence is very similar to the one above.     □

**Theorem 2.** *If $A_1 \cong A_1'$ and $A_2 \cong A_2'$ then we have $(A_1 \vee A_2) \cong (A_1' \vee A_2')$, $(A_1 + A_2) \cong (A_1' + A_2')$, $(A_1 - A_2) \cong (A_1' - A_2')$ and $(A_1;A_2) \cong (A_1';A_2')$.*

*Proof.* This follows trivially from Lemma 1 and Definition 10.     □

Using the weak equivalence, we can formulate a number of algebraic laws.

**Theorem 3.** *For any event expressions A, B and C, the following laws hold:*

$$
\begin{array}{rrcl}
R1: & A \vee B & \cong & B \vee A \\
R2: & A \vee A & \cong & A \\
R3: & A \vee (B \vee C) & \cong & (A \vee B) \vee C \\
R4: & A; (B; C) & \cong & (A; B); C \\
R5: & A + B & \cong & B + A \\
R6: & A + A & \cong & A \\
R7: & A + (B + C) & \cong & (A + B) + C \\
R8: & A + (B \vee C) & \cong & (A + B) \vee (A + C) \\
R9: & (A \vee B) + C & \cong & (A + C) \vee (B + C) \\
R10: & (A \vee B) - C & \cong & (A - C) \vee (B - C) \\
R11: & (A - B) - B & \cong & A - B \\
R12: & A - (B \vee C) & \cong & (A - B) - C
\end{array}
$$

*Proof.* $R1$, $R2$ and $R3$ follow trivially from Definitions 10 and 5 and the definition of *remsim*. For $R4$, we first take an arbitrary $d \in [\![A; (B; C)]\!]$. Using Theorem 1 it is straightforward to show that $d \in [\![(A; B); C]\!]$ which implies that there is some $d' \in [\![(A; B); C]\!]$ with $start(d) \leq start(d')$ and $end(d') \leq end(d)$. In the same way, this implies that there is some $d'' \in [\![A; (B; C)]\!]$ with $start(d') \leq start(d'')$ and $end(d'') \leq end(d')$. The minimality condition in the definition of *restrict* means that we must in fact have $d \cong d''$, which implies $d \cong d'$. Thus, for an arbitrary $d \in [\![A; (B; C)]\!]$ there is a $d' \in [\![(A; B); C]\!]$ such that $d \cong d'$. In the same way we can show that for an arbitrary $d \in [\![(A; B); C]\!]$ there is a $d' \in [\![A; (B; C)]\!]$ with $d \cong d'$.

$R5$ and $R6$ follow trivially from Definitions 10 and 5 and the definition of *remsim*. The proofs of $R7$ and $R8$ are very similar to that of $R4$. $R9$ follows trivially from $R5$ and $R8$.

For $R10$, we take an arbitrary $d \in [\![(A \vee B) - C]\!]$. This means that $d \in [\![A \vee B]\!]$ and $empty([\![C]\!], d)$. Thus either $d \in [\![A]\!]$ or $d \in [\![B]\!]$, which means that $d \in [\![A - C]\!]$ or $d \in [\![B - C]\!]$, but in both cases we have $d \in dissem([\![A - C]\!], [\![B - C]\!])$. Thus there is some $d' \in [\![(A - C) \vee (B - C)]\!]$ with $start(d) \leq start(d')$ and $end(d) = end(d')$. Since $d' \in dissem([\![A]\!], [\![B]\!])$, by minimality of *remsim* we must have $d \cong d'$. In a similar way we can show that any $d \in [\![(A - C) \vee (B - C)]\!]$ implies the existence of an $d' \in [\![(A \vee B) - C]\!]$ such that $d \cong d'$.

$R11$ follows trivially from Definitions 10 and 5. For $R12$, if $a \in [\![A - (B \vee C)]\!]$ we have $a \in [\![A]\!]$ and $empty([\![B \vee C]\!], a)$. By Theorem 1, we must have $empty([\![B \vee C]\!], a)$ and thus, $empty([\![B]\!], a)$ and $empty([\![C]\!], a)$. Then $a \in [\![A - B]\!]$, and $a \in [\![(A - B) - C]\!]$. Starting instead with an $a \in [\![(A - B) - C]\!]$, this means $a \in [\![A]\!]$, $empty([\![B]\!], a)$ and $empty([\![C]\!], a)$. Then $empty([\![B \vee C]\!], a)$ and thus $a \in [\![A - (B \vee C)]\!]$. $\square$

## 5    Event Detection Algorithm

For the detection algorithm, we let 1 denote the first time point at which events may occur, using 0 only when referring to the time of system initialisation.

To simplify the algorithm presentation, we use the following auxiliary functions (*match* is not well-defined, but any function that meets the condition can be used).

$$
\begin{aligned}
get(A, \mathcal{S}, \tau) &= \begin{cases} \{\langle v, \tau \rangle\} \text{ if } \{\langle v, \tau \rangle\} \in \mathcal{S}(A) \\ \langle \rangle \text{ if no such instance exists in } \mathcal{S}(A) \end{cases} \\
match(y, q) &= \begin{cases} y \cup \text{ an element in } filter(y, q) \text{ with maximum start time} \\ \langle \rangle \text{ if } filter(y, q) \text{ is empty} \end{cases} \\
filter(y, q) &= \{e \mid e \in q \wedge end(e) < start(y)\}
\end{aligned}
$$

The symbol $\langle \rangle$ is used to represent a non-occurrence, and we use the symbol $\tau^c$ when referring to the current time in the algorithm. Since each operator occurrence in the expression requires its own state variables, we simplify the presentation by using variables that are indexed with subexpressions. Thus, for each subexpression $A$, $v_A$ denotes the $v$ variable of $A$. An equivalent method would be to number each subexpression, and use ordinary integer indexed variables.

### 5.1    Algorithm Description

Figure 1 presents the algorithm for detecting an event expression $E$. The algorithm is presented in a meta format that can be instantiated for any fixed expression. The top level conditionals can be evaluated statically, which permits statically unrolling the foreach statement. All indices can also be evaluated statically. A concrete example of this is given in Example 2.

In the initial state, at time 0, let $w_A = z_A = \langle \rangle$, $t_A = 0$ and $q_A = \emptyset$ for every subexpression $A$ in $E$. Each time instant, the algorithm takes as input the current instances of primitive events (provided by the $get$ function) and computes the current instance of $E$, if there is one. The following theorem formalises the output of the algorithm.

**Theorem 4.** *For any subexpression $A$ in $E$, after executing the algorithm at time instants 1 to $\tau$, $v_A = a$ if there is an instance $a \in [\![A]\!]$ with $end(a) = \tau$. If there is no such instance in $[\![A]\!]$, $v_A = \langle \rangle$.*

*Proof.* We only outline very informally the core of the correctness proof, providing some intuition to the relation between the algorithm and the formal semantics. When processing a subexpression $A$ on the form $B \vee C$, $v_B$ and $v_C$ already contain the current instances of $B$ and $C$, respectively, since the original expression is processed bottom-up. The algorithm assigns to $v_A$ the one with latest start time, which according to the definition of *remsim* is the current instance of $A$.

Conjunctions are handled by storing in $w$ and $z$ the instances with latest start time from $B$ and $C$, respectively. If there is a current instance of $B$ or $C$, the

For each subexpression $A$ in $E$, in bottom-up order, do the following:

if $A \in \mathcal{I}$ then $v_A := get(A, \mathcal{S}, \tau^c)$

if $A$ is $B \vee C$ then
  if $v_B = \langle \rangle$ or $(v_C \neq \langle \rangle$ and $start(v_B) \leq start(v_C))$
  then $v_A := v_C$
  else $v_A := v_B$

if $A$ is $B + C$ then
  if $v_B \neq \langle \rangle$ and $(w_A = \langle \rangle$ or $start(w_A) < start(v_B))$ then $w_A := v_B$
  if $v_C \neq \langle \rangle$ and $(z_A = \langle \rangle$ or $start(z_A) < start(v_C))$ then $z_A := v_C$
  if $v_B \neq \langle \rangle$ and $((v_C = \langle \rangle$ and $z_A \neq \langle \rangle)$ or
              $(v_C \neq \langle \rangle$ and $start(v_C) \leq start(v_B)))$
  then $v_A := v_B \cup z_A$
  if $v_C \neq \langle \rangle$ and $((v_B = \langle \rangle$ and $w_A \neq \langle \rangle)$ or
              $(v_B \neq \langle \rangle$ and $start(v_B) < start(v_C)))$
  then $v_A := w_A \cup v_C$

if $A$ is $B - C$ then
  if $v_C \neq \langle \rangle$ and $t_A < start(v_C)$ then $t_A := start(v_C)$
  if $v_B \neq \langle \rangle$ and $t_A < start(v_B)$ then $v_A := v_B$

if $A$ is $B; C$ then
  if $v_C = \langle \rangle$ then $v_A := \langle \rangle$
    else $v_A := match(v_C, q_A);$    $q_A := q_A - filter(v_C, q_A)$
  if $v_B \neq \langle \rangle$ and $t_A < start(v_B)$
  then $q_A := q_A \cup \{v_B\};$    $t_A := start(v_B)$

**Fig. 1.** Meta-algorithm for the detection of $E$ under the interpretation $\mathcal{S}$

current instance of $A$ is formed by combining instances from $B$ and $C$ such that at least one is a current instance, and such that the start time of the combination is as late as possible.

For negations, the variable $t$ contains the latest start time of all instances of $C$ that has occurred until now. The current instance of $B$ becomes the current instance of $A$ if it starts later than $t$, which conforms to the definition of *negsem*.

To deal with sequences, the variable $q$ stores instances of $B$ that has not yet been possible to match with any instance of $C$. In addition, the variable $t$ is used to ensure that no instances in $q$ are fully overlapping. If there is a current instance of $C$, it is combined with the best matching instance in $q$ to form the current instance of $A$. Also, the definition of *restrict* dictates that instances of $B$ that end before the start time of the current instance of $C$, may not be used to form future instances of $A$. Hence, these are removed from $q$.                    □

$$
\begin{aligned}
&v_1 := get(A, \mathcal{S}, \tau^c) \\
&v_2 := get(B, \mathcal{S}, \tau^c) \\
&v_3 := get(C, \mathcal{S}, \tau^c) \\
&\text{if } v_1 = \langle\rangle \text{ or } (v_2 \neq \langle\rangle \text{ and } start(v_1) \leq start(v_2)) \\
&\quad \text{then } v_4 := v_2 \\
&\quad \text{else } v_4 := v_1 \\
&\text{if } v_3 \neq \langle\rangle \text{ and } t_5 < start(v_3) \text{ then } t_5 := start(v_3) \\
&\text{if } v_4 \neq \langle\rangle \text{ and } t_5 < start(v_4) \text{ then } v_5 := v_4
\end{aligned}
$$

**Fig. 2.** Instantiated algorithm for detecting $(A \vee B) - C$

*Example 2.* Assume we are detecting the event $(A \vee B) - C$. After instantiating the meta-algorithm, we can unroll the foreach statement and statically evaluate the top-level conditionals. We also instantiate the subexpression indicies with corresponding integers. The resulting algorithm is presented in Figure 2.

## 6   Conclusions and Future Work

We have developed an interval based algebra for detection of complex events. The algebra includes a strong restriction policy in order to comply with the resource requirements of embedded or real-time applications. The restriction policy is justified by a theorem stating that it never adds instances, compared to the unrestricted semantics. Also, every removed instance leaves some trace in the restricted version, as the interval between the start and end time of the removed instance must still contain at least one instance.

An event detection algorithm that implements the proposed algebra was presented. In this algorithm, each disjunction, conjunction and negation in the event expression requires a constant amount of storage, and contributes with a constant factor to the computation time. For the sequence $A; B$, on the other hand, a set of instances must be stored and the computation time is proportional to the size of this set. This is a result of Theorem 1, since it is not enough to store a single best instance of $A$ (i.e., the one with latest start time). Once an instance of $B$ occurs, it must be combined with the best *allowed* instance of $A$. This might not be the best instance of $A$ that has occurred so far, if the interval of $B$ is long.

This is clearly a weakness, but as it follows from one of the desired properties of the restriction, we have to look for other ways to ensure limited resource demands. The maximum size of the storage set for $A; B$ depends on the relative frequence of occurrences in $A$ and $B$. Roughly, if no more that $n$ instances of $A$ can occur during the longest possible interval in which no $B$ occurs, $n$ is the maximum size of the storage set.

We are currently formalising this idea, including how to calculate frequence bounds for complex events from frequence bounds of the primitive events. This seems to be possible for all expressions except negations, so there is still a problem with expressions like $A; (B - C)$. If $C$ and $B$ occur together, $B - C$ never occurs at all, so every instance of $A$ is stored forever.

Additional future work includes finishing the formal proof of Theorem 4. We are also considering extending the algebra with additional operations, especially time limited versions of sequence and conjunction.

# References

[1] R. Adaikkalavan and S. Chakravarthy. Event operators: Formalization algorithms, and implementation using interval-based semantics. Technical Report CSE-2002-3, University of Texas at Arlington, Department of Computer Science and Engineering, June 2002. 123

[2] J. F. Allen and G. Ferguson. Actions and events in interval temporal logic. Journal of Logic and Computation, 4(5):531–579, October 1994. 122

[3] S. Chakravarthy, V. Krishnaprasad, E. Anwar, and S.-K. Kim. Composite events for active databases: Semantics, contexts and detection. In Jorge B. Bocca, Matthias Jarke, and Carlo Zaniolo, editors, 20th International Conference on Very Large Data Bases, September 12-15, 1994, Santiago, Chile proceedings, pages 606–617, Los Altos, CA 94022, USA, 1994. Morgan Kaufmann Publishers. 122, 123

[4] A. Galton and J. C. Augusto. Two approaches to event definition. In R. Cicchetti, A. Hameurlain, and R. Traunmüller, editors, Proc. of Database and Expert Systems Applications 13th International Conference (DEXA '02), volume 2453 of Lecture Notes in Computer Science, pages 547–556, Aix-en-Provence, France, 2–6 September 2002. Springer-Verlag. 123

[5] S. Gatziu and K. R. Dittrich. Events in an Active Object-Oriented Database System. In N. W. Paton and H. W. Williams, editors, Proc. 1st Intl. Workshop on Rules in Database Systems (RIDS), Edinburgh, UK, September 1993. Springer-Verlag, Workshops in Computing. 122

[6] N. Gehani, H. V. Jagadish, and O. Shmueli. COMPOSE: A system for composite specification and detection. In Advanced Database Systems, volume 759 of Lecture Notes in Computer Science. Springer, 1993. 122

[7] A. Hinze and A. Voisard. A parameterized algebra for event notification services. In Proceedings of the 9th International Symposium on Temporal Representation and Reasoning (TIME 2002), Manchester, UK, 2002. 122, 123

[8] R. Kowalski and M. Sergot. A logic-based calculus of events. In J. W. Schmidt and C. Thanos, editors, Foundations of Knowledge Base Management: Contributions from Logic, Databases, and Artificial Intelligence, pages 23–55. Springer, Berlin, Heidelberg, 1989. 122

[9] G. Liu, A. Mok, and P. Konana. A unified approach for specifying timing constraints and composite events in active real-time database systems. In 4th IEEE Real-Time Technology and Applications Symposium (RTAS '98), pages 199–209, Washington - Brussels - Tokyo, June 1998. IEEE. 122

[10] J. Mellin and S. F. Adler. A formalized schema for event composition. In Proc. 8th Int. Conf on Real-Time Computing Systems and Applications (RTCSA 2002), pages 201–210, Tokyo, Japan, 18–20 March 2002. 122, 123

[11] D. Zimmer and R. Unland. On the Semantics of Complex Events in Active Database Management Systems. In Proceedings of the 15th International Conference on Data Engineering, pages 392–399. IEEE Computer Society Press, 1999. 123

# PARS: A Process Algebra with Resources and Schedulers

MohammadReza Mousavi, Michel Reniers, Twan Basten, and Michel Chaudron

Eindhoven University of Technology
Post Box 513
NL-5600 MB, Eindhoven, The Netherlands
{m.r.mousavi,m.a.reniers,a.a.basten,m.r.v.chaudron}@tue.nl

**Abstract.** In this paper, we introduce a dense time process algebraic formalism with support for specification of (shared) resource requirements and resource schedulers. The goal of this approach is to facilitate and formalize introduction of scheduling concepts into process algebraic specification using separate specifications for resource requiring processes, schedulers and systems composing the two. The benefits of this research are twofold. Firstly, it allows for formal investigation of scheduling strategies. Secondly, it provides the basis for an extension of schedulability analysis techniques to the formal verification process, facilitating the modelling of real-time systems in a process algebraic manner using the rich background of research in scheduling theory.

## 1 Introduction

Scheduling theory has a rich history of research in computer science dating back to the 60's and early 70's. Process algebras have been studied as a formal theory of system design and verification since about the same time. These theories have remained separate until recently some connections have been investigated. However, combining scheduling theory in a process algebraic design still involves many theoretical and practical complications. In this paper, building upon previous attempts in this direction, we propose a process algebra for the design of scheduled real-time systems called *PARS* (for Process Algebra with Resources and Schedulers). Previous attempts to incorporate scheduling algorithms in process algebra either did not have an explicit notion of schedulers such as that of [3, 12, 13] (thus, coding the scheduling policy in the process specification) or scheduling is treated for restricted cases such as those of [4, 10] (that only support single-processor scheduling).

Our approach to modelling scheduled systems is depicted in Figure 1. Process specification (including aspects such as causal relations of actions, their timing and resource requirements) is separated from specification of schedulers. System level specification consists of applying schedulers to process specifications, on the one hand, and composing scheduled systems, on the other hand. A distinguishing feature of our process algebra is the possibility of specifying schedulers as process terms (similar to resource-consuming processes). Another advantage

K.G. Larsen and P. Niebert (Eds.): FORMATS 2003, LNCS 2791, pp. 134–150, 2004.
© Springer-Verlag Berlin Heidelberg 2004

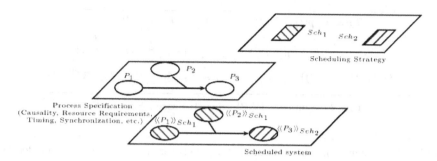

**Fig. 1.** Schematic view of the *PARS* approach

of the proposed approach is the separation between process specification and scheduler specification that provides a separation of concerns, allows for specifying generic scheduling strategies and makes it possible to apply schedulers to systems at different levels of abstraction. Common to most process algebraic frameworks for resources, the proposed framework provides the possibility of extending standard schedulability analysis to the formal verification process.

The paper is organized as follows. We define the syntax and semantics of *PARS* in three parts. In Section 2, we build a process algebra with asynchronous relative dense time (i.e., with the possibility of interleaving timing transitions) for process specification that has a notion of resource consumption. In Section 3, a similar process algebraic theory is developed for schedulers as resource providers. Section 4 defines application of a scheduler to a process. In each section, we first give the formal syntax and semantics of our language and then explain its usage using different aspects of one or more examples. In Section 5, we compare our approach to several recent extensions of process algebra with resources and finally, Section 6 concludes the results and presents future research directions. Due to space restrictions, in this paper, we leave out a few details of the theory and some definitions. We give informal explanation for the eliminated parts and refer the interested reader to [14] for a detailed version of this paper.

## 2   Process Specification

The first part of specification in *PARS* consists of process specification which represents the behavior of the system with the resource requirements of its basic actions. In our framework, resources are represented by a set $R$. The amount of resources required by a basic durational action is modeled by a function $\rho : R \to \mathbb{R}^{\geq 0}$ (indicating required quantity of each resource). We assume the resource demand to be constant during execution of basic actions. The resources provided by schedulers are modeled using a function $\overline{\rho} : R \to \mathbb{R}^{\leq 0}$. Active tasks (actions currently being executed) that require or provide resources are represented by multisets of such tasks in the semantics. As a notational convention, we refer to the set of all multisets as $M$. (We assume that the type of elements in the

$$P ::= \delta \mid p(t) \mid P \; ; \; P \mid P \parallel P \mid P \parallel\parallel P \mid P + P \mid$$

$$\sigma_t(P) \mid \mu X.P(X) \mid \int_{x_t \in T} P(x_t) \mid \partial_{Act}(P) \mid id : P$$

$$p \in (A \times (R \to \mathbf{R}^{\geq 0})) \cup \{\epsilon\}, \; t \in \mathbf{R}^{\geq 0}, \; \text{Act} \subseteq A, \; x_t \in V_t, T \subseteq \mathbf{R}^{\geq 0}, id \in \mathbf{N}$$

**Fig. 2.** Syntax of PARS, Part 1: Process Specification

multiset is clear from the context.) The operator $+$ and $-$ are overloaded to represent addition and subtraction of multisets.

The syntax of process specification in *PARS* is presented in Figure 2. It resembles a relative dense time process algebra (such as relative dense time ACP of [2]) with empty process ($\epsilon(0)$) and deadlock ($\delta$). The main difference with such a theory is the attachment of resource requirements to basic actions (most process algebras abstract from resource requirements by assuming abundant availability of shared resources) and our interpretation of time as duration of action execution. Basic action $\epsilon(t)$ represents idling which lasts for $t$ time and does not require any resource. Other basic actions $(a, \rho)(t)$ are pairs of actions from the set $A$ together with the respective resource requirement function $\rho$ and the timing $t$ during which the resource requirement should be provided to the action. Thus, the time annotation $t$ should be interpreted as a duration, corresponding to the time duration which action $a$ is to be executed; in standard timed process algebras, time annotations are usually interpreted as (absolute or relative) points in time corresponding to the occurrence or completion of an action. Terms $P \; ; \; P$, $P \parallel P$, $P \parallel\parallel P$, $P + P$ represent sequential composition, abstract, and strict parallel composition, and nondeterministic choice, respectively. Abstract parallel composition refers to cases where the ordering (and possible preemption) of actions has to be decided by a scheduling strategy. Strict parallel composition is similar to standard parallel composition in timed process algebra in that it forces concurrent execution of the two operands. The deadline operator applied to process $P$ in $\sigma_t(P)$ specifies that process $P$ should terminate within $t$ units of time or it will deadlock. Recursion is specified explicitly using the expression $\mu X.P(X)$ where free variable $X$ may occur in process $P$ and is bound by $\mu X$. The term $\int_{x_t \in T} P(x_t)$ specifies continuous choice of $x_t$ (from the set of timing variables $V_t$) over set $T$. Similar to recursion, variable $x_t$ is bound in term $P$ by operator $\int_{x_t \in T}$. In this paper, we are only concerned with closed terms (processes that do not have free recursion or timing variables). To prevent process $P$ from performing particular actions in some given set $Act$ (in particular, to force synchronization among two parallel processes, see e.g., [2]), the encapsulation expression $\partial_{Act}(P)$ is used. Using the $id :$ construct, process terms are decorated with identifiers (natural numbers, following the idea of [4]) which serve to group processes for scheduling purposes. Note that an atomic action is neither required to have an identifier, nor its identifier needs to be unique. Later on, in the semantics, a process identifier is augmented with a few estimations of performance measures of processes, namely relative deadline and worst-case execution time. Such a semantic identifer, in turn, is referenced by the scheduler specification domain in order to devise scheduling strategies. Precedence of binding among binary composition operators is ordered as $;, |||, ||, +$ where $;$ binds the strongest and $+$ the weakest. Unary operators are followed by a pair of parentheses or they bind to the smallest possible term.

The operational semantics of process specification is given in Figure 3. States are process terms and the semantics has two types of transitions. First, time passage (by spending time on resources or idling) $\overset{M,t}{\rightarrow}$ where $M$ is the multiset that represents the amount of resources required by the actions participating in the transition. Elements of $M$ are of the form $(ids, \rho)$, where $ids$ is the set of identifiers related to the action having resource requirements $\rho$. The second type of transitions, $\overset{a}{\rightarrow}$, represent the completion of actions. These transitions occur when an action has used the resources it requires for the specified amount of time. We do not combine resource requirements of different actions, but keep them separate in a multiset, since they may be provided by different scheduling policies (based on their respective process identifiers). We use $\overset{X}{\rightarrow}$ as a shorthand for either of the two transitions. Predicate $P\surd$ refers to the possibility of successful termination of $P$. The semantics of process specification is the smallest transition relation (union of the time and action transition relations) and the smallest termination predicate satisfying the rules of Figure 3.

Rules **(I0)** and **(I1)** specify the transitions and termination options of idling processes. In rule **(I1)**, $\overline{0}$ is a shorthand for the function mapping all resources to zero. Rules **(A0)** and **(A1)** specify how an atomic action can spend time on its resources and after that commit. The semantics of sequential composition is captured by **(S0)-(S2)**. Abstract parallel composition is specified by **(P0)-(P4)** and strict parallel composition by **(SP0)-(SP3)**. In rule **(P0)**, $t \gg Q$ uses an auxiliary unary operator (called deadline shift) specifying that $Q$ is getting $t$ units of time closer to its deadlines. Semantics of this operator is as follows:

$$\textbf{(Sh0)} \frac{p(t') \overset{X}{\rightarrow} P'}{t \gg p(t') \overset{X}{\rightarrow} P'} \qquad \textbf{(Sh1)} \frac{t \le t' \quad \sigma_{t'-t}(t \gg P) \overset{X}{\rightarrow} P'}{t \gg (\sigma_{t'}(P)) \overset{X}{\rightarrow} P'}$$

$$\textbf{(Sh2)} \frac{t \gg P \overset{X}{\rightarrow} P'}{t \gg (P \,;\, Q) \overset{X}{\rightarrow} P' \,;\, Q} \qquad \textbf{(Sh3)} \frac{P\surd \quad t \gg Q \overset{X}{\rightarrow} Q'}{t \gg (P \,;\, Q) \overset{X}{\rightarrow} Q'}$$

$$\textbf{(Sh4)} \frac{(t \gg P) \, Op_1 \, (t \gg Q) \overset{X}{\rightarrow} P'}{t \gg (P \, Op_1 \, Q) \overset{X}{\rightarrow} P'} \qquad \textbf{(Sh5)} \frac{Op_2 \, (t \gg P) \overset{X}{\rightarrow} P'}{t \gg Op_2 \, (P) \overset{X}{\rightarrow} P'}$$

$$\textbf{(Sh6)} \frac{t+t' \gg P \overset{X}{\rightarrow} P'}{t' \gg t \gg P \overset{X}{\rightarrow} P'} \qquad \textbf{(Sh7)} \frac{P\surd}{t \gg P\surd}$$

$$Op_1 \in \{||, |||, +\}, t, t' \in \mathbb{R}^{\ge 0} \qquad Op_2 \in \{\mu X., \int_{x_t \in T}, \partial_{Act}, id :\}$$

In the above semantics, the rules for sequential composition ((**Sh2**) and (**Sh3**)) are in line with the intuition that in scheduling theory only *ready* actions can take part in scheduling and other actions have to wait for their causal predecessors to commit. Function $\gamma(a, b)$ in rules **(P3)** and **(SP2)** specifies the

$$\textbf{(I0)}\frac{}{\epsilon(0)\sqrt{}} \quad \textbf{(I1)}\frac{t' \le t}{\epsilon(t) \overset{[(\emptyset,\bar{0})],t'}{\longrightarrow} \epsilon(t-t')}$$

$$\textbf{(A0)}\frac{t' \le t}{(a,\rho)(t) \overset{[(\emptyset,\rho)],t'}{\longrightarrow} (a,\rho)(t-t')} \quad \textbf{(A1)}\frac{}{(a,\rho)(0) \overset{a}{\to} \epsilon(0)}$$

$$\textbf{(S0)}\frac{P \overset{\chi}{\to} P'}{P\,;Q \overset{\chi}{\to} P'\,;Q} \quad \textbf{(S1)}\frac{P\sqrt{} \quad Q \overset{\chi}{\to} Q'}{P\,;Q \overset{\chi}{\to} Q'} \quad \textbf{(S2)}\frac{P\sqrt{} \quad Q\sqrt{}}{P\,;Q\sqrt{}}$$

$$\textbf{(P0)}\frac{P \overset{M,t}{\to} P'}{P \parallel Q \overset{M,t}{\to} P' \parallel t \gg Q} \quad \textbf{(P1)}\frac{P \overset{a}{\to} P'}{P \parallel Q \overset{a}{\to} P' \parallel Q} \quad \textbf{(P2)}\frac{P \overset{M,t}{\to} P' \quad Q \overset{M',t}{\to} Q'}{P \parallel Q \overset{M+M',t}{\to} P' \parallel Q'}$$

$$Q \parallel P \overset{M,t}{\to} t \gg Q \parallel P' \qquad Q \parallel P \overset{a}{\to} Q \parallel P'$$

$$\textbf{(P3)}\frac{P \overset{a}{\to} P' \quad Q \overset{b}{\to} Q' \quad \gamma(a,b)=c}{P \parallel Q \overset{c}{\to} P' \parallel Q'} \quad \textbf{(P4)}\frac{P\sqrt{} \quad Q\sqrt{}}{P \parallel Q\sqrt{}}$$

$$\textbf{(SP0)}\frac{P \overset{M,t}{\to} P' \quad Q \overset{M',t}{\to} Q'}{P \parallel\parallel Q \overset{M+M',t}{\to} P' \parallel\parallel Q'} \quad \textbf{(SP1)}\frac{P \overset{a}{\to} P'}{P \parallel\parallel Q \overset{a}{\to} P' \parallel\parallel Q}$$

$$Q \parallel\parallel P \overset{a}{\to} Q \parallel\parallel P'$$

$$\textbf{(SP2)}\frac{P \overset{a}{\to} P' \quad Q \overset{b}{\to} Q' \quad \gamma(a,b)=c}{P \parallel\parallel Q \overset{c}{\to} P' \parallel\parallel Q'} \quad \textbf{(SP3)}\frac{P\sqrt{} \quad Q\sqrt{}}{P \parallel\parallel Q\sqrt{}}$$

$$\textbf{(C0)}\frac{P \overset{\chi}{\to} P'}{P + Q \overset{\chi}{\to} P'} \quad \textbf{(C1)}\frac{P\sqrt{}}{P + Q\sqrt{}}$$

$$Q + P \overset{\chi}{\to} P' \qquad Q + P\sqrt{}$$

$$\textbf{(D0)}\frac{P \overset{M,t}{\to} P' \quad t \le t_0}{\sigma_{t_0}(P) \overset{M,t}{\to} \sigma_{t_0-t}(P')} \quad \textbf{(D1)}\frac{P \overset{a}{\to} P'}{\sigma_{t_0}(P) \overset{a}{\to} \sigma_{t_0}(P')} \quad \textbf{(D2)}\frac{P\sqrt{}}{\sigma_{t_0}(P)\sqrt{}}$$

$$\textbf{(E0)}\frac{P \overset{a}{\to} P' \quad a \notin Act}{\partial_{Act}(P) \overset{a}{\to} \partial_{Act}(P')} \quad \textbf{(E1)}\frac{P \overset{M,t}{\to} P'}{\partial_{Act}(P) \overset{M,t}{\to} \partial_{Act}(P')} \quad \textbf{(E2)}\frac{P\sqrt{}}{\partial_{Act}(P)\sqrt{}}$$

$$\textbf{(R0)}\frac{P[\mu X.P(X)/X] \overset{\chi}{\to} P'}{\mu X.P(X) \overset{\chi}{\to} P'} \quad \textbf{(R1)}\frac{P[\mu X.P(X)/X]\sqrt{}}{\mu X.P(X)\sqrt{}}$$

$$\textbf{(CC0)}\frac{t_0 \in T \quad P(t_0) \overset{\chi}{\to} P'}{\int_{x_t \in T} P(x_t) \overset{\chi}{\to} P'} \quad \textbf{(CC1)}\frac{t_0 \in T \quad P(t_0)\sqrt{}}{\int_{x_t \in T} P(x_t)\sqrt{}}$$

$$\textbf{(Id0)}\frac{P \overset{M,t}{\to} P'}{id:P \overset{M\oplus \tilde{id},t}{\to} id:P'} \quad \textbf{(Id1)}\frac{P \overset{a}{\to} P'}{id:P \overset{a}{\to} id:P'} \quad \textbf{(Id2)}\frac{P\sqrt{}}{id:P\sqrt{}}$$

$$a,b,c \in A,\ t,t' \in \mathbb{R}^{>0}, t_0 \in \mathbb{R}^{\ge 0}, \chi \in (\mathbb{M} \times \mathbb{R}^{>0}) \cup A$$

**Fig. 3.** Semantics of *PARS*, Part 1: Process Specification

result of a synchronized communication between $a$ and $b$. The semantics of abstract parallel composition deviates from standard semantics of parallelism in timed process algebras in that it allows for asynchronous spending of time by the two parties (rule **(P0)**). This reflects that depending on availability of resources and due to scheduling, concurrent execution of tasks can be preempted and serialized at any moment of time. Components not spending time on resources do not participate (actively) in a time transition. Rules **(C0)-(C1)** provide a semantics for nondeterministic choice. Given our interpretation of time, the choice operator does not have the property of time-determinism (which states that passage of time cannot determine choices): Starting to spend time on an action reveals the choice in the same way executing an action determines the choice in untimed process algebras. The deadline operator is defined by **(D0)-(D2)**. There is no rule for $\sigma_0(P)$ when $P$ can only do a time step. This means that this process deadlocks (i.e., missing a deadline results in deadlock). Encapsulation is defined in rules **(E0)-(E2)** stating that the encapsulation operator prevents process $P$ from performing actions in $Act$. Rules **(R0)-(R1)** and **(CC0)-(CC1)** specify the semantics of recursion and continuous choice. Note that in the semantics of continuous choice, the choice is made as soon as the process term starts making a transition. Rules **(Id0)-(Id2)** specify the semantics of $id$ by adding the semantic identifier $\widetilde{id}$ to the multiset in the transition, where $\widetilde{id}$ is the tuple $(id, Dl(P), WCET(P))$ consisting of the syntactic $id$, (an estimation of) the deadline, and the worst-case execution time of $P$. We omit detailed definitions of $Dl$ and $WCET$. They are defined in [14] using structural induction on process terms. Other performance measures can extend or replace this notion of semantic identifier. In semantic rule **(Id0)**, $\oplus$ stands for adding a semantic identifier to the set of identifiers of each resource-requirement function in the multiset.

The standard notion of strong bisimulation is not a congruence with respect to the operators defined in the process language. The problem lies particularly in the interaction between deadlines and abstract parallel composition. In [14], it is shown that strong bisimulation is a congruence with respect to a restricted subset of the language without the deadline operator. Also, there we define a notion of deadline-sensitive bisimulation that is a congruence. To show how the process specification language is to be used, next we specify a few common patterns from scheduling literature [5].

**Example 1** (Periodic and Aperiodic Tasks) First, we specify a periodic task, consisting of an atomic action $a$ requiring a single CPU and 100 units of memory during its computation time of $t$, and with period of $t'$.

$$P_1 = \mu X.((a, \{CPU \mapsto 1, Mem \mapsto 100\})(t) \,|||\, \epsilon(t') \,;\, X)$$

Note that the computation time of the periodic task may be larger than the period (which means that any feasible scheduler must allow for task parallelism). Now, suppose that the exact computation time of $a$ is not known. However, we know that the computation time is within a (possibly infinite) interval $I$, then the periodic task is specified as follows.

$$P_2 = \mu X.(\int_{x_t \in I} (a, \{CPU \mapsto 1, Mem \mapsto 100\})(x_t) \;|||\; \epsilon(t') \; ; \; X)$$

In the remainder, we use syntactic shorthand $p(I)$ instead of $\int_{x_t \in I} p(x_t)$.

Aperiodic tasks follow a similar pattern with the difference that instead of computation time, their period of arrival is not known:

$$S = \mu X.((b, \{CPU \mapsto 1\})(t) \;|||\; \epsilon([0, \infty)) \; ; \; X)$$

If the process specification of the system consists of periodic user level tasks and aperiodic system level tasks (e.g., system interrupts) that are to be scheduled with different policies, the specification goes as follows:

$$SysProc = System : (S) \;||\; User : (P_2)$$

where $System$ and $User$ are distinct integer id's for these two types of tasks.

**Example 2** (Portable Tasks) Suppose that the task $a$ can run on different platforms, either on a dual-processor machine on which it will take 2 units of time and 100 units of memory (during those 2 time units) or on a single processor for which it will require 4 units of time and 70 units of memory (over the 4 time units). Then it is specified as follows:

$$P = (a, \{CPU \mapsto 2, Mem \mapsto 100\})(2) + (a, \{CPU \mapsto 1, Mem \mapsto 70\})(4)$$

## 3   Scheduler Specification

The second part of system specification in $PARS$ is about scheduler specification. In this part, we model availability of resources and the strategy to grant these resources to processes requiring them. This is done by using predicates referring to properties of processes eligible for receiving the resources. The syntax of scheduler specification $(Sc)$ is similar to process specification and is specified in Figure 4. Basic actions of schedulers are predicates $(Pred)$ mentioning appropriate processes to be provided with resources and the amount of resources available $(\bar{\rho})$ during the specified time $(t)$. The predicate can refer to the syntactic identifier, deadline or worst-case execution time of processes. In the syntax of $Pred$, $Id$ is a variable from the set of semantic identifiers $V_i$ (with a distinguished member $\underline{Id}$ and typical members $Id_0, Id_1, etc.$). To refer to the specific process receiving the provided resources, we use $\underline{Id}$ and to refer to the processes in its context we use other members of $V_i$. Following the structure of a semantic identifier, $Id$ is a tuple containing syntactic identifier $(Id.id)$, deadline $(Id.Dl)$ and execution time $(Id.WCET)$. As in the process language, the language for predicates can be extended to other metrics of processes.

$$Sc \quad ::= \delta \mid s(t) \mid Sc \, ; Sc \mid Sc \parallel Sc \mid Sc \mid\mid\mid Sc \mid Sc + Sc \mid$$

$$\int_{x_t \in T} Sc(x_t) \mid Sc \triangleright Sc \mid Sc \triangleright^n Sc \mid \big\rangle_{x_t \in T} Sc(x_t) \mid \big\rangle^n_{x_t \in T} Sc(x_t) \mid \mu X.Sc(X)$$

$$Pred ::= Id.id \; Op_1 \; Num \mid Id.D1Op_1 \; time \mid Id.W \; CET \; Op_1 \; time \mid Pred \; Op_2 \; Pred$$

$$Op_1 ::= < \; \mid \; = \; \mid \; > \qquad Op_2 ::= \wedge \mid \vee$$

$$s \in (Pred \times (R \to \mathbf{R}^{\leq 0})) \cup \{\epsilon\}, t \in \mathbf{R}^{\geq 0}, x_t \in V_t, time \in V_t \cup \mathbf{R}^{\geq 0}, Id \in V_i, Num \in \mathbb{N}$$

**Fig. 4.** Syntax of *PARS*, Part 2: Scheduler Specification

A couple of new operators are added to the ones in the process specification language. The preemptive precedence operator $\triangleright$ gives precedence to the right-hand-side term (with the possibility of the right-hand side taking over the execution of the left-hand side at any point). Continuous preemptive precedence $\big\rangle_{x_t \in T}$ gives precedence to the least possible matching of $x_t \in T$. To be more precise, a continuous precedence operator generates a symbolic transition system with all possible $t \in T$, but, when confronted with a process, allows a transition with a particular $t'$ if the processes confronted with cannot make a transition with $t'' \in T \wedge t'' < t'$. The non-preemptive counter-parts of precedence operators $\triangleright^n$ and $\big\rangle^n_{x_t \in T}$ have the same intuition but they do not allow taking over of one side if the other side has already decided to start. Timing variables bound by continuous choice or continuous precedence operators can be used in predicates (as timing constants) and in process timings.

The semantics of schedulers is presented in Figure 5. It induces a symbolic transition system that has predicates indicating resource grants on its labels. At this level, we assume no information about resource requiring processes that the scheduler is to be confronted with. Thus, the resource grant predicates specify the criteria that processes receiving resources should satisfy and the criteria they should falsify. The latter can be used to state that a process should not be able to perform higher priority transitions. The transition relation is of the form $\overset{M,t}{\to}$, where $M$ is a multiset containing predicates about processes that can receive a certain amount of resources during time $t$. Elements of $M$ are of the form $(pred, npred, \overline{\rho})$ where $pred$ is the positive predicate that the process receiving resources should satisfy, $npred$ is the negative predicate that it should falsify and $\overline{\rho}$ is the function representing the amount of different resources offered. Rules (**ScA0**) and (**ScA1**) specify the semantics of atomic scheduler actions. Rule (**ScA1**) shows that a scheduler can provide its resources if the requiring process satisfies its predicate. The negative predicate of a basic scheduler is set to false (which is by default falsified). Rules (**Pr0**)-(**Pr2**) specify the semantics for the precedence operator. In these rules, $M \vee_{neg} pred$ stands for adding $pred$ as a disjunction to all negative predicates in $M$. Enabledness of a process term is used as a negative predicate to assure that a lower priority process cannot

$$(\text{ScA0})\frac{}{(p,\rho)(0)\sqrt{}} \qquad (\text{ScA1})\frac{t' \le t}{(p,\rho)(t)\overset{[(p,false,\rho)],t'}{\to}(p,\rho)(t-t')}$$

$$(\text{Pr0})\frac{P\overset{M,t}{\to}P'}{P\rhd Q\overset{M\vee_{\text{neg}}en(Q),t}{\to}P'\rhd Q} \qquad (\text{Pr1})\frac{Q\overset{M,t}{\to}Q'}{P\rhd Q\overset{M,t}{\to}P\rhd Q'} \qquad (\text{Pr2})\frac{P\sqrt{}\quad Q\sqrt{}}{P\rhd Q\sqrt{}}$$

$$(\text{CPr0})\frac{t'\in T \quad P(t')\overset{M,t}{\to}P'(t')}{\wr_{x_t\in T}P(x_t)\overset{M\vee_{\text{neg}}(en(P(x_t))\wedge x_t\in\lfloor T\rfloor_{t'}),t}{\to}\wr_{x_t\in T}P'(x_t)} \qquad (\text{CPr1})\frac{t'\in T \quad P(t')\sqrt{}}{\wr_{x_t\in T}P(x_t)\sqrt{}}$$

$$(\text{NPr0})\frac{P\overset{M,t}{\to}P'}{P\rhd^{\text{n}}Q\overset{M\vee_{\text{neg}}en(Q),t}{\to}P'} \qquad (\text{NPr1})\frac{Q\overset{M,t}{\to}Q'}{P\rhd^{\text{n}}Q\overset{M,t}{\to}Q'} \qquad (\text{NPr2})\frac{P\sqrt{}\quad Q\sqrt{}}{P\rhd^{\text{n}}Q\sqrt{}}$$

$$(\text{NCPr0})\frac{t'\in T \quad P(t')\overset{M,t}{\to}P'}{\wr^{\text{n}}_{x_t\in T}P(x_t)\overset{M\vee_{\text{neg}}(en(P(x_t))\wedge x_t\in\lfloor T\rfloor_{t'}),t}{\to}P'} \qquad (\text{NCPr1})\frac{t'\in T \quad P(t')\sqrt{}}{\wr^{\text{n}}_{x_t\in T}P(x_t)\sqrt{}}$$

$$t\in \mathbf{R}^{>0},\ t'\in \mathbf{R}^{\ge 0}$$

**Fig. 5.** Semantics of $PARS$, Part 2: Scheduler Specification

take over a higher priority one. The notion of enabledness is defined as follows. ($P\to$ stands for the possibility of performing a transition $P\overset{\chi}{\to}P'$ for some $P'$ and $\chi$. Moreover $P\nrightarrow$ stands for its negation.)

$$en((pred,\rho)(t)) \doteq pred$$

$$en(P\ ;\ Q) \doteq \begin{cases} en(P)\vee en(Q) & if\ P\sqrt{}\wedge P\to \\ en(P) & if\ \neg(P\sqrt{}) \\ en(Q) & if\ P\sqrt{}\wedge P\nrightarrow \end{cases}$$

$$en(P\ ||\ Q) \doteq en(P+Q) \doteq en(P\rhd Q) \doteq en(P\rhd^{\text{n}}Q) \doteq en(P)\vee en(Q)$$

$$en(\textstyle\int_{x_t\in T}P(x_t)) \doteq en(\wr_{x_t\in T}P(x_t)) \doteq en(\wr^{\text{n}}_{x_t\in T}P(x_t)) \doteq x_t\in T\wedge en(P(x_t))$$

Rules **(CPr0)-(CPr1)** present the semantic rules for the continuous precedence operators. In rule **(CPr0)**, expression $\lfloor T\rfloor_t$ is defined as $\{t'|t'\in T\wedge t' < t\}$. Note that in both preemptive precedence operators, the possibility of other options (lower or higher priority processes) always remains after making a transition. This allows for preempting or changing the resource provision at any point of time based on the processes that the scheduler is confronted with. Rules **(NPr0)-(NPr2)** and **(NCPr0)-(NCPr1)** specify the semantics of non-preemptive precedence operators. We omit the semantic rules for operators shared with process specification since they are analogous to those specified in the process specification semantics. Apart from action transition rules such as **(P1)** and **(P3)** that are absent in the semantics of schedulers, the rest of the rules in Figure 3 remain intact for this semantics.

**Example 3** Consider the process specification of Example 1, where the system consists of two types of processes: User processes and system processes. Suppose that our execution platform can provide two processors and 200 units of memory. System processes have priority over user processes (in using CPUs). The following scheduler is the first attempt to specify our scheduling strategy:

$$Sch_{Mem} \quad = (true, \{Mem \mapsto -200\})([0, \infty))$$
$$PrSch_{CPU0} = (\underline{Id}.id = User, \{CPU \mapsto -2\})([0, \infty)) \rhd$$
$$(\underline{Id}.id = System, \{CPU \mapsto -2\})([0, \infty))$$
$$Sch_0 \quad = Sch_{Mem} \ ||| \ PrSch_{CPU0}$$

The above specification generates a transition system that allows arbitrary time transitions providing both CPUs and 200 units of memory with negative predicate $false$ to system processes (meaning that that there is no process that can take over a system process). However, according to rule **(Pr0)**, for transitions providing CPU to user processes, the predicate $t \in [0, \infty) \wedge \underline{Id}.id = System$ is added as a negative predicate. Intuitively, this should mean that CPUs are provided to a user process if no system process is able to take that transition. However, this would prevent the user process from gaining access to its CPU requirement even if only a single CPU is used by a system process (thus, one CPU can be wasted without any reason). The following scheduler specification solves this problem by separating the scheduling process of the two CPUs:

$$PrSch_{CPU1} = (\underline{Id}.id = User, \{CPU \mapsto -1\})([0, \infty)) \rhd$$
$$(\underline{Id}.id = System, \{CPU \mapsto -1\})([0, \infty))$$
$$Sch_1 \quad = PrSch_{CPU1} \ ||| \ PrSch_{CPU1} \ ||| \ Sch_{Mem}$$

Part of the symbolic transition system of scheduler $Sch_1$ is depicted in Figure 6. Of course, part of the intuitive explanation given above remains to be formalized by the semantics of applying schedulers to processes where resources are provided to actual tasks (i.e., the symbolic transition of a scheduler is matched with an actual transition of a process).

**Example 4** (Specifying Scheduling Strategies) To illustrate the scheduler specification language, we specify a few generic single-processor scheduling strategies. *Non-preemptive Round-Robin Scheduling*: Consider a scheduling strategy where a single processor is going to be granted to processes non-preemptively in an increasing order of process identifiers (from 0 to $n$). The following scheduler specifies the round-robin strategy.

$$Sch_{NP-RR} = \mu X.((\underline{Id} = n, \{CPU \mapsto 1\})[0, \infty) \rhd^n \ldots \rhd^n$$
$$((\underline{Id} = 1, \{CPU \mapsto 1\})[0, \infty) \rhd^n (\underline{Id} = 0, \{CPU \mapsto 1\})[0, \infty))) \ ; X$$

*Monotonic Scheduling*: Consider the following process specifications of several periodic tasks:

$$SysProc = P_0 \parallel P_1 \parallel \ldots \parallel P_n$$
$$P_i = \mu X.(2i+1):((a_i, \rho_i)(t)) \parallel\parallel (2i):(\epsilon(t')) ; X$$

The following scheduler specifies the preemptive rate monotonic strategy, where processes with the shortest period (the highest rate) have priority:

$$RMSch(i, x_t) =$$
$$(\underline{Id}.id = 2i+1 \wedge Id_0 = 2i \wedge Id_0.WCET = x_t, \{CPU \mapsto 1\})([0, \infty))$$
$$RMSch = \mathord{\int}_{x_t \in I\!R^{\geq 0}} RMSch(0, x_t) + \ldots + RMSch(n, x_t)$$

Scheduler process $RMSch(i, t)$ specifies that the process receiving CPU should have an odd identifier (thus, being an action) and its corresponding period should have worst-case execution time $t$. Process $RMSch$ states that the processes with the least period have precedence over the others.

# 4  Applying Schedulers to Processes

Scheduled systems are processes resulting from application of a scheduler to processes. Syntax of scheduled systems is presented in Figure 7. In this syntax, $P$ and $Sc$ refer to the syntactic classes of processes and schedulers presented in the previous sections. Term $\langle\!\langle Sys \rangle\!\rangle_{Sc}$ denotes applying scheduler $Sc$ to the system $Sys$ and $\partial_{Res}(Sys)$ is used to close a system specification and prevent it from requiring resources in $Res$.

The semantics of the new operators for scheduled systems is defined in Figure 8. The type of labels in the transition relation is the same as that of the transition relation in the process specification semantics of Figure 3 (hence, multisets in time transitions are resource requirement multisets). Since a process is a system by definition, all semantic rules of Figure 3 carry over to the semantics of systems. It should be understood that the variables ranging over

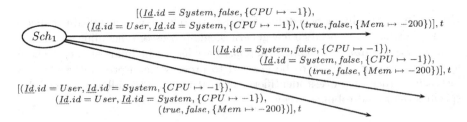

**Fig. 6.** Part of the transition system of scheduler $Sch_1$ in Example 3

$$Sys ::= P \mid \langle\langle Sys \rangle\rangle_{Sc} \mid Sys \mathbin{;} Sys \mid Sys \parallel Sys \mid Sys \parallel\parallel Sys \mid Sys + Sys \mid$$

$$\partial_{Res}(Sys) \mid \sigma_t(Sys) \mid \mu X.Sys(X) \mid id : Sys$$

**Fig. 7.** Syntax of *PARS*, Part 3: Syntax of Scheduled Systems

process terms in Figure 3, are in this case ranging over the more general class of system terms.

The application operator $\langle\langle Sys \rangle\rangle_{Sc}$ is defined by semantic rules **(Sys0)**-**(Sys2)**. In **(Sys0)**, the operator $apply : Sys \times \mathbb{M} \times \mathbb{M} \to \mathbb{P}(\mathbb{M})$ is meant to apply a multiset of resource providing predicates (third parameter) to a multiset of resource requiring tasks (second parameter) originated from a system (first parameter). The formal definition of this operator is the smallest function satisfying the following constraint:

$$\forall_{m \in M'} \; \forall_{M'' \in applyTask(S,M,m,\emptyset)} \, apply(S, M'', M' - [m]) \subseteq apply(S, M, M')$$

In this statement, *applyTask* (defined below) is meant to provide the set of possible outcomes of applying a single resource providing task $m$ to the resource requiring multiset $M$ (the forth parameter of *applyTask* is used to keep track of resource requiring tasks checked so far to receive the provided resource). This statement means that the application of a scheduler task to a multiset of process tasks is done by taking an arbitrary scheduler task and applying it to the multiset of process tasks and starting over with the rest of scheduler tasks. The function *applyTask* is the smallest function satisfying the following constraints:

$$applyTask(S, \emptyset, m, M) \doteq \{\emptyset\}$$
$$N + N' \in applyTask(S, [(ids, \rho)] + M, (pred, npred, \overline{p}), M'),$$

where

$$
\begin{cases}
\begin{aligned}
&if \;\; pred(ids, M + M' + [(ids, \rho)]) \wedge \\
&\quad \neg engage(S, M + M' + [(ids, \rho)], (pred, npred, \overline{p})) \\
&\qquad N = [(ids, max(\overline{0}, \rho + \overline{p})] \wedge \\
&\qquad\quad N' \in applyTask(S, M, (pred, npred, min(\overline{0}, \overline{p} + \rho)), M' + [(ids, \rho)]) \\
&otherwise \\
&\qquad N = [(ids, \rho)] \\
&\qquad\quad N' \in applyTask(S, M, (pred, npred, \overline{p}), M' + [(ids, \rho)])
\end{aligned}
\end{cases}
$$

The above expression states that if we pick a resource requiring task $(ids, \rho)$ which satisfies the positive predicate (specified by $pred(ids, M + M' + [(ids, \rho)])$)

$$(\textbf{Sys0}) \frac{Sys \overset{M,t}{\to} Sys' \quad Sc \overset{M',t}{\to} Sc' \quad M'' \in apply(Sys, M, M')}{\langle\!\langle Sys \rangle\!\rangle_{Sc} \overset{M'',t}{\to} \langle\!\langle Sys' \rangle\!\rangle_{Sc'}}$$

$$(\textbf{Sys1}) \frac{Sys \overset{a}{\to} Sys'}{\langle\!\langle Sys \rangle\!\rangle_{Sc} \overset{a}{\to} \langle\!\langle Sys' \rangle\!\rangle_{Sc}} \qquad (\textbf{Sys2}) \frac{Sys\surd}{\langle\!\langle Sys \rangle\!\rangle_{Sc}\surd}$$

$$(\textbf{ER0}) \frac{Sys \overset{M,t}{\to} Sys' \quad \forall_{(ids,\rho)\in M, r\in Res}\rho(r) = 0}{\partial_{Res}(Sys) \overset{M,t}{\to} \partial_{Res}(Sys')}$$

$$(\textbf{ER1}) \frac{Sys \overset{a}{\to} Sys'}{\partial_{Res}(Sys) \overset{a}{\to} \partial_{Res}(Sys')} \qquad (\textbf{ER2}) \frac{Sys\surd}{\partial_{Res}(Sys)\surd}$$

$$t \in \mathbb{R}^{>0}, \; a \in A, \; Res \subseteq R$$

**Fig. 8.** Semantics of *PARS*, Part 3: Applying Schedulers to Processes

and the tasks in its context (including the picked task itself) cannot satisfy the negative predicate ($\neg engage(S, M + M' + [(ids, \rho)], (pred, npred, \overline{p}))$) then we can grant the resources to this task and continue feeding the remaining tasks with the remaining resources. Otherwise, we leave this resource requiring task and proceed with the remaining tasks. In this expression, $min(\overline{0}, \overline{p} + \rho)$ and $max(\overline{0}, \rho + \overline{p})$ are point-wise minimum and maximum of $\overline{p}(r) + \rho(r)$ with 0, respectively. The predicate $pred(\widetilde{id}, M)$ means that there exists a mapping from variables in $V_i$ (set of id variables) to the semantic identifiers in $M$ (particularly mapping $\underline{Id}$ to a member of $ids$) which satisfies $pred$. The predicate $engage$ is formally defined as follows:

$$engage(S, M, (pred, npred, \overline{p})) \doteq \exists_{M',t,S',ids,\rho,\widetilde{id}} S \overset{M',t}{\to} S' \wedge M \subseteq M' \wedge$$
$$(ids, \rho) \in M' \wedge \widetilde{id} \in ids \wedge npred(\widetilde{id}, M') \wedge \exists_{r\in R}\rho(r) > 0 \wedge \overline{p}(r) < 0$$

This predicate checks if $S$ can perform a transition with a resource requiring multiset $M'$ that firstly, contains $M$ (thus, extending the same group of tasks), secondly, there exists a task identifier $\widetilde{id}$ in it that can satisfy the negative predicate ($npred(\widetilde{id}, M')$, defined in the same way as $pred(\widetilde{id}, \ldots)$) and the corresponding task can potentially use the resources offered by $\overline{p}$. To summarize, it checks for existence of a higher priority task in the context that can possibly consume the resources offered by the scheduler.

Note that application of a scheduler to a system does not necessarily satisfy all resource requirements of the system. Since the transition system of a scheduled system is itself a process specification transition system, several schedulers can be applied to a system in a distributed (using parallel composition of several schedulers) or hierarchical (using several levels of application operator) fashion in order to satisfy all its requirements.

Rules **(ER0)-(ER2)** represent preventing the system from requiring resources of a certain type by using an encapsulation operator on a given set of resources (similar to the encapsulation construct for actions).

**Example 5** Consider the following process specification and the two different Earliest Deadline First (EDF) schedulers:

$$Proc \; = 1 : (\sigma_1(a, \{CPU \mapsto 1, Mem \mapsto 100\}(1)))$$
$$2 : (\sigma_2(b, \{CPU \mapsto 1, Mem \mapsto 100\}(2)))$$
$$EDF_1 = \mu X.(\big\rangle_{x_t \in I\!\!R^{\geq 0}}(\underline{Id}.Dl = x_t)\{CPU \mapsto -2, Mem \mapsto -200\}(2)) \, ; X$$
$$EDF_2 = \mu X.(\big\rangle_{x_t \in I\!\!R^{\geq 0}}((\underline{Id}.Dl = x_t)\{CPU \mapsto -1, Mem \mapsto -100\}(2)) \;|||$$
$$\big\rangle_{x_t \in I\!\!R^{\geq 0}}((\underline{Id}.Dl = x_t)\{CPU \mapsto -1, Mem \mapsto -100\}(2))) \, ; X$$

In the system $\partial_{\{CPU,Mem\}}(\langle\!\langle Proc \rangle\!\rangle_{EDF_1})$, the scheduler should start providing all available resources to task 1 for one unit of time, thus wasting one CPU and 100 units of memory. After that, available resources will be given to process 2. However, the process misses its deadline, since it needs 2 units of time to compute while its deadline has been shifted to 1 already. In contrast, system $\partial_{\{CPU,Mem\}}(\langle\!\langle Proc \rangle\!\rangle_{EDF_2})$ allows for a successful run. In this case, at the first time unit each of the two processes can receive a CPU and 100 units of memory. This is due to the fact that after providing the required resources of process 1 by one of the basic schedulers, the other scheduler may assign its resources to process 2. It follows from the semantics that after applying one resource offer to process 1 the whole process cannot engage in a resource interaction with a deadline of less than 2 and thus process 2 can receive its required resources.

This example helps us to realize that although scheduling policies such as earliest deadline first are assumed to be well-defined scheduling policies, formalizing their definition shows that different flavors of them may exist in practice (especially with respect to multiple resources), some of which may perform better than others for different systems.

## 5   Related Work

Several theories of process algebra with resources have been proposed recently. Our approach is mainly based on dense time ACSR of [3]. ACSR [13, 12] is a process algebra enriched with priorities and specification of resources. Several extensions to ACSR have been proposed over time for which [13] provides a summary. The main shortcoming of this process algebra is the absence of an explicit scheduling concept. In ACSR, scheduling strategy is coded by means of priorities inside the process specification domain. Due to lack of a resource provision model, some other restrictions are also imposed on resource demands of processes. For example, two parallel processes are not allowed to call for one resource or they deadlock.

Our work has also been inspired by [4]. There, a process algebraic approach to resource modelling is presented and application of scheduling to process terms is investigated. This approach has an advantage over that of ACSR in that scheduling is separated from the process specification domain. However, firstly, there is no structure or guideline to define schedulers in this language (as [13] puts it, the approach looks like defining a new language semantics for each scheduling strategy) and secondly, the scheduling is restricted to a single resource (single CPU) concept.

Scheduling algebra of [17] defines a process algebra that has processes with interval timing. Computing the possible start time of tasks (so-called *anchor points*) is the only aspect of scheduling that is taken into account and it abstracts from resource requirements/provisions.

RTSL of [10] defines a discrete-time process algebra for scheduling analysis of single processor systems. The only shared resource in this process algebra is the single CPU. The restriction of tasks, in this approach, to sequential processes makes the language less expressive than ours (for example, in the process language a periodic task whose computation time may be larger than its period cannot be specified). Also, coding the scheduling policy in terms of a priority function may make specification of scheduling more cumbersome (similar to [4]).

Timed automata, as a well-known specification method for timed systems, has been extended to cover the notion of resources and scheduling as well (see [9], for example). Papers [16] and [11] are examples of an extension of untimed models with resources.

Asynchrony in timed parallel composition (interleaving of relative timed-transitions) has been of little interest in timed process algebras. Semantics of parallel composition in ATP [15] and different versions of timed-ACP [2], timed-CCS [6, 7] and timed-CSP [8] all enforce synchronization of timed transitions such that both parallel components evolve concurrently in time. The *cIPA* of [1] is among the few timed process algebras that contain a notion of timed asynchrony. In this process algebra, non-synchronizing actions are forced to make asynchronous (interleaving) time transitions and synchronizing actions are specified to perform synchronous (concurrent) time transition. This distinction is not necessary in our framework, since non-synchronizing actions may find enough resources to execute in true concurrency and synchronizing actions may be forced to make interleaving time transitions due to the use of shared resources (e.g., scheduling two synchronizing actions on a single CPU).

## 6   Conclusion

In this paper, we propose an approach to integrate the separate specifications of real-time behavior (including aspects such as duration of actions, causal dependencies, synchronization) and scheduling strategy in an integrated and uniform process algebraic formalism. This allows for formalizing scheduling algorithms and benefiting from them in process algebraic design of systems as independent specification entities that influence the real-time behavior of the system. Our technical contribution to the current real-time and/or resource-based process algebraic formalisms can be summarized as defining a dense and asynchronous timed process algebra for resource consuming processes, providing a (similar) process algebraic language with basic constructs for defining resource providing processes (schedulers with multiple resources) and defining application of schedulers to processes in an algebraic fashion.

The theory presented in this paper can be completed/extended in several ways. Among those, axiomatizing *PARS* is one of the most important ones. As it can be seen in this paper, the three phases of specifications share a major part of the semantics; thus bringing the three levels of specification closer (for example, allowing for multi-level scheduling of a resource or allowing resource consuming schedulers) can be beneficial. Furthermore, applying the proposed theory in practice calls for simplification (e.g., to discrete time), optimization for implementation, tooling and experimenting in the future.

# References

[1] L. Aceto and D. Murphy. Timing and causality in process algebra. *Acta Informatica*, 33(4):317–350, 1996. 148

[2] J. C. M. Baeten and C. A. Middelburg. *Process Algebra with Timing*. EATCS Monographs. Springer-Verlag, Berlin, Germany, 2002. 136, 148

[3] P. Brémond-Grégoire and I. Lee. A process algebra of communicating shared resources with dense time and priorities. *Theoretical Computer Science*, 189(1–2):179–219, 1997. 134, 147

[4] M. Buchholtz, J. Andersen, and H. H. Loevengreen. Towards a process algebra for shared processors. In *Proceedings of MTCS '01, Electronic Notes in Theoretical Computer Science*, 52(3), 2002. 134, 136, 148

[5] G. C. Buttazzo. *Hard Real-Time Computing Systems*. Kluwer Academic Publishers, Boston, MA, USA, 1997. 139

[6] F. Corradini, D. D'Ortenzio, and P. Inverardi. On the relationships among four timed process algebras. *Fundamenta Informaticae*, 38(4):377–395, 1999. 148

[7] M. Daniels. Modelling real-time behavior with an interval time calculus. In J. Vytopil, editor, *Proceedings of FTRTF '91*, volume 571 of *Lecture Notes in Computer Science*, pages 53–71. Springer-Verlag, Berlin, Germany, 1991. 148

[8] J. Davies and S. Schneider. A brief history of Timed CSP. *Theoretical Computer Science*, 138(2):243–271, Feb. 1995. 148

[9] E. Fersman, P. Pettersson, and W. Yi. Timed automata with asynchronous processes: Schedulability and decidability. In *Proceedings of TACAS 2002*, volume 2280 of *Lecture Notes in Computer Science*, pages 67–82. Springer-Verlag, Berlin, Germany, 2002. 148

[10] A. N. Fredette and R. Cleaveland. RTSL: A language for real-time schedulability analysis. In *Proceedings of the Real-Time Systems Symposium*, pages 274–283. IEEE Computer Society Press, Los Alamitos, CA, USA, 1993. 134, 148

[11] P. Gastin and M. W. Mislove. A truly concurrent semantics for a process algebra using resource pomsets. *Theoretical Computer Science*, 281:369–421, 2002. 148

[12] I. Lee, J.-Y. Choi, H. H. Kwak, A. Philippou, and O. Sokolsky. A family of resource-bound real-time process algebras. In Proceedings of FORTE '01, pages 443–458. Kluwer Academic Publishers, Aug. 2001. 134, 147

[13] I. Lee, A. Philippou, and O. Sokolsky. A general resource framework for real-time systems. In Proceedings of the Monterey Workshop, Venice, Italy, 2002. 134, 147, 148

[14] M. Mousavi, M. Reniers, T. Basten and M. Chaudron. *PARS*: A process algebra with resource and schedulers. Technical report, Department of Computer Science, Eindhoven University of Technology, 2003. To appear. 135, 139

[15] X. Nicollin and J. Sifakis. The algebra of timed processes ATP: theory and application. Information and Computation, 114(1):131–178, Oct. 1994. 148

[16] M. Núñez and I. Rodríguez. PAMR: A process algebra for the management of resources in concurrent systems. In Proceedings of FORTE '01, pages 169–185. Kluwer Academic Publishers, 2001. 148

[17] R. van Glabbeek and P. Rittgen. Scheduling algebra. In A. M. Haeberer, editor, Proceedings of AMAST '99, volume 1548 of Lecture Notes in Computer Science, pages 278–292. Springer-Verlag, Berlin, Germany, 1999. 148

# Formal Semantics of Hybrid Chi

R.R.H. Schiffelers, D.A. van Beek, K.L. Man, M.A. Reniers, and J.E. Rooda

Department of Mechanical Engineering
and Department of Mathematics and Computer Science
Eindhoven University of Technology
P.O.Box 513, 5600 MB Eindhoven, The Netherlands
{R.R.H.Schiffelers,D.A.v.Beek,K.L.Man,M.A.Reniers,J.E.Rooda}@tue.nl

**Abstract.** The verification formalism / modeling and simulation language hybrid Chi is defined. The semantics of hybrid Chi is formally specified using Structured Operational Semantics (SOS) and a number of associated functions. The $\chi$ syntax and semantics can also deal with local scoping of variables and/or channels, implicit differential algebraic equations, such as higher index systems, and they are very well suited for specification of pure discrete event systems.

## 1 Introduction

The hybrid $\chi$ (Chi) language was originally designed as a modeling and simulation language for specification of discrete-event (DE), continuous time (CT) or combined DE/CT models (so-called hybrid models). The language and simulator have been successfully applied to a large number of industrial cases, such as an integrated circuit manufacturing plant, a brewery, and process industry plants [1]. For the purpose of verification, the discrete-event part of the language was mapped onto the $\chi_\sigma$ process algebra, for which a structured operational semantics was defined, bisimulation relations were derived, and a model checker was built [2]. In this way, verification of DE $\chi$ models was made possible [3].

One of the goals of our research is the development of a hybrid verification formalism / modeling and simulation language with associated verification and simulation tools. The recent formalization of the $\chi$ language, including the continuous part, resulted in the $\chi_{\sigma_h}$ process algebra, described in this paper, and in a more elegant $\chi$ modeling language. The $\chi$ language now has the same operators, with the same semantics, as the $\chi_{\sigma_h}$ formal language. The $\chi$ modeling language extends $\chi_{\sigma_h}$ with, among others, parameterized process and experiment definitions and instantiations. A straightforward syntactical translation of $\chi$ to $\chi_{\sigma_h}$ is described in [4].

The $\chi_{\sigma_h}$ language is a hybrid process algebra, and is thus related to other hybrid process algebras, such as HyPa [5], the $\phi$-Calculus [6], and hybrid formalisms based on CSP [7], [8]. It is also related to hybrid Petri nets [9], hybrid I/O automata [10], hybrid automata [11], and to work derived from hybrid automata, such as Charon [12] and Masaccio [13]. The main difference between

K.G. Larsen and P. Niebert (Eds.): FORMATS 2003, LNCS 2791, pp. 151–165, 2004.
© Springer-Verlag Berlin Heidelberg 2004

the $\chi$ formalism and these other formalisms is that $\chi$ is overall a more expressive formalism. Higher expressivity means either that certain phenomena can be modeled in $\chi$ whereas they cannot be modeled in some other formalisms, or that certain phenomena can be modeled more concisely or more intuitively in $\chi$. The higher expressivity is a result of:

1. The relatively large number of operators dedicated to modeling of discrete-event behavior. This makes it easy to abstract from continuous behavior and specify pure discrete-event models, without any continuous variables. In this respect, $\chi$ has much in common with the $\phi$-Calculus [6], and the hybrid formalisms based on CSP [7], [8].
2. The division of continuous variables into three subclasses. This allows for specification of steady state initialization, initialization of algebraic variables, consistent initialization of higher index systems, mode switches accompanied by index changes [14], and variables changing dynamically from differential to algebraic. In HyPa [5], such phenomena can in principle also be specified. HyPa, however requires a categorization of variables attached to every equation, whereas in $\chi$ this can be specified once, by means of a scope operator.
3. The scope operator combined with parameterized process definition and instantiation that enable hierarchical composition of processes. In this respect, the $\chi$ language is related to Charon [12], that allows components to be defined and instantiated. Local variables and variable abstraction are present in many formalisms. In $\chi$, however, the concepts of variable abstraction and channel abstraction (comparable with action abstraction in other formalisms) are integrated in the scope operator, which also provides a local scope for the three classes of continuous variables and for recursive process definitions.

Section 2 describes the syntax of the $\chi_{\sigma_h}$ language. In Section 3, the semantics of $\chi_{\sigma_h}$ is formally specified using a structured operational semantics (SOS) and a number of associated functions. Examples in Section 4 are used to illustrate the language.

## 2   Syntax of the $\chi_{\sigma_h}$ Language

A $\chi_{\sigma_h}$ process is a triple $\langle p, \sigma, E \rangle$, where $p$ denotes a process term, $\sigma$ denotes a valuation, and $E$ denotes an environment. A valuation is a partial function from variables to values (constants). Syntactically, a valuation is denoted by a set of pairs $\{x_0 \mapsto c_0, \ldots, x_n \mapsto c_n\}$, where $x_i$ denotes a variable and $c_i$ its value. An environment is a five-tuple $(E_\Gamma, E_J, E_F, E_C, E_R)$, where $E_\Gamma, E_J, E_F$ denote sets of "normal" continuous variables, jumping continuous variables, and fixed continuous variables, respectively. In most models, the normal continuous variables are used. The behavior of these variables depends on the way they occur in equations: a normal continuous variable that occurs differentiated or algebraic (not differentiated) behaves as a fixed continuous variable or as a jumping continuous variable, respectively (see the semantics of function $\Omega$ in Section 3). All variables must be in the domain of $\sigma$. The variables that are not in any of the sets $E_\Gamma, E_J, E_F$ are discrete. In the environment, $E_C$ denotes a set of channel labels, and $E_R$ denotes a recursive process definition. A recursive process definition is a partial function from recursion variables to process terms. Syntactically, a recursive process definition is denoted by a set of pairs $\{X_0 \mapsto p_0, \ldots, X_m \mapsto p_m\}$, where $X_i$ denotes a recursion variable and $p_i$ the process term defining it. Process terms $P$ in $\chi_{\sigma_h}$ are built from atomic process terms $(AP)$ using operators for combining them:

$$AP ::= \text{skip} \mid x := e \mid \quad m!e \quad \mid \quad m?x \mid u \mid \Delta e_n$$
$$P ::= AP \mid X \mid b \to P \mid P \rhd P \mid P; P \mid P \oplus P$$
$$\mid P \llbracket P \mid P \parallel P \mid \llbracket \sigma, E \mid P \rrbracket \mid \partial(P) \mid \pi(P)$$

An informal (concise) explanation of this syntax is given below. Section 3 gives a more detailed account of their meaning.

The process term skip represents an internal action. The value of variables can be changed instantaneously through assignments. An assignment is a process term of the form $x := e$ with $x$ a variable and $e$ an expression. In principle, the continuous variables change arbitrarily over time. Predicates $(u)$ are used to control these changes, i.e., a predicate restricts the allowed behavior of the continuous variables. In $\chi$ two types of predicates over continuous and discrete variables are allowed: (1) differential equations of the form $rde_1 = rde_2$ where $rde_1$ and $rde_2$ are real-valued expressions in which the derivative operator may be used (e.g., $\dot{x} = -x + y$), and (2) predicates in which the derivative operator may not be used (e.g., $x \geq 0$, $y = 2x + 2$, true).

More complex process terms can be obtained by composing process terms by means of among others sequential composition (;), choice ($\oplus$), alternative composition ($\llbracket$), parallel composition ($\parallel$) and guarding a process term $p$ by a boolean expression $b$: $b \to p$. The process term $b \to p$ denotes the process term that behaves as process term $p$ in case the boolean expression $b$ evaluates to true and deadlocks otherwise.

Processes interact either through the use of shared variables or by synchronous point-to-point communication over a channel. By means of $m!e$, the value of expression $e$ is sent over channel $m$. By means of $m?x$ a value is received from channel $m$ in variable $x$. The acts of sending and receiving a value have to take place at the same moment in time. The encapsulation operator $\partial$ is introduced to block internal send and receive events in order to assure that only their synchronous execution takes place.

Some of the atomic process terms in $\chi_{\sigma_h}$ are delay-able (sending and receiving), others are not delay-able (skip, assignments). By means of the delay process term $\Delta e_n$ a process can be forced to delay for the amount of time units specified by the value of numerical expression $e_n$. By means of the maximal progress operator $\pi$, execution of actions can be given priority over passage of time.

The disrupt operator ($\rhd$) is used for describing that a process is allowed to take over execution from another process even if that process is not finished yet (this in contrast with sequential composition). This is useful for describing mode switches and interrupts/disrupts.

In $\chi$, two operators can be used for the purpose of describing alternative behaviors; the choice operator ($\oplus$) and the alternative composition operator ($[\![]\!]$). The choice operator allows choice between different kinds of continuous behavior of a process, where the choice depends on the initial state of the continuous-time or hybrid process. The alternative composition operator allows choice between different actions/events of a process, usually between time-events, state-events or communication events of a discrete-event controller. In such a case, time-passing should not make a choice. The choice is delayed until the first action is possible.

A scope process term $[\![\ \sigma, E \mid p\ ]\!]$ is used to declare a local scope. Here $\sigma$ denotes a valuation of local variables, and $E$ denotes a local environment as defined in the beginning of this section.

The operators are listed in descending order of their binding strength as follows $\{;, \rightarrow, \triangleright\}, \{\oplus, [\!], \| \}$. The operators inside the braces have equal binding strength. In addition, operators of equal binding strength associate to the left, and parentheses may be used to group expressions.

# 3   Semantics of the $\chi_{\sigma_h}$ Language

In this section, the structured operational semantics (SOS) of $\chi_{\sigma_h}$ is presented. It associates a hybrid transition system [15] with a $\chi_{\sigma_h}$ process.

## 3.1   General Description of the SOS

The main purpose of such an SOS is to define the behavior of $\chi_{\sigma_h}$ processes at a certain chosen level of abstraction. The meaning of a $\chi_{\sigma_h}$ process depends on the values of the variables and on the environment. A set $V$ of variables, and a set $\mathbb{C}$ of channel labels that may be used in $\chi_{\sigma_h}$ specifications are assumed. The values of the variables at a specific moment in time are captured by means of a valuation, i.e., a partial function from the variables to the union of the set of values $\Lambda$ (containing at least the booleans $\mathbb{B}$, and the reals $\mathbb{R}$) and a "value" $\perp$ (indicating undefinedness). The set of all valuations is denoted $\Sigma$: $\Sigma = V \mapsto (\Lambda \cup \{\perp\})$. The set $T$ is used to represent points in time; usually $T = \mathbb{R}_{\geq 0}$. The set of environments $ES$ is defined as $ES = \mathcal{P}(V) \times \mathcal{P}(V) \times \mathcal{P}(V) \times \mathcal{P}(\mathbb{C}) \times RS$, where $\mathcal{P}$ denotes the powerset function and $RS = XS \mapsto P$ denotes the set of all partial functions of recursion variables $XS$ to process terms $P$. The elements of an environment $E \in ES$ can be obtained by means of five functions: $\Gamma, \mathcal{J}, \mathcal{F} \in ES \rightarrow \mathcal{P}(V)$, $\mathcal{C} \in ES \rightarrow \mathcal{P}(\mathbb{C})$, and $\mathcal{R} \in ES \rightarrow RS$. The function $\Gamma$ is defined as $\Gamma(E_\Gamma, E_J, E_F, E_C, E_R) = E_\Gamma$. The functions $\mathcal{J}$, $\mathcal{F}$, $\mathcal{C}$ and $\mathcal{R}$ are defined in a similar way to the function $\Gamma$. The SOS is chosen to represent the following:

1. instantaneous execution of discrete transitions:
   (a) $\_ \overset{\cdot}{\rightarrow} \_ \subseteq (P \times \Sigma \times ES) \times (A_\tau \times \Sigma) \times (P \times \Sigma \times ES)$, where $A_\tau$ denotes the actions, and is defined as $A_\tau = \{\alpha(m, c) \mid \alpha \in \{isa, ira, ca\}, m \in \mathbb{C}, c \in$

$\Lambda\} \cup \{\tau\}$. Here, $isa, ira, ca$ denote action labels for internal send action, internal receive action and communication action respectively, $m \in \mathbb{C}$ denotes a channel, $c \in \Lambda$ denotes a value, and $\tau$ is the internal action.

The intuition of a transition $\langle p, \sigma, E \rangle \xrightarrow{a, \sigma'} \langle p', \sigma', E \rangle$ is that the process $\langle p, \sigma, E \rangle$ executes the discrete action $a \in A_\tau$ and thereby transforms into the process $\langle p', \sigma', E \rangle$, where $\sigma'$ denotes the accompanying valuation of the process term $p'$ after the discrete action $a$ is executed.

(b) $\_ \xrightarrow{\cdot} \langle \checkmark, \_, \_ \rangle \subseteq (P \times \Sigma \times ES) \times (A_\tau \times \Sigma) \times (\Sigma \times ES)$. The intuition of a transition $\langle p, \sigma, E \rangle \xrightarrow{a, \sigma'} \langle \checkmark, \sigma', E \rangle$ is that the process $\langle p, \sigma', E \rangle$ executes the discrete action $a$ and thereby transforms into the terminated process $\langle \checkmark, \sigma', E \rangle$.

2. continuous behavior: $\_ \overset{\cdot}{\rightsquigarrow} \_ \subseteq (P \times \Sigma \times ES) \times ((T \mapsto \Sigma) \times T) \times (P \times \Sigma \times ES)$.

The intuition of a transition $\langle p, \sigma, E \rangle \overset{\varsigma, t}{\rightsquigarrow} \langle p', \varsigma(t), E \rangle$ is that the variables in $\text{dom}(\sigma)$ behave (continuously) according to the trajectories in $\varsigma$ until (and including) time $t$ and then result in the process $\langle p', \varsigma(t), E \rangle$, where $\varsigma(t) \in \Sigma$ is the valuation at the end point $t$ of the trajectory $\varsigma$.

These relations and predicates are defined through so-called deduction rules. A deduction rule is of the form $\frac{H}{r}$, where $H$ is a number of hypotheses separated by commas and $r$ is the result of the rule. The result of a deduction rule can be derived if all of its hypotheses are derived. In case the set of hypotheses is empty, the deduction rule is called a deduction axiom. The notation $\frac{H}{R}$, where $R$ is a number of results separated by commas, is a shorthand for a deduction rule $\frac{H}{r}$ for each result $r \in R$. In order to increase the readability of the $\chi_{\sigma_\text{h}}$ deduction rules, some abbreviations are used. The notation

$$\frac{\langle p_1, \sigma_1, E_1 \rangle \xrightarrow{a_1, \sigma_1'} \left\langle \begin{matrix} q_{10} \\ \vdots \\ q_{1n} \end{matrix}, \sigma_1', E_1 \right\rangle, \quad \cdots, \quad \langle p_m, \sigma_m, E_m \rangle \xrightarrow{a_m, \sigma_m'} \left\langle \begin{matrix} q_{m0} \\ \vdots \\ q_{mn} \end{matrix}, \sigma_m', E_m \right\rangle, \quad C}{\langle r, \sigma, E \rangle \xrightarrow{b, \sigma'} \left\langle \begin{matrix} s_0 \\ \vdots \\ s_n \end{matrix}, \sigma', E \right\rangle}$$

where $q_{j_i}, s_i \in P \cup \{\checkmark\}$ and $C$ denotes an optional hypothesis that must be satisfied in the deduction rule, is an abbreviation for the following rules (one for each $i$):

$$\frac{\langle p_1, \sigma_1, E_1 \rangle \xrightarrow{a_1, \sigma_1'} \langle q_{1i}, \sigma_1', E_1 \rangle, \quad \cdots, \quad \langle p_m, \sigma_m, E_m \rangle \xrightarrow{a_m, \sigma_m'} \langle q_{mi}, \sigma_m', E_m \rangle, \quad C}{\langle r, \sigma, E \rangle \xrightarrow{b, \sigma'} \langle s_i, \sigma', E \rangle}$$

Based on [10] we use the following definitions of operators $\cup$, $\upharpoonright$, and $\downarrow$ applied on functions. If $f$ is a function, $\text{dom}(f)$ and $\text{range}(f)$ denote the domain and range of $f$, respectively. If $S$ is a set, $f \upharpoonright S$ denotes the restriction of $f$ to $S$, that is, the function $g$ with $\text{dom}(g) = \text{dom}(f) \cap S$, such that $g(c) = f(c)$ for each $c \in \text{dom}(g)$.

If $f$ and $g$ are functions with $\mathrm{dom}(f) \cap \mathrm{dom}(g) = \emptyset$, then $f \cup g$ denotes the unique function $h$ with $\mathrm{dom}(h) = \mathrm{dom}(f) \cup \mathrm{dom}(g)$ satisfying the condition: for each $c \in \mathrm{dom}(h)$, if $c \in \mathrm{dom}(f)$ then $h(c) = f(c)$, and $h(c) = g(c)$ otherwise.

If $f$ is a function whose range is a set of functions and $S$ is a set, then $f \downarrow S$ denotes the function $g$ with $\mathrm{dom}(g) = \mathrm{dom}(f)$ such that $g(c) = f(c) \restriction S$ for each $c \in \mathrm{dom}(g)$. If $f$ is a function whose range is a set of functions, all of which have a particular element $d$ in their domain, then $f \downarrow d$ denotes the function $g$ with $\mathrm{dom}(g) = \mathrm{dom}(f)$ such that $g(c) = f(c)(d)$ for each $c \in \mathrm{dom}(g)$.

## 3.2   Deduction Rules

**Atomic Process Terms:** For the deduction rules of the atomic process terms, it is assumed that $\Gamma(E), \mathcal{J}(E), \mathcal{F}(E) \subseteq \mathrm{dom}(\sigma)$, $m \in \mathcal{C}(E)$, $x \in \mathrm{dom}(\sigma)$, and $\bar{\sigma}(e), \bar{\sigma}(e_n), c \in \Lambda$.

Rule 1 states that the skip process term performs the $\tau$ action to the terminated process $\checkmark$ and has no effect on the valuation or environment.

The execution of the assignment process term $x := e$ (see Rule 2) leads to a new valuation where all variables are unchanged except for variable $x$. $\sigma[\bar{\sigma}(e)/x]$ denotes the update of valuation $\sigma$ such that the new value of variable $x$ is $\bar{\sigma}(e)$, which denotes the value of $e$ with respect to $\sigma$. Internal send and receive process terms are intended to be used in parallel composition (see Rule 25). The value of expression $e$ which is sent via channel $m$ is evaluated in valuation $\sigma$ (see Rule 3). The receive process term $m?x$ can receive any value $c$ (see Rule 4).

$$\frac{}{\langle \mathrm{skip}, \sigma, E \rangle \xrightarrow{\tau,\sigma} \langle \checkmark, \sigma, E \rangle}\ 1 \qquad \frac{}{\langle x := e, \sigma, E \rangle \xrightarrow{\tau,\sigma[\bar{\sigma}(e)/x]} \langle \checkmark, \sigma[\bar{\sigma}(e)/x], E \rangle}\ 2$$

$$\frac{}{\langle m!e, \sigma, E \rangle \xrightarrow{isa(m,\bar{\sigma}(e)),\sigma} \langle \checkmark, \sigma, E \rangle}\ 3 \qquad \frac{}{\langle m?x, \sigma, E \rangle \xrightarrow{ira(m,c),\sigma[c/x]} \langle \checkmark, \sigma[c/x], E \rangle}\ 4$$

The predicate process term can perform a time transition for all trajectories $\varsigma$ for predicate $u$ as defined by Rule 5.

$$\frac{\varsigma \in \Omega(\sigma, \Gamma(E), \mathcal{J}(E), \mathcal{F}(E), u, t)}{\langle u, \sigma, E \rangle \xrightarrow{\varsigma,t} \langle u, \varsigma(t), E \rangle}\ 5$$

Function $\Omega \in \Sigma \times \mathcal{P}(V) \times \mathcal{P}(V) \times \mathcal{P}(V) \times U \times T \to \mathcal{P}(T \mapsto \Sigma)$, where $U$ denotes the set of all predicates , returns a set of trajectories from time to a valuation for the variables, given a valuation representing the current values of the variables, a set of normal continuous variables, a set of jumping continuous variables, a set of fixed continuous variables, a predicate and a time point that denotes the duration of the trajectory. Formally, the function $\Omega$ is defined as:

$\Omega(\sigma, \Gamma_E, J_E, F_E, u, d) =$
$\{\, \varsigma' \downarrow \operatorname{dom}(\sigma)$
$\mid \varsigma' \in T \mapsto ((V \cup V') \mapsto \Lambda)$
$, \operatorname{dom}(\varsigma') = [0, d],\ d > 0$
$, \forall_{\sigma' \in \operatorname{range}(\varsigma')}$  $\quad\operatorname{dom}(\sigma') = \operatorname{dom}(\sigma) \cup \{x' \mid x \in \mathcal{D}(u)\}$
$, \forall_{0 \le t \le d,\ x \in \operatorname{dom}(\sigma) \setminus (\Gamma_E \cup J_E \cup F_E)}$  $(\varsigma' \downarrow x)(t) = \sigma(x)$
$, \forall_{x \in (\mathcal{D}(u) \setminus J_E) \cup F_E}$  $(\varsigma' \downarrow x)(0) = \sigma(x)$
$, \forall_{x \in (\Gamma_E \cup J_E \cup F_E) \setminus \mathcal{D}(u)}$  $\varsigma' \downarrow x$ is a bounded function that is continuous almost everywhere.
$, \forall_{x \in \mathcal{D}(u)}$  $\varsigma' \downarrow x'$ is a bounded function that is continuous almost everywhere.
$, \forall_{0 \le t \le d}$  $\varsigma'(t) \models \mathcal{T}_u(u)$
$, \forall_{0 \le t \le d,\ x \in \mathcal{D}(u)}$  $(\varsigma' \downarrow x)(t) = (\varsigma' \downarrow x)(0) + \int_0^t (\varsigma' \downarrow x')(s)ds$
$\}$

In lines 5 and 6 of the body of function $\Omega$, it is assumed that the value of $x$ is defined ($\sigma(x) \in \Lambda$). Function $\mathcal{D} \in U \to \mathcal{P}(V)$ extracts the differential variables from a predicate. E.g. $\mathcal{D}(x = \dot{y}) = \{y\}$. Function $\mathcal{T}_u \in U \to U'$ replaces every occurrence of the derivative $\dot{x}$ of a variable with name $x$ in a predicate $u \in U$ by a fresh variable $x' \in V'$ that has the same name as $x$ postfixed with the prime character. The set $V'$ is defined as $V' = \{x' \mid x \in V\}$, and $U'$ denotes the set of predicates on variables $x \in V$ and $x' \in V'$. For example, the application of function $\mathcal{T}_u$ to the equation $\dot{x} = -y + \dot{z}$ gives the equation $x' = -y + z'$.

The behavior of each variable $x$ is described by a function of time $\varsigma' \downarrow x$. The behavior of the discrete variables $x \in \operatorname{dom}(\sigma) \setminus (\Gamma_E \cup J_E \cup F_E)$ is specified by constant functions ($\forall_{0 \le t \le d} (\varsigma' \downarrow x)(t) = \sigma(x)$). The initial conditions of the non-jumping differential variables $x \in \mathcal{D}(u) \setminus J_E$ and the fixed continuous variables $x \in F_E$ are specified by $(\varsigma' \downarrow x)(0) = \sigma(x)$. The behavior $\varsigma' \downarrow x$ of the algebraic variables $x \in (\Gamma_E \cup J_E \cup F_E) \setminus \mathcal{D}(u)$ and the behavior $\varsigma' \downarrow x'$ of the derivatives ($x'$ such that $x \in \mathcal{D}(u)$) is a bounded function (not set-valued) that is continuous almost everywhere (except for a set of measure zero). The trajectory $\varsigma'$ satisfies the predicate for all time points of its domain ($\forall_{0 \le t \le d}\ \varsigma'(t) \models \mathcal{T}_u(u)$). The function $\varsigma' \downarrow x$ of a differential variable $x \in \mathcal{D}(u)$ is the integral of the function $\varsigma' \downarrow x'$ of its derivative.

For a normal continuous variable $x \notin (J_E \cup F_E)$, its occurrence in $u$ as differential (occurring differentiated in $u$) or algebraic (not occurring differentiated in $u$), determines the behavior of the variable at the beginning of a time transition. I.e. in time transitions at $t = 0$, differential variables may not behave discontinuously (i.e. may not jump so that $(\varsigma' \downarrow x)(0) = \sigma(x)$). Algebraic variables, on the other hand, may show discontinuous behavior at $t = 0$, so that for these variables $(\varsigma' \downarrow x)(0)$ may be different from $\sigma(x)$. In some cases, differential variables may jump. This is, for example, the case in steady state initializations ($\dot{x} = 0$). E.g. in $\dot{x} = -x + 1 \parallel \dot{x} = 0$, where $x \in J_E$, $(\varsigma' \downarrow x)(0)$ jumps to 1, independently of $\sigma(x)$. The set of fixed continuous variables $F_E$ is needed in cases where algebraic variables need to be initialized. For example consider $\dot{x} = f(x, y, z) \parallel \dot{y} = g(x, y, z) \parallel h(x, y, z) = 0$. Normally, $x$ and $y$ are initialized, and the value of $z$ is then determined by the equations. If, for example, the modeler would prefer to initialize variables $x$ and $z$, so that the value of $y$ is then determined by the equations, the sets $F_E$ and $J_E$ should be such that $z \in F_E$ and $y \in J_E$. Such initializations are common in, for instance, chemical systems.

$$\frac{\varsigma \in \Omega(\sigma, \Gamma(E), \mathcal{J}(E), \mathcal{F}(E), \text{true}, t)}{\langle m!e, \sigma, E \rangle \stackrel{\varsigma,t}{\leadsto} \langle m!e, \varsigma(t), E \rangle} \; 6 \qquad \frac{\varsigma \in \Omega(\sigma, \Gamma(E), \mathcal{J}(E), \mathcal{F}(E), \text{true}, t)}{\langle m?x, \sigma, E \rangle \stackrel{\varsigma,t}{\leadsto} \langle m?x, \varsigma(t), E \rangle} \; 7$$

$$\frac{\bar{\sigma}(e_n) = 0}{\langle \Delta e_n, \sigma, E \rangle \stackrel{\tau,\sigma}{\longrightarrow} \langle \checkmark, \sigma, E \rangle} \; 8 \qquad \frac{0 < t \leq \bar{\sigma}(e_n), \; \varsigma \in \Omega(\sigma, \Gamma(E), \mathcal{J}(E), \mathcal{F}(E), \text{true}, t)}{\langle \Delta e_n, \sigma, E \rangle \stackrel{\varsigma,t}{\leadsto} \langle \Delta \bar{\sigma}(e_n) - t, \varsigma(t), E \rangle} \; 9$$

Rules 6 and 7 state that $m!e$ and $m?x$ can perform any time transition $\varsigma, t$ that satisfies $\varsigma \in \Omega(\sigma, \Gamma(E), \mathcal{J}(E), \mathcal{F}(E), \text{true}, t)$. The predicate true does not restrict the continuous behavior of the (continuous) variables.

The delay process term specifies a certain amount of delay. The full amount of delay does not have to be performed in one transition (see Rule 9). Note that $\bar{\sigma}(e_n)$ denotes the value of expression $e_n$ with respect to valuation $\sigma$ before the delay. In case that the amount of delay is zero, the delay process term terminates with an internal action as defined by Rule 8. Since there are no rules for the case that the amount of delay is negative, such a delay leads to deadlock.

**Recursion Variable:** Recursion is used among others to model repetition. The recursion variable $X$ simply behaves as the process term given by $\mathcal{R}(E)(X)$. Here $\mathcal{R}(E)(X)$ is the process term that is defined for recursion variable $X$ in recursive process definition $\mathcal{R}(E)$. It is assumed that $X \in \text{dom}(\mathcal{R}(E))$.

$$\frac{\langle \mathcal{R}(E)(X), \sigma, E \rangle \stackrel{a,\sigma'}{\longrightarrow} \langle \stackrel{\checkmark}{p'}, \sigma', E \rangle}{\langle X, \sigma, E \rangle \stackrel{a,\sigma'}{\longrightarrow} \langle \stackrel{\checkmark}{p'}, \sigma', E \rangle} \; 10 \qquad \frac{\langle \mathcal{R}(E)(X), \sigma, E \rangle \stackrel{\varsigma,t}{\leadsto} \langle p', \varsigma(t), E \rangle}{\langle X, \sigma, E \rangle \stackrel{\varsigma,t}{\leadsto} \langle p', \varsigma(t), E \rangle} \; 11$$

**Guard Operator:** In case that the guard $b$ evaluates to false (i.e. $\sigma \models \neg b$), there are no transitions. In case that the guard evaluates to true (i.e. $\sigma \models b$), the guarded process term simply behaves as $p$.

$$\frac{\langle p, \sigma, E \rangle \stackrel{a,\sigma'}{\longrightarrow} \langle \stackrel{\checkmark}{p'}, \sigma', E \rangle, \; \sigma \models b}{\langle b \to p, \sigma, E \rangle \stackrel{a,\sigma'}{\longrightarrow} \langle \stackrel{\checkmark}{p'}, \sigma', E \rangle} \; 12 \qquad \frac{\langle p, \sigma, E \rangle \stackrel{\varsigma,t}{\leadsto} \langle p', \varsigma(t), E \rangle, \; \sigma \models b}{\langle b \to p, \sigma, E \rangle \stackrel{\varsigma,t}{\leadsto} \langle p', \varsigma(t), E \rangle} \; 13$$

**Sequential Composition Operator:** The sequential composition of the process terms $p$ and $q$ behaves as process term $p$ until $p$ terminates, and then continues to behave as process term $q$.

$$\frac{\langle p, \sigma, E \rangle \stackrel{a,\sigma'}{\longrightarrow} \langle \stackrel{\checkmark}{p'}, \sigma', E \rangle}{\langle p; q, \sigma, E \rangle \stackrel{a,\sigma'}{\longrightarrow} \langle \stackrel{q}{p';q}, \sigma', E \rangle} \; 14 \qquad \frac{\langle p, \sigma, E \rangle \stackrel{\varsigma,t}{\leadsto} \langle p', \varsigma(t), E \rangle}{\langle p; q, \sigma, E \rangle \stackrel{\varsigma,t}{\leadsto} \langle p'; q, \varsigma(t), E \rangle} \; 15$$

**Disrupt Operator:** The disrupt operator $p \rhd q$ is introduced to model a kind of sequential composition, where the process term $q$ may take over execution from process term $p$ at any moment, without waiting for its termination.

$$\frac{\langle p, \sigma, E \rangle \xrightarrow{a,\sigma'} \langle \overset{\checkmark}{p'}, \sigma', E \rangle}{\langle p \rhd q, \sigma, E \rangle \xrightarrow{a,\sigma'} \langle \overset{\checkmark}{p' \rhd q}, \sigma', E \rangle} \; 16 \qquad \frac{\langle p, \sigma, E \rangle \overset{\varsigma,t}{\leadsto} \langle p', \varsigma(t), E \rangle}{\langle p \rhd q, \sigma, E \rangle \overset{\varsigma,t}{\leadsto} \langle p' \rhd q, \varsigma(t), E \rangle} \; 17$$

$$\frac{\langle q, \sigma, E \rangle \xrightarrow{a,\sigma'} \langle \overset{\checkmark}{q'}, \sigma', E \rangle}{\langle p \rhd q, \sigma, E \rangle \xrightarrow{a,\sigma'} \langle \overset{\checkmark}{q'}, \sigma', E \rangle} \; 18 \qquad \frac{\langle q, \sigma, E \rangle \overset{\varsigma,t}{\leadsto} \langle q', \varsigma(t), E \rangle}{\langle p \rhd q, \sigma, E \rangle \overset{\varsigma,t}{\leadsto} \langle q', \varsigma(t), E \rangle} \; 19$$

**Choice Operator:** The effect of applying the choice operator to the process terms $p$ and $q$ is that the execution of a transition by either one of them results in a definite choice.

$$\frac{\langle p, \sigma, E \rangle \xrightarrow{a,\sigma'} \langle \overset{\checkmark}{p'}, \sigma', E \rangle}{\langle p \oplus q, \sigma, E \rangle \xrightarrow{a,\sigma'} \langle \overset{\checkmark}{p'}, \sigma', E \rangle, \; \langle q \oplus p, \sigma, E \rangle \xrightarrow{a,\sigma'} \langle \overset{\checkmark}{p'}, \sigma', E \rangle} \; 20$$

$$\frac{\langle p, \sigma, E \rangle \overset{\varsigma,t}{\leadsto} \langle p', \varsigma(t), E \rangle}{\langle p \oplus q, \sigma, E \rangle \overset{\varsigma,t}{\leadsto} \langle p', \varsigma(t), E \rangle, \; \langle q \oplus p, \sigma, E \rangle \overset{\varsigma,t}{\leadsto} \langle p', \varsigma(t), E \rangle} \; 21$$

**Alternative Composition Operator:** The action behavior of the alternative composition operator is equal to that of the choice operator (see Rule 22). The weak time-determinism principle is adopted for the time transitions. This principle means that the passage of time by itself cannot result in making a choice between two alternatives that can perform that time transition with the same trajectory $\varsigma$ and the same time step $t$. This is captured in Rule 24. Rule 23 states that if one of the two process terms $p$ and $q$ can perform a time transition and the other cannot, then the alternative composition can also perform that time transition, but loses the alternative that could not perform a time transition.

$$\frac{\langle p, \sigma, E \rangle \xrightarrow{a,\sigma'} \langle \overset{\checkmark}{p'}, \sigma', E \rangle}{\langle p \, [\!] \, q, \sigma, E \rangle \xrightarrow{a,\sigma'} \langle \overset{\checkmark}{p'}, \sigma', E \rangle, \; \langle q \, [\!] \, p, \sigma, E \rangle \xrightarrow{a,\sigma'} \langle \overset{\checkmark}{p'}, \sigma', E \rangle} \; 22$$

$$\frac{\langle p, \sigma, E \rangle \overset{\varsigma,t}{\leadsto} \langle p', \varsigma(t), E \rangle, \; \langle q, \sigma, E \rangle \not\leadsto}{\langle p \, [\!] \, q, \sigma, E \rangle \overset{\varsigma,t}{\leadsto} \langle p', \varsigma(t), E \rangle, \; \langle q \, [\!] \, p, \sigma, E \rangle \overset{\varsigma,t}{\leadsto} \langle p', \varsigma(t), E \rangle} \; 23$$

$$\frac{\langle p, \sigma, E \rangle \overset{\varsigma,t}{\leadsto} \langle p', \varsigma(t), E \rangle, \; \langle q, \sigma, E \rangle \overset{\varsigma,t}{\leadsto} \langle q', \varsigma(t), E \rangle}{\langle p \, [\!] \, q, \sigma, E \rangle \overset{\varsigma,t}{\leadsto} \langle p' \, [\!] \, q', \varsigma(t), E \rangle} \; 24$$

**Parallel Composition Operator:** The parallel composition of the processes $p$ and $q$ has as its behavior with respect to action transitions the interleaving of the behaviors of $p$ and $q$ (see Rule 26). The time transitions of the parallel composition of two process terms have to synchronize to obtain the time transition (with same trajectory $\varsigma$ and the same time step $t$) of their parallel composition as defined by Rule 27. The parallel composition allows the synchronization of matching send and receive actions. A send action $isa(m,c)$ and a receive action $ira(m',c')$ match iff $m = m'$ and $c = c'$ (i.e. the channels used for sending and receiving are same, and also the value sent and the value received are identical). The result of the synchronization is a communication action $ca(m,c)$ as defined by Rule 25.

$$\frac{\langle p,\sigma,E\rangle \xrightarrow{isa(m,c),\sigma'} \langle \overset{\checkmark}{\underset{p'}{p'}},\sigma',E\rangle,\ \langle q,\sigma',E\rangle \xrightarrow{ira(m,c),\sigma''} \langle \overset{\checkmark}{\underset{q'}{\checkmark}},\sigma'',E\rangle}{\langle p \parallel q,\sigma,E\rangle \xrightarrow{ca(m,c),\sigma''} \langle \overset{\checkmark}{\underset{p'\parallel q'}{p'}},\sigma'',E\rangle,\ \langle q \parallel p,\sigma,E\rangle \xrightarrow{ca(m,c),\sigma''} \langle \overset{\checkmark}{\underset{q'\parallel p'}{p'}},\sigma'',E\rangle} \quad 25$$

$$\frac{\langle p,\sigma,E\rangle \xrightarrow{a,\sigma'} \langle \overset{\checkmark}{p'},\sigma',E\rangle}{\langle p \parallel q,\sigma,E\rangle \xrightarrow{a,\sigma'} \langle \underset{p'\parallel q}{q},\sigma',E\rangle,\ \langle q \parallel p,\sigma,E\rangle \xrightarrow{a,\sigma'} \langle \underset{q\parallel p'}{q},\sigma',E\rangle} \quad 26$$

$$\frac{\langle p,\sigma,E\rangle \overset{\varsigma,t}{\rightsquigarrow} \langle p',\varsigma(t),E\rangle,\ \langle q,\sigma,E\rangle \overset{\varsigma,t}{\rightsquigarrow} \langle q',\varsigma(t),E\rangle}{\langle p \parallel q,\sigma,E\rangle \overset{\varsigma,t}{\rightsquigarrow} \langle p' \parallel q',\varsigma(t),E\rangle} \quad 27$$

**Scope Operator:** By means of the scope operator, local variables (optionally with an initial value) and a local environment can be introduced in a $\chi_{\sigma_h}$ process. The application of the scope operator to a process $p$ results in the behavior of the process $p$ after the addition of the local variables (in fact the valuation for the local variables) to the global valuation ($\mu(\sigma,\sigma_s)$), and the addition of the local environment to the global environment ($\mu_E(E,E_s)$). Function $\mu \in \Sigma \times \Sigma \to \Sigma$ merges two valuations. If $\sigma,\ \sigma' \in \Sigma$, $\mu(\sigma,\sigma')$ denotes the valuation $\sigma''$ with $\mathrm{dom}(\sigma'') = \mathrm{dom}(\sigma) \cup \mathrm{dom}(\sigma')$, such that $\forall_{x\in\mathrm{dom}(\sigma')}\ \sigma''(x) = \sigma'(x)$ and $\forall_{x\in\mathrm{dom}(\sigma)\backslash\mathrm{dom}(\sigma')}\ \sigma''(x) = \sigma(x)$. Function $\mu_E \in ES \times ES \to ES$ merges two environments. It is defined as $\mu_E(E,E_s) = (\Gamma(E)\cup\Gamma(E_s), \mathcal{J}(E)\cup\mathcal{J}(E_s), \mathcal{F}(E)\cup\mathcal{F}(E_s), \mathcal{C}(E)\cup\mathcal{C}(E_s), \mu_R(\mathcal{R}(E),\mathcal{R}(E_s)))$. Function $\mu_R \in RS \times RS \to RS$ merges two recursive process definitions. If $R, R' \in RS$, $\mu_R(R,R')$ denotes the recursive process definition $R''$, with $\mathrm{dom}(R'') = \mathrm{dom}(R) \cup \mathrm{dom}(R')$ such that $\forall_{x\in\mathrm{dom}(R')}\ R''(x) = R'(x)$ and $\forall_{x\in\mathrm{dom}(R)\backslash\mathrm{dom}(R')}\ R''(x) = R(x)$.

The scope operator is also used for abstraction: action abstraction and data abstraction. The skip and assignment actions are internal ($\tau$) actions already. The internal send and receive actions on a local channel are encapsulated (blocked). Therefore, they need not be abstracted. The only action that needs to be abstracted by substitution of a $\tau$ action (action abstraction) is the communication action $ca(m,c)$ via a local channel $m \in \mathcal{C}(E_s)$ (see Rule 28). Function $\mathrm{ch} \in A_\tau \to \mathbb{C} \cup \{\bot\}$ extracts the channel label from an action. It is defined as $\mathrm{ch}(\alpha(m,c)) = m$ and $\mathrm{ch}(\tau) = \bot$.

The changes of local variables are abstracted (made invisible) outside the scope operator, by removing them from the transition arrow. For action transitions, data abstraction is defined using $\sigma_{\mu s}$, where $\sigma_{\mu s}$ denotes $\mu(\sigma, \sigma' \upharpoonright (\mathrm{dom}(\sigma) \setminus \mathrm{dom}(\sigma_s)))$, as shown in rules 28 and 29. The changed valuation of local variables is stored in the local valuation ($\sigma' \upharpoonright \mathrm{dom}(\sigma_s)$).

For time transitions, data abstraction is defined using $\varsigma_\sigma$, where $\varsigma_\sigma$ denotes $\varsigma \downarrow (\mathrm{dom}(\sigma) \setminus \mathrm{dom}(\sigma_s)) \cup \varsigma_{\mathrm{corr}}$. The correction function $\varsigma_{\mathrm{corr}}$ specifies the continuous behavior of the variables in the start valuation that were redefined in the local valuation $\sigma_s$. It is defined as $\varsigma_{\mathrm{corr}} \in \Omega(\sigma \upharpoonright \mathrm{dom}(\sigma_s), \Gamma(E) \cap \mathrm{dom}(\sigma_s), \mathcal{J}(E) \cap \mathrm{dom}(\sigma_s), \mathcal{F}(E) \cap \mathrm{dom}(\sigma_s), \mathrm{true}, t)$ (see Rule 30).

$$\frac{\langle p, \mu(\sigma, \sigma_s), \mu_E(E, E_s) \rangle \xrightarrow{ca(m,c),\sigma'} \langle \overset{\checkmark}{p'}, \sigma', \mu_E(E, E_s) \rangle, \ m \in \mathcal{C}(E_s)}{\langle [\![ \ \sigma_s, E_s \mid p \ ]\!], \sigma, E \rangle \xrightarrow{\tau,\sigma_{\mu s}} \langle [\![ \ \overset{\checkmark}{\sigma' \upharpoonright \mathrm{dom}(\sigma_s), E_s \mid p'} \ ]\!], \sigma_{\mu s}, E \rangle} \ 28$$

$$\frac{\langle p, \mu(\sigma, \sigma_s), \mu_E(E, E_s) \rangle \xrightarrow{a,\sigma'} \langle \overset{\checkmark}{p'}, \sigma', \mu_E(E, E_s) \rangle, \ \mathrm{ch}(a) \notin \mathcal{C}(E_s)}{\langle [\![ \ \sigma_s, E_s \mid p \ ]\!], \sigma, E \rangle \xrightarrow{a,\sigma_{\mu s}} \langle [\![ \ \overset{\checkmark}{\sigma' \upharpoonright \mathrm{dom}(\sigma_s), E_s \mid p'} \ ]\!], \sigma_{\mu s}, E \rangle} \ 29$$

$$\frac{\langle p, \mu(\sigma, \sigma_s), \mu_E(E, E_s) \rangle \overset{\varsigma,t}{\leadsto} \langle p', \varsigma(t), \mu_E(E, E_s) \rangle}{\langle [\![ \ \sigma_s, E_s \mid p \ ]\!], \sigma, E \rangle \overset{\varsigma_\sigma,t}{\leadsto} \langle [\![ \ \varsigma(t) \upharpoonright \mathrm{dom}(\sigma_s), E_s \mid p' \ ]\!], \varsigma_\sigma(t), E \rangle} \ 30$$

**Encapsulation Operator:** The behavior of the encapsulation of a process $\partial(p)$ is the same as the behavior of the process argument $p$ with the restriction that only actions from the set $A_x = \{ca(m,c) \mid m \in \mathbb{C}, c \in \Lambda\} \cup \{\tau\}$ can be executed (see Rule 31). In this way, internal send actions $isa(m,c)$ and internal receive actions $ira(m,c)$ are blocked, and only communication actions $ca(m,c)$ and $\tau$ actions are allowed. Encapsulation has no effect on time transitions, as defined by Rule 32.

$$\frac{\langle p, \sigma, E \rangle \xrightarrow{a_x,\sigma'} \langle \overset{\checkmark}{p'}, \sigma', E \rangle, \ a_x \in A_x}{\langle \partial(p), \sigma, E \rangle \xrightarrow{a_x,\sigma'} \langle \overset{\checkmark}{\partial(p')}, \sigma', E \rangle} \ 31 \qquad \frac{\langle p, \sigma, E \rangle \overset{\varsigma,t}{\leadsto} \langle p', \varsigma(t), E \rangle}{\langle \partial(p), \sigma, E \rangle \overset{\varsigma,t}{\leadsto} \langle \partial(p'), \varsigma(t), E \rangle} \ 32$$

**Maximal Progress Operator:** The maximal progress operator gives action transitions a higher priority than time transitions. Rule 33 states that action

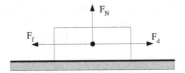

**Fig. 1.** Dry friction

behavior is not affected by maximal progress. Time transitions are allowed only if it is not possible to perform any action transitions as defined by Rule 34.

$$\frac{\langle p, \sigma, E \rangle \xrightarrow{a,\sigma'} \langle \overset{\checkmark}{p'}, \sigma', E \rangle}{\langle \pi(p), \sigma, E \rangle \xrightarrow{a,\sigma'} \langle \overset{\checkmark}{\pi(p')}, \sigma', E \rangle} \ 33 \qquad \frac{\langle p, \sigma, E \rangle \overset{\varsigma,t}{\leadsto} \langle p', \varsigma(t), E \rangle, \ \langle p, \sigma, E \rangle \nrightarrow}{\langle \pi(p), \sigma, E \rangle \overset{\varsigma,t}{\leadsto} \langle \pi(p'), \varsigma(t), E \rangle} \ 34$$

For all $\chi_{\sigma_h}$ operators, strong (state-based) bisimulation has been proven to be a congruence.

## 4    Examples

The two examples in this section are related to the kind of hybrid systems that can be modeled by means of hybrid automata and related formalisms. This makes it easier to become familiar with $\chi_{\sigma_h}$ specifications. In practice, however, a modeler would specify models in the $\chi$ language, which has a more user-friendly syntax for the scope operator.

### 4.1    Dry Friction Phenomenon

A driving force $F_d$ is applied to a body on a flat surface with frictional force $F_f$ (Figure 1). When the body is moving with positive velocity $v$, the frictional force is given by $F_f = \mu F_N$, where $F_N = mg$. When the velocity of the body is zero and $|F_d| < \mu_0 F_N$, the frictional force neutralizes the applied driving force. Instead of locations (hybrid automaton), $\chi$ uses recursion variables to specify the modes "neg", "stop", and "pos". The mode "stop" requires that $v$ is initially 0. The mode "stop" is maintained for as long as the parallel composition ($v = 0 \rightarrow v = 0 \parallel -\mu_0 F_N \leq F_d \leq \mu_0 F_N$) can delay. Otherwise, the process term ($F_d \leq -\mu_0 F_N \rightarrow$ neg $\oplus$ $F_d \geq \mu_0 F_N \rightarrow$ pos) after the disrupt operator $\triangleright$ takes over. The choice operator $\oplus$ specifies that either process term $F_d \leq -\mu_0 F_N \rightarrow$ neg or $F_d \geq \mu_0 F_N \rightarrow$ pos is executed. Therefore, depending on the value of $F_d$, either the process term specified by recursion variable (mode) neg or pos is executed. The mode "pos" is maintained until condition $v \leq 0 \wedge F_d < \mu_0 F_N$ becomes true. In $\chi$, action transitions have priority over time transitions. Therefore, when $v \leq 0$ and $F_d < \mu_0 F_N$, the process term skip is enabled and is immediately executed. Subsequently the mode "stop" is executed again. Symbols $m$, $F_N$, $\mu_0$, $\mu$, $x_0$ and $v_0$ are constants.

$\langle\ \pi([\![\ \emptyset,\ (\ \emptyset,\emptyset,\emptyset,\emptyset$
$\quad\quad,\ \{\,\text{stop}\ \mapsto\ (v=0\rightarrow v=0\ \|\ -\mu_0 F_{\mathrm{N}}\le F_{\mathrm{d}}\le\mu_0 F_{\mathrm{N}})$
$\quad\quad\quad\quad\quad\quad\quad\quad \rhd(\ F_{\mathrm{d}}\le-\mu_0 F_{\mathrm{N}}\rightarrow\text{neg}\oplus F_{\mathrm{d}}\ge\mu_0 F_{\mathrm{N}}\rightarrow\text{pos}\ )$
$\quad\quad\quad,\ \text{pos}\ \mapsto\ (v\ge 0\ \|\ m\dot{v}=F_{\mathrm{d}}-\mu F_{\mathrm{N}})$
$\quad\quad\quad\quad\quad\quad\quad\ \rhd\ (v\le 0\wedge F_{\mathrm{d}}<\mu_0 F_{\mathrm{N}}\rightarrow\text{skip};\ \text{stop})$
$\quad\quad\quad,\ \text{neg}\ \mapsto\ (v\le 0\ \|\ m\dot{v}=F_{\mathrm{d}}+\mu F_{\mathrm{N}})$
$\quad\quad\quad\quad\quad\quad\quad\ \rhd\ (v\ge 0\wedge F_{\mathrm{d}}>-\mu_0 F_{\mathrm{N}}\rightarrow\text{skip};\ \text{stop})$
$\quad\quad\quad\}$
$\quad\quad )$
$\quad\ |\ F_{\mathrm{d}}=\sin(t)\ \|\ \dot{t}=1\ \|\ \dot{x}=v\ \|\ (\text{neg}\oplus\text{stop}\oplus\text{pos})$
$\quad ]\!])$
$,\ \{t\mapsto 0,x\mapsto x_0,v\mapsto v_0,F_{\mathrm{d}}\mapsto\perp\},\ (\{t,x,v,F_{\mathrm{d}}\},\emptyset,\emptyset,\emptyset,\emptyset)$
$\rangle$

## 4.2  Railroad Gate Controller

In [16] a railroad gate controller is modeled using a hybrid automaton. When a train approaches the gate the controller must close the gate. The controller has a reaction delay of $\alpha$ time units. After the train has passed the gate the controller must open the gate. The purpose of the model is to determine the value of $\alpha$, to ensure that the gate is always fully closed when the train is at a certain distance from the gate.

A formal specification of the railroad gate controller using $\chi_{\sigma_{\mathrm{h}}}$ is given below. Channels *approach*, *exit*, *open* and *close* are used for pure synchronization, no data is communicated. The train, gate and controller are modeled using different scopes. The scope process term modeling the train consists of a parallel composition of an infinite loop ($*(\ldots)$) and an equation ($\dot{x}=v$). The velocity of the train can be any function of time between 40 and 50. The process waits until the train has reached position $x=1000$ and then synchronizes with the controller (*approach*!). The train is now approaching the gate. If the train has reached the exit position $x=2100$, the train synchronizes with the controller, the position $x$ of the train is reset to zero ($x:=0$), and the loop is re-executed. The train is now past the gate. The scope process term modeling the gate consists of a parallel composition of an infinite loop and an equation ($\dot{\phi}=n$). The infinite loop is an alternative composition of four process terms. The first process term waits until the gate is closed ($\phi=0$) and then turns off the gate. The second process term waits until the gate is open ($\phi=90$). The third and fourth process term wait for synchronization with the controller in order to open or close the gate (*open*? and *close*? respectively). The four process terms delay in parallel until one of the four events ($\nabla\phi\le 0$, $\nabla\phi\ge 90$, *open*?, *close*?) takes place. The controller consists of an infinite loop. It tries to synchronize with the train, in order to open or close the gate (*approach*? and *exit*? respectively). The constant $\alpha$ is used to model the reaction delay in the controller. After $\alpha$ time units ($\Delta\alpha$) the controller synchronizes with the gate, and the loop is re-executed. In the specification, some abbreviations are used which are listed in the table below.

| Abbreviation | Meaning |
|---|---|
| $* p$ | $[\![ \; \emptyset, (\emptyset, \emptyset, \emptyset, \emptyset, \{X \mapsto p; X\}) \mid X \; ]\!]$ |
| $\nabla x \geq e$ | $x \leq e \rhd (x \geq e \to \text{skip})$ |
| $\nabla x \leq e$ | $x \geq e \rhd (x \leq e \to \text{skip})$ |
| $m!$ | $m!\text{true}$ |
| $m?$ | $[\![ \; \{x \mapsto \perp\}, (\emptyset, \emptyset, \emptyset, \emptyset, \emptyset) \mid m?x \; ]\!]$ |

$\langle \; \pi(\partial( \; [\![ \; \{v \mapsto \perp\}, (\{v\}, \emptyset, \emptyset, \emptyset, \emptyset)$
$\qquad\qquad \mid \dot{x} = v \parallel *( \; (40 \leq v \leq 50 \; [\!] \; \nabla x \geq 1000); \; \text{approach}\,!$
$\qquad\qquad\qquad\quad ; \; (30 \leq v \leq 50 \; [\!] \; \nabla x \geq 2100); \; \text{exit}\,!; \; x := 0$
$\qquad\qquad\qquad\quad )$

$\qquad ]\!]$
$\quad \parallel [\![ \; \{n \mapsto 0\}, (\emptyset, \emptyset, \emptyset, \emptyset, \emptyset)$
$\qquad\qquad \mid \dot{\phi} = n \parallel *( \; n < 0 \to (\nabla\phi \leq 0; \; n := 0)$
$\qquad\qquad\qquad\qquad\quad [\!] \; n > 0 \to (\nabla\phi \geq 90; \; n := 0)$
$\qquad\qquad\qquad\qquad\quad [\!] \; \text{open}\,?; \; n := 9$
$\qquad\qquad\qquad\qquad\quad [\!] \; \text{close}\,?; \; n := -9$
$\qquad\qquad\qquad\qquad\quad )$

$\qquad ]\!]$
$\quad \parallel *( \; \text{approach}\,?; \; \Delta\alpha; \; \text{close}\,! \; [\!] \; \text{exit}?; \; \Delta\alpha; \; \text{open}\,! \; )$
$\quad ))$
$, \{x \mapsto 0, \phi \mapsto 90\}, (\{x, \phi\}, \emptyset, \emptyset, \{\text{approach}, \text{exit}, \text{open}, \text{close}\}, \emptyset)$
$\rangle$

# 5   Conclusions and Future Research

The semantics of the hybrid $\chi$ language has been formally specified using a relatively small set of deduction rules and associated functions. The language is highly expressive and can be used to specify a wide range of systems, including pure discrete-event systems, systems with local scoping of variables and/or channels, and systems of implicit algebraic differential equations. Future work entails the extension of the discrete-event $\chi$ verification tool to enable verification of hybrid models. Furthermore, the hybrid $\chi$ simulator will be redesigned to correspond to the new syntax and formal semantics.

# Acknowledgements

The authors would like to thank Pieter Cuijpers, Albert Hofkamp, Niek Jansen, Erjen Lefeber, Sasha Pogromsky, Frits Vaandrager, and the anonymous referees for their helpful comments on drafts of this paper.

# References

[1] van Beek, D.A., van den Ham, A., Rooda, J.E.: Modelling and control of process industry batch production systems. In: 15th Triennial World Congress of the International Federation of Automatic Control, Barcelona (2002) CD-ROM  151

[2] Bos, V., Kleijn, J.J.T.: Formal Specification and Analysis of Industrial Systems. PhD thesis, Eindhoven University of Technology (2002)  151

[3] Bos, V., Kleijn, J.J.T.: Automatic verification of a manufacturing system. Robotics and Computer Integrated Manufacturing **17** (2000) 185–198  151

[4] Schiffelers, R.R.H., van Beek, D.A., Man, K.L., Reniers, M.A., Rooda, J.E.: A hybrid language for modeling, simulation and verification. In Engell, S., Guéguen, H., Zaytoon, J., eds.: IFAC Conference on Analysis and Design of Hybrid Systems, Saint-Malo, Brittany, France (2003) 235–240  151

[5] Cuijpers, P.J.L., Reniers, M.A.: Hybrid process algebra. Technical Report Computer Science Reports 03-07, Eindhoven University of Technology, Department of Computer Science, The Netherlands (2003)  151, 152

[6] Rounds, W.C., Song, H.: The $\phi$-calculus - a hybrid extension of the $\pi$-calculus to embedded systems. Technical Report CSE 458-02, University of Michigan, USA (2002)  151, 152

[7] Jifeng, H.: From CSP to hybrid systems. In Roscoe, A.W., ed.: A Classical Mind, Essays in Honour of C.A.R. Hoare. Prentice Hall (1994) 171–189  151, 152

[8] Chaochen, Z., Ji, W., Ravn, A.P.: A formal description of hybrid systems. In Alur, R., Henzinger, T.A., Sonntag, E.D., eds.: Hybrid Systems III - Verification and Control. Lecture Notes in Computer Science 1066. Springer (1996) 511–530  151, 152

[9] David, R., Alla, H.: On hybrid Petri nets. Discrete Event Dynamic Systems: Theory & Applications **11** (2001) 9–40  151

[10] Lynch, N.A., Segala, R., Vaandrager, F.W.: Hybrid I/O automata. Technical Report MIT-LCS-TR-827d, MIT Laboratory for Computer Science, Cambridge, MA 02139 (2003) to appear in Information and Computation  151, 155

[11] Alur, R., Courcoubetis, C., Halbwachs, N., Henzinger, T.A., Ho, P.H., Nicollin, X., Olivero, A., Sifakis, J., Yovine, S.: The algorithmic analysis of hybrid systems. In: Theoretical Computer Science 138. Springer (1995) 3–34  151

[12] Alur, R., Dang, T., Esposito, J., Hur, Y., Ivancic, F., Kumar, V., Lee, I., Mishra, P., Pappas, G.J., Sokolsky, O.: Hierarchical modeling and analysis of embedded systems. Proceedings of the IEEE **91** (2003) 11–28  151, 152

[13] Henzinger, T.A.: Masaccio: A formal model for embedded components. In: First IFIP International Conference on Theoretical Computer Science (TCS). Lecture Notes in Computer Science 1872. Springer (2000) 549–563  151

[14] Mosterman, P.J., Ciolfi, J.E.: Embedded code generation for efficient reinitialization. In: 15th Triennial World Congress of the International Federation of Automatic Control. (2002) CD-ROM  152

[15] Cuijpers, P.J.L., Reniers, M.A., Heemels, W.P.M.H.: Hybrid transition systems. Technical Report Computer Science Reports 02-12, Eindhoven University of Technology, Department of Computer Science, The Netherlands (2002)  154

[16] Alur, R., Henzinger, T.A., Ho, P.H.: Automatic symbolic verification of embedded systems. IEEE Transactions on Software Engineering **22** (1996) 102–119  163

# Run-Time Guarantees for Real-Time Systems*

Reinhard Wilhelm

Informatik, Saarland University
Saarbruecken, Germany

**Abstract.** Hard Real-Time systems are subject to stringent timing constraints, which result from the interaction with the surrounding physical environment. The provider of the system has to guarantee that all timing constraints will be met. Such a guarantee is typically given by successfully executing a schedulability analysis. A schedulability analysis of a set of tasks requires the worst case execution times (WCET) of the tasks to be known. Since in general the problem of computing WCETs is not decidable, estimations of the WCET in form of upper bounds have to be calculated. The upper bounds always exist, since real-time programs don't allow unbounded iteration or recursion. These upper bounds are still called the worst case execution times of the tasks. The estimations have to be safe, i.e., they may never underestimate the real execution time. Furthermore, they should be tight, i.e., the overestimation should be as small as possible.

In modern processor architectures, caches, pipelines, and different kinds of speculative execution are key features for improving performance. Unfortunately, they make the prediction of the behaviour of instructions very difficult since this behaviour now depends on the execution history. Therefore, most classical approaches to worst case execution time prediction are not directly applicable or lead to results exceeding the real execution time by orders of magnitude.

We split the analysis into a set of subtasks: Value Analysis, Cache and Pipeline Analysis, and Worst-Case Path Determination. Value analysis attempts to determine the values in registers for each program point in order to statically compute Effective Addresses normally known only at execution time. Effective addresses are needed for the data cache analysis. Cache Analysis predicts the instruction and data cache behaviour of the program, and Pipeline Analysis predicts the pipeline behaviour. These three analyses are all done by Abstract Interpretation.

The essential idea is the following: The execution of an instruction or even a single memory access or a pipeline phase during the execution of an instruction can contribute different costs to the program's execution time depending on the execution history. All non-optimal executions of an instruction or part of an instruction we will consider as T im e A ccidents. We then regard Safety Properties being the absence of time accidents at individual instructions. Abstract Interpretation is then used to verify as many of such safety properties as possible. Any verified safety property allows the reduction of the WCET.

* Work supported by project IST-2001-34820, Advanced Real-Time Systems (ARTIST)

K.G. Larsen and P. Niebert (Eds.): FORMATS 2003, LNCS 2791, pp. 166–167, 2004.

The final step of the run-time prediction is Worst-case Path Analysis. It solves an Integer Linear Program (ILP) expressing the program control flow and taking into account the predicted maximum number of machine cycles for each Basic Block of the program. Maximizing an objective function expressing the total number of machine cycles for each program path yields an upper bound of the program's execution times.

WCET tools have been implemented for several processors [2, 3, 1] and are now being used in the aeronautics and the automotive industries [4]. Benchmarks have shown that very tight bounds on the execution times can be derived by the techniques mentioned above.

# References

[1] K. Engblom. Processor Pipelines and Static Worst-Case Execution Time Analysis. PhD thesis, Uppsala University, 2002.  167

[2] C. Ferdinand, R. Heckmann, M. Langenbach, F. Martin, M. Schmidt, H. Theiling, S. Thesing, and R. Wilhelm. WCET Determination for a Real-Life Processor. In T.A. Henzinger and C. Kirsch, editors, Embedded Software, volume 2211 of Lecture Notes in Computer Science, pages 469 – 485. Springer, 2001.  167

[3] M. Langenbach, S. Thesing, and R. Heckmann. Pipeline Modelling for Timing Analysis. In Static Analysis Symposium, volume 2274 of LNCS, pages 294–309. Springer Verlag, 2002.  167

[4] Stephan Thesing, Jean Souyris, Reinhold Heckmann, Famantanantsoa Randimbivololona, Marc Langenbach, Reinhard Wilhelm, and Christian Ferdinand. An abstract interpretation-based timing validation of hard real-time avionics software systems. In Proceedings of the Performance and Dependability Symposium, San Francisco, CA, June 2003.  167

# A Nonarchimedian Discretization for Timed Languages*

Cătălin Dima

ENSEIRB
1, avenue du dr. Albert Schweitzer,
Domaine Universitaire - BP 99, 33402 Talence Cedex, France

**Abstract.** We give a new discretization of behaviors of timed automata. In this discretization, timed languages are represented as sets of words containing action symbols, a clock tick symbol 1, and two delay symbols $\delta^-$ (negative delay) and $\delta^+$ (positive delay). Unlike the region construction, our discretization commutes with intersection. We show that discretizations of timed automata are, in general, context-sensitive languages over $\Sigma \cup \{1, \delta^+, \delta^-\}$, and give a class of automata that equals the class of languages that are discretizations of timed automata, and show that their emptiness problem is decidable.

## 1 Introduction

Timed automata [AD94] are a successful and widely used extension of finite automata for modeling real-time systems. They are finite automata endowed with clocks that measure time passage. Clocks are real-valued variables and evolve synchronously at rate 1. Transitions are guarded by arithmetic conditions on the clocks, and some transitions might reset some clocks to 0.

The essential construction for timed automata is the *region construction*, a finite abstraction of the infinite state-space of the automaton. This abstraction preserves emptiness and is the basis of both theoretic results concerning monoid recognizability, and of tools for model-checking timed systems [Yov98, LPY97]. Recently, researchers have shown interest in laying the foundations of a formal timed language theory. Some important progress has been reported on adapting regular expressions to timed automata [BP99, BP02, ACM02, AD03], on adapting monoid recognizability to timed languages [BPT01], and on pumping lemmas for timed languages [Her98, Bea98].

The research reported here has started from the observation that the region construction is not compositional w.r.t. intersection. That is, the region automaton of an intersection of languages of two timed automata is not the intersection of the region automata of the two timed automata. E.g., take the timed languages

$$L_1 = \{t_1 a\ t_2 \mid t_1, t_2 \in ]0, 1[\} = \left\| \langle \underline{t} \rangle_{]0,1[} \langle a \underline{t} \rangle_{]0,1[} \right\|,$$
$$L_2 = \{t_1 a\ t_2 \mid t_1 + t_2 \in ]1, 2[\} = \left\| \langle \underline{t} a\ \underline{t} \rangle_{]1,2[} \right\|$$

---

* This work has been partially supported by the Action Spécifique STIC-CNRS no. 93 "Automates Distribués et Temporisés"

K.G. Larsen and P. Niebert (Eds.): FORMATS 2003, LNCS 2791, pp. 168–181, 2004.
© Springer-Verlag Berlin Heidelberg 2004

Here we have used timed regular expressions similar to [ACM02], in which $\underline{t}$ represents time passage and the *time binding* construction $\langle E \rangle_I$ requires that each behavior generated by expression $E$ be observed within the interval $I$. The following figure gives the region constructions for these languages:

The classical "intersection construction" applied to $\mathcal{A}_1$ and $\mathcal{A}_2$ does not yield a region automaton for the timed language $L_1 \cap L_2$. Rather, we need to construct first a timed automaton for $L_1 \cap L_2$ and then to discretize it into a region automaton.

We present here a different discretization for timed automata, as (sets of) words over an alphabet containing action symbols, "clock tick" symbols and two special symbols $\delta^-$ and $\delta^+$ denoting positive, resp. negative delay. The delay symbols give some information about how to "skew" a sequence of clock ticks with some positive or negative delay, in order to obtain dense-time behaviors from a discrete-time information. They have a "nonarchimedian" semantics, in the sense that the semantics of any number of positive delays is smaller than one clock tick. In other words, clock ticks are used for quantitative timing information, while delay symbols give a qualitative timing information, as they only "alter" the information provided by clock ticks. We show that this discretization produces context-sensitive languages, fact which may give another explanation for the undecidability of the universality problem for timed automata.

We then investigate a class of automata that is suitable for representing discretizations of languages of timed automata. Our automata, called here *oscillator automata*, have the ability to add positive and/or negative delays, and to say whether the result is a positive or a negative delay. However they are not able to count delays. We show that this class of automata has a decidable emptiness problem, and that the class of languages that are discretizations of timed automata equals the class of languages accepted by oscillator automata.

The paper is divided as follows: in the next section we remind the notions of timed automata and timed regular expressions. In the third section we give the semantics of our discretization, and several basic properties they have. In the fourth section we introduce oscillator automata and show the decidability of their emptiness problem. The fifth section gives the results concerning the connection between oscillator automata and timed automata.

## 2   Preliminaries

Behaviors of timed systems can be modeled by **timed words** over a set of symbols $\Sigma$. Two presentations can be given to (most of the) timed words, one being as sequences of symbols from $\Sigma$ with time stamps, the other as mixed sequences of symbols from $\Sigma$ and real numbers. For example, $(a, 3)(b, 3.1)$ and

$12\,a\,0.1\,b$ represent the same timed word. A special case occurs when time elapses after the last action – it is the case of the timed word $3\,a\,1\,b\,1$. We will then associate to this the time stamp sequence $(a,3)(b,4)(\varepsilon,5)$, hence accepting that the last tuple in a time stamp sequence be labeled with the empty word. The *untiming* of a timed word is the sequence of actions in it. *Renamings* can be defined on timed words in the straightforward way: for example, $[a \mapsto b](a,3)(b,3.1) = (b,3)(b,3.1)$. Finally, the *length* of a timed word $w$ is the time stamp of the last symbol in $w$, or, equivalently, the sum of the real numbers in it. Hence, $\ell((a,3)(b,4)(\varepsilon,5)) = \ell(3\,a\,1\,b\,1) = 5$.

A **timed automaton** [AD94] is a tuple $\mathcal{A} = (Q, \mathcal{X}, \Sigma, \delta, Q_0, Q_f)$ where $Q$ is a finite set of *states*, $\mathcal{X}$ is a finite set of *clocks*, $\Sigma$ is a finite set of *action symbols*, $Q_0, Q_f \subseteq Q$ are sets of *initial*, resp. *final* states, and $\delta$ is a finite set of tuples (*transitions*) $(q, C, a, X, r)$ where $q, r \in Q$, $X \subseteq \mathcal{X}$, $a \in \Sigma \cup \{\varepsilon\}$ and $C$ is a finite conjunction of *clock constraints* of the form $x \in I$, where $x \in \mathcal{X}$ and $I \subseteq [0, \infty[$ is an interval with integer (or infinite) bounds. For each transition $(q, C, a, X, r) \in \delta$, the component $C$ is called the *guard* of the transition, $a$ is called the *action label* of the transition, and $X$ is called the *reset component* of the transition. We consider $\mathcal{X} = \{x_1, \ldots, x_n\}$, and identify each reset component $X$ with $\{i \mid i \in [1 \ldots n], x_i \in X\}$. Here $[1 \ldots n]$ stands for $\{1, \ldots, n\}$, and, in general, we denote $[i \ldots j] = \{i, i+1, \ldots, j\}$ for any $i, j \in \mathbb{Z}$.

The semantics of a timed automaton is given in terms of a *timed transition system* $\mathcal{T}(\mathcal{A}) = (\mathcal{Q}, \theta, \mathcal{Q}_0, \mathcal{Q}_f)$ where $\mathcal{Q} = Q \times \mathbb{R}_{\geq 0}^n$, $\mathcal{Q}_0 = Q_0 \times \{\mathbf{0}_n\}$, $\mathcal{Q}_f = Q_f \times \mathbb{R}_{\geq 0}^n$ and

$$\theta = \left\{ (q, v) \xrightarrow{t} (q, v') \mid v_i' = v_i + t, \ \forall i \in [1 \ldots n] \right\} \cup$$
$$\left\{ (q, v) \xrightarrow{a} (q', v') \mid \exists (q, C, a, X, q') \in \delta \text{ s.t. } v \models C \text{ and } \forall i \in [1 \ldots n], \right.$$
$$\left. \text{if } i \in X \text{ then } v_i' = 0 \text{ and if } i \notin X \text{ then } v_i' = v_i \right\}$$

A *run* in $\mathcal{T}(\mathcal{A})$ is a *finite* chain $(q^0, v^0) \xrightarrow{\xi_1} (q^1, v^1) \xrightarrow{\xi_2} \ldots \xrightarrow{\xi_k} (q^k, v^k)$ of transitions from $\theta$. An *accepting run* in $\mathcal{T}(\mathcal{A})$ is a run which starts in $\mathcal{Q}_0$ and ends in $\mathcal{Q}_f$. The *accepted language* of $\mathcal{A}$ is the set of timed words which label some accepting run of $\mathcal{T}(\mathcal{A})$. Two timed automata are called *equivalent* iff they have the same language. Note that we work here with finite runs.

The class of **timed regular expressions** is built using the following grammar:

$$E ::= 0 \mid \varepsilon \mid \underline{t}z \mid E + E \mid E \cdot E \mid E^* \mid \langle E \rangle_I \mid E \wedge E \mid [a \mapsto z]E, \tag{1}$$

where $z \in \Sigma \cup \{\varepsilon\}$ and $I$ is an interval. Their semantics is as follows:

$$\|\underline{t}z\| = \{tz \mid t \in \mathbb{R}_{\geq 0}\} \qquad \|E_1 + E_2\| = \|E_1\| \cup \|E_2\|$$
$$\|E_1 \cdot E_2\| = \|E_1\| \cdot \|E_2\| \qquad \|\langle E \rangle_I\| = \{w \in \|E\| \mid \ell(w) \in I\}$$
$$\|E^*\| = \|E\|^* \qquad \|0\| = \emptyset, \qquad \|\varepsilon\| = \{\varepsilon\}$$
$$\|E_1 \wedge E_2\| = \|E_1\| \cap \|E_2\| \qquad \|[a \mapsto z]E\| = \{[a \mapsto z]w \mid w \in \|E\|\}$$

**Theorem 1** ([ACM02]). *The class of timed languages accepted by timed automata equals the class of timed languages which are the semantics of some timed regular expression.*

## 3    Nonarchimedian Delays

The basic idea that we exploit is that timed words represented by timed automata can be grouped into *regions* (the name shows the similarity to the regions of [AD94]). An $n$-dimensional region is a *nonempty* subset of $\mathbb{R}_{\geq 0}{}^n$ which represents the solution to a constraint of the form $\bigwedge_{0 \leq i < j \leq n} x_j - x_i \in I_{ij}$, where $I_{ij}$ is either a point interval or an open unit interval, and $x_0 = 0$.

*Remark 1.* [Dim02] The constraint $\bigwedge_{0 \leq i < j \leq n} x_j - x_i \in I_{ij}$ has a nonempty semantics iff for all $i < j < k$, $I_{ij} + I_{jk} \supseteq I_{ik}$. This set of inclusions is called the *triangle property.*

Hence, the set of all timed words of length $n$ can be decomposed into an infinite but countable family of languages of the form $L = \{(a_1, t_1) \ldots (a_n, t_n) \mid \forall 0 \leq i < j \leq n, t_j - t_i \in I_{ij}, t_0 = 0\}$. We will call each such set as a *region timed language*. The $n$-dimensional subset of reals $\{(t_1, \ldots, t_n) \mid \forall 0 \leq i < j \leq n, t_j - t_i \in I_{ij}, t_0 = 0\}$ will be called the *underlying region* of $L$ and denoted $\mathcal{U}(L)$. Our aim is then to code all countable families like $L$ as a set of words over an extended set of symbols.

The set of symbols we use is $\Delta = \Sigma \cup \{1, \delta^+, \delta^-\}$ with $1, \delta^+, \delta^- \notin \Sigma$. $\delta^+$ will be called the *positive delay symbol* while $\delta^-$ will be called the *negative delay symbol*. For each $w \in \Delta^*$ and $x \in \Sigma \cup \{1, \delta^+, \delta^-\}$ we denote $|w|_x$ the *number of occurrences* of the symbol $x$ in $w$. Words over $\Delta$ will be called $\delta$-**words** while languages in $\Delta$ will be called as $\delta$-**languages**. Our aim is to code *timed languages* with $\delta$-*languages*.

Consider first the "point region" timed language $L_0 = \{t_1 a\, b \mid t_1 = 1\} = \|\langle \underline{t}a\, b\rangle_1\|$. Obviously, the symbol 1 can be used with the meaning of *clock tick*, hence we may code this language as $1ab$. Observe that time does not elapse between $a$ and $b$.

Consider now the region timed language $L_1 = \{t_1 a\, t_2 b \mid t_1 + t_2 = 1, t_1, t_2 > 0\} = \|\langle\, \underline{t}a\, \underline{t}b\rangle_1\|$. We would like to reuse the encoding $1ab$ by adding some information that should say that clock ticks must be "adjusted" with positive or negative delays. This will be the role of the delay symbols: $\delta^+=positive\ delay$, that is, the clock tick is "skewed positively", while $\delta^-=negative\ delay$, that is, the clock tick is "skewed negatively". The following $\delta$-word encodes $L_1$: $\omega_1 = 1\delta^- a\, \delta^+ b$:

- Before the action $a$, the time elapses with at most one clock tick,
- In between $a$ and $b$ there must be a positive delay and
- The duration of the whole behavior is exactly one clock tick, hence the number of positive delays equals the number of negative delays in the whole word.

As another example, the language $L_2 = \{t_1 a\, t_2 a\, t_3 \mid t_i \in ]0,1[\,, t_1 + t_2 + t_3 = 1\} = \|\langle\langle\underline{t}a\rangle_{]0,1[}\langle\underline{t}a\rangle_{]0,1[}\langle\underline{t}\rangle_{]0,1[}\rangle_1\|$ can be coded by the $\delta$-word $\omega_2 = 1\delta^- \delta^- a\ \delta^+ a\ \delta^+$ because we need *two* negative delay symbols before $a$ to counterbalance the two positive delays. Observe that, in the $\delta$-word $\omega_1$ above, the sub-language $\langle\underline{t}a\rangle_{]0,1[}$ was coded by $1\delta^- a$, while in the $\delta$-word $\omega_2$ it is coded by $1\delta^- \delta^- a$.

As mentioned in the introduction, the delay symbols have a *nonarchimedian* semantics, that is, the semantics of any number of $\delta^+$ is always less than one clock tick. Any number of delay symbols modifies clock tick durations by some positive or negative amount, an amount which may never be larger than one time unit.

**Definition 1.** *Consider a $\delta$-word $\omega = a_0 \eta_1 a_1 \eta_2 \ldots \eta_k a_k$, with $a_i \in \Sigma$ for all $1 \le i \le k-1$, $a_0, a_k \in \Sigma \cup \{\varepsilon\}$ and $\eta_i \in \{1, \delta^+, \delta^-\}^*$ for each $1 \le i \le k$. Then, for each $1 \le i < j \le k+1$, denote $p_{ij} = |\eta_i \ldots \eta_{j-1}|_{\delta^+}$, the number of $\delta^+$'s in the word $\eta_i \ldots \eta_{j-1}$, and similarly $n_{ij} = |\eta_i \ldots \eta_{j-1}|_{\delta^-}$ and $v_{ij} = |\eta_i \ldots \eta_{j-1}|_1$. The **semantics** of $\omega$ is the set of timed words $w = a_0 t_1 a_1 t_2 \ldots t_k a_k$ for which, for each $1 \le i < j \le k$,*

- *If $p_{ij} > n_{ij}$ then $t_i + \ldots + t_j \in \,]v_{ij}, v_{ij}+1[$*
- *If $p_{ij} = n_{ij}$ then $t_i + \ldots + t_j = v_{ij}$*
- *If $p_{ij} < n_{ij}$ then $t_i + \ldots + t_j \in \,]v_{ij}-1, v_{ij}[$*

We also say that $\omega$ is the (**nonarchimedian**) **discretization** of its semantics $\|\omega\|$. *Nonarchimedianity* amounts in fact to the property that for all $n \in \mathbb{N}, \|(\delta^+)^n\| < 1$ and $\|1(\delta^-)^n\| > 0$, where the inequality signs are the extensions of $<$ and $>$ on sets of reals.

*Remark 2.* Note that not all $\delta$-words have nonempty semantics: $\|\delta^-\| = \emptyset$ because we cannot have timed words with negative durations. However:

**Proposition 1.** *For each $\delta$-word $\omega = a_0 \eta_1 a_1 \eta_2 \ldots \eta_k a_k$ as in the definition 1, we have $\|\omega\| \ne \emptyset$ if and only if for all $i \in [1 \ldots k]$, if $|\eta_i|_1 = 0$ then $|\eta_i|_{\delta^-} \le |\eta_i|_{\delta^+}$.*

*Remark 3.* Note that semantics of $\delta$-words is not compositional w.r.t. concatenation: $\|\delta^+ a\| = \|\delta^+ \delta^+ a\|$, but $\|\delta^+ a \cdot \delta^- 1\| \ne \|\delta^+ \delta^+ a\| \cdot \|\delta^- 1\|$. Therefore, no choice of a "representative $\delta$-word" for a class of timed language is "good" w.r.t. concatenation.

Let us then consider sets of $\delta$-words that represent a given timed language, that is, equivalence classes w.r.t. semantics. We will call a $\delta$-language $D$ as *saturated* iff, whenever $\omega \in D$ and $\|\omega\| = \|\omega'\|$ for some $\omega, \omega' \in \Delta^*$, then $\omega' \in D$. If $D$ is not saturated, we denote $\mathrm{sat}(D)$ the least saturated $\delta$-language that contains $D$. Let us see first the structure of the saturated $\delta$-language discretizing a region timed language $L = \{(a_1, t_1) \ldots (a_n, t_n) \mid \forall 0 \le i < j \le n, t_j - t_i \in I_{ij}, t_0 = 0\}$, in which $a_n = \varepsilon$:

Denote $P$ the underlying region for $L$ and take $P'$ one of the point regions which is "closest" to $P$. ($P'$ may be equal to $P$ if $P$ is a point region). The constraint corresponding to $P'$ can be built by choosing, for each $0 \le i < j \le n$,

$I'_{ij}$ as either $\{\inf I_{ij}\}$ or $\{\sup I_{ij}\}$, and by checking whether the family $(I_{ij})_{i,j}$ satisfies the triangle property. $I'_{i-1,i}$ will give the number of clock ticks in between the $(i - 1)$-th action symbol and the $i$-th action symbol, with the convention that the 0 action symbol is the empty word. We will consider its coding in unary, as $1^{\alpha_{i-1,i}} = \underbrace{11 \ldots 1}_{\alpha_{i-1,i}}$. The addition of positive and negative delay symbols is then done as follows: for each $0 \leq i < j \leq n$, we define $J_{ij}$ as the point $\{0\}$, when $I'_{ij} = I_{ij}$, or $J_{ij} = ]0, \infty[$ when $I'_{ij} = \inf I_{ij} \neq I_{ij}$, or $J_{ij} = -\infty, 0[$ when $I'_{ij} = \sup I_{ij} \neq I_{ij}$. In [Dim02] we have shown that the family $(J_{ij})_{i,j}$ also satisfies the triangle property. $J_{ij}$ gives the constraint for the difference between the number of $\delta^+$'s and $\delta^-$'s in between the $i$-th and the $j$-th action. Then, the saturated set of discretizations of $L$ is the $\delta$-language $\xi_1 a_1 \xi_2 a_2 \ldots a_{n-1} \xi_n$ in which $\xi_i$ is a "shuffle" of $1^{\alpha_{i-1,i}}$ with a word $\zeta_i \in \{\delta^-, \delta^+\}^*$ with $|\zeta_i|_{\delta^+} - |\zeta_i|_{\delta^-} \in J_{i-1,i}$.

**Proposition 2.** *Given two saturated $\delta$-languages $W_1, W_2 \subseteq \Delta^*$, if $\|W_1\| \cap \|W_2\| \neq \emptyset$ then $W_1 \cap W_2 \neq \emptyset$. Moreover, if $z \notin \Sigma$, then $W_1 z W_2$ is saturated and $\|W_1 z\| \cdot \|W_2\| = \|W_1 \cdot z \cdot W_2\|$.*

*Proof.* The first part follows easily from the definition of saturated sets. We prove the second part for $W_1$ and $W_2$ being equivalence classes w.r.t. the semantic equivalence.

Take $w_1 \in W_1$ and $w_2 \in W_2$. For the left-to-right inclusion, consider $\sigma \in \|w_1 z\| \|w_2\|$ and denote $L$ the region timed language to which $\sigma$ belongs. By the discussion above on the construction of saturated sets, we may construct a $\delta$-word $w$ with $\|w\| = L$. It is then easy to decompose $w$ as $w = w'_1 z w'_2$, and $\sigma$ as $\sigma = \sigma_1 z \sigma_2$, and to observe that $\sigma_i \in \|w_i\| \cap \|w'_i\| = \|w_i\| = \|w'_i\| = L$. The right-to-left inclusion is straightforward, as is the property of $W_1 z W_2$ being saturated.                                                                                         $\square$

*Remark 4.* In general, concatenation of saturated $\delta$-languages does not give a saturated $\delta$-language. E.g., consider $L = \|\langle \mathbf{t} \rangle_{]0,1[}\|$. The saturated discretization of $L \cdot L$ contains $\omega = 1\delta^+\delta^-1$. But clearly this $\delta$-word cannot be decomposed as $\omega = \omega_1 \cdot \omega_2$ such that $\|\omega_1\| = \|\omega_2\| = L$. Therefore, if $D$ is the saturated $\delta$-language discretizing $L$, then $D \cdot D$ does not contain $1\delta^+\delta^-1$ which is in the saturated $\delta$-language discretizing $L \cdot L$.

On the other hand, the intersection of two non-saturated $\delta$-languages might be empty, even when the intersection of their semantics is nonempty. But our aim is to construct discretizations of timed languages by union, concatenation, star and intersection applied to some basic $\delta$-languages. Without solving this non-compositionality of intersection w.r.t. non-saturation, we might get an empty $\delta$-language as a discretization of a nonempty timed language.

The solution is to work with images of saturated $\delta$-languages under *deletion morphisms*. A $\delta$-language $D \subseteq \Delta^*$ is **weakly saturated** iff there exists a set of symbols $\Gamma$ with $\Gamma \cap \Sigma = \emptyset$ and a saturated $\delta$-language $D' \subseteq (\Gamma \cup \Sigma \cup \{1\delta^+, \delta^-\})^*$ such that $D = [\Gamma \mapsto \varepsilon]D'$. Here, $[\Gamma \mapsto \varepsilon]$ is the morphism which deletes any occurrence of a symbol from $\Gamma$. The following proposition ensures the correctness of this approach:

**Proposition 3.** *Given two weakly saturated δ-languages $W_1, W_2$, we have that*

1. *If $\|W_1\| \cap \|W_2\| \neq \emptyset$ then $W_1 \cap W_2 \neq \emptyset$ and $W_1 \cap W_2$ is weakly saturated.*
2. *$\|W_1 \cdot W_2\| = \|W_1\| \cdot \|W_2\|$ and $W_1 \cdot W_2$ is weakly saturated.*

*Proof.* This proposition is a corollary of the following lemma:

**Lemma 1.** *Symbol deletion in δ-words commutes with semantics: for each $\omega \in \Delta^*$ and $a \in \Sigma$, $\|[a \mapsto \varepsilon]\omega\| = [a \mapsto \varepsilon]\|\omega\|$. Moreover, $\|[a \mapsto \varepsilon]\omega\|$ is a region timed language.*

*Proof.* We will prove the result for $\|\omega\|_a = 1$, the general case following from the straightforward fact that renaming commutes with semantics.

For the left-to-right inclusion, suppose that $a_k = a$ for some $1 \leq k \leq n$. What is then required to prove is that the following $(n-1)$-dimensional region:

$$R' = \{(t_1, \ldots, t_{k-1}, t_{k+1}, \ldots, t_n) \mid \forall 0 \leq i < j \leq n \text{ with } i, j \neq k, t_j - t_i \in I_{ij}, t_0 = 0\}$$

is such that $R' \subseteq \{(t_1, \ldots, t_{k-1}, t_{k+1}, \ldots, t_n) \mid \exists t_k \in \mathbb{R}_{\geq 0} \text{ s.t.}$
$(t_1, \ldots, t_{k-1}, t_k, t_{k+1}, \ldots, t_n) \in \mathcal{U}(\|\omega\|)\}$, where $\mathcal{U}(\|\omega\|)$ is the underlying region for the region language $\|\omega\|$. But this is a well-known property for conjunctions of difference constraints satisfying the triangle property, see [Dim01] for a proof. The right-to-left proof is straightforward, as well as the second conclusion.    □

It remains to identify a class of saturated languages over $\Delta^*$ that contains exactly the languages whose semantics equals the timed language of some timed automaton. This is the subject of the next section. But before that, we will show here that, in general, discretizations of languages of timed automata are context-sensitive languages over $\Delta$. To see that, consider the following discretization of the (semantics of the) expression $\langle \underline{t}a_1\underline{t}a_2 \rangle_{]2,3[}\underline{t} \wedge \underline{t}a_1\langle \underline{t}a_2\underline{t}\rangle_{]1,2[}$:

$$D = \{\eta_1 a_1 \eta_2 a_2 \eta_3 \mid |\eta_1\eta_2|_{\delta^+} > |\eta_1\eta_2|_{\delta^-}, |\eta_2\eta_3|_{\delta^+} < |\eta_2\eta_3|_{\delta^-}, |\eta_1|_1 = |\eta_2|_1 = |\eta_3|_1 = 1\}$$

It is easy to see that this language is context-sensitive: its intersection with the regular language $(\delta^+)^*(\delta^-)^*1a_1(\delta^+)^*(\delta^-)^*1a_2(\delta^+)^*(\delta^-)^*1a_3$ is

$$L_4 = \{(\delta^+)^m(\delta^-)^n 1a_1(\delta^+)^p(\delta^-)^q 1a_2(\delta^+)^r(\delta^-)^s 1a_3 \mid m-n > p-q, p-q < r-s\}.$$

## 4    Oscillator Automata

An appropriate class of automata for accepting saturated sets of δ-words must have the ability to compare numbers of $\delta^+$'s and $\delta^-$'s in infixes of δ-words, but, for decidability reasons, it must be unable to *count* them. The formalization of these leads to the notion of *oscillator automata*, that we present here.

Before that, let us give some notations used in this section. We will use the symbol $\forall$ as the symbol of the total binary relation on the integers, that is, $x \forall y$ for all $x, y \in \mathbb{Z}$. We denote $\mathcal{R} = \{<, =, >, \}$ and $\mathcal{R}_\forall = \mathcal{R} \cup \{\forall\}$. $R_\forall^n$ denotes the $n$-tuple $R_\forall^n = (\forall, \ldots, \forall)$. For $c \in \mathbb{Z}^n$ and $X \subseteq [1 \ldots n]$, we denote by $c[X := 0]$ the tuple whose components are $c[X := 0]_i = c_i$ for $i \notin X$ and $c[X := 0]_i = 0$ for $i \in X$. The unit tuple is denoted $\mathbf{1} = (1, \ldots, 1)$, and the zero tuple is denoted $\mathbf{0}$. We will also use addition and subtraction on tuples. Hence, $(c + 1)_i = c_i + 1$ and $(c - 1)_i = c_i - 1$ for all $i \in [1 \ldots n]$.

**Definition 2.** *An **oscillator automaton of degree** $n$ (or $n$-**oscillator automaton**) is a tuple $\mathcal{A} = (Q, \theta, Q_0, Q_f)$ where $Q$ is a finite set of states, $Q_0, Q_f \subseteq Q$ are the sets of* initial, *resp.* final *states, while $\theta \subseteq Q \times \Delta \times \mathcal{R}_\forall^n \times \mathcal{P}([1 \ldots n]) \times Q$ is the transition relation. Additionally, it is required that $(q, \delta^+, R, \emptyset, q') \in \theta$ if and only if $(q, \delta^-, R, \emptyset, q') \in \theta$, and in this case $q = q'$ and $R = R_\forall^n$.*

A *configuration* of an oscillator automaton is a tuple $(q, c_1, \ldots, c_n) \subseteq Q \times \mathbb{Z}^n$ consisting of a state of the automaton and an $n$-tuple of *counters*. We denote the set of configurations of $\mathcal{A}$ as $\mathcal{C}(\mathcal{A})$. The set of *initial configurations* is $Q_0 \times \{(0, 0, \ldots, 0)\}$, while the set of *final configurations* is $Q_f \times \mathbb{Z}^n$. Configurations are connected by transitions as follows: for any $x \in \Delta$ for which $(q, x, R, X, q') \in \theta$, if we denote $X = \{i \in [1 \ldots n] \mid R_i \neq \forall\}$, then $(q, c_1, \ldots, c_n) \vdash^x (q', c_1', \ldots, c_n')$ if $\forall i \in [1 \ldots n]$, $c_i R_i 0$ and

- If $x \notin \{\delta^-, \delta^+\}$ then $c' = c[X := 0]$.
- If $x = \delta^+$ then $c' = c + 1$, and if $x = \delta^-$ then $c' = c - 1$.

Hence, in a tuple $(q, x, R, X, q') \in \theta$, the $R$ component gives the *precondition* that the counters must meet in order to take the transition. Moreover, only $\delta^+$-transitions increment counters, while only $\delta^-$-transitions decrement counters.

A *run* in $\mathcal{A}$ is a sequence $\rho = \left((q_{i-1}, c_1^{i-1}, \ldots, c_n^{i-1}) \vdash^{x_i} (q_i, c_1^i, \ldots, c_n^i)\right)_{i \in [0 \ldots k]}$ of transitions between configurations. We say that the $\delta$-word $x_1 \ldots x_n$ is *associated* to the run $\rho$. A run is *accepting* if it starts with an initial configuration and ends in a final configuration. The $\delta$-*language accepted* by $\mathcal{A}$ is the set of $\delta$-words $\omega = x_1 \ldots x_k$ associated to an accepting run, and is denoted $D(\mathcal{A})$. The *timed language accepted* by $\mathcal{A}$, denoted $L(\mathcal{A})$, is $\|D(\mathcal{A})\|$

As an example, the 2-oscillator automaton shown on the right accepts the discretization of the timed language $\|\langle \mathtt{tat} \rangle_{]0,1[}\|$. The dotted arrows represent the $\delta^+$-transitions, while the dashed arrows represent the $\delta^-$-transitions.

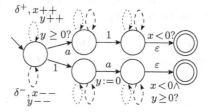

Note that the counter $y$ is necessary for ensuring that, e.g., the $\delta$-word $\delta^- a 1$ (which has an empty semantics) is not accepted.

**Theorem 2.** *The emptiness problem is decidable for oscillator automata.*

*Proof.* The proof idea is that, for checking emptiness, we only need to keep the sign of each counter, together with the sign of the difference between each pair of counters. Essentially, this information is similar to the *ordering of the fractional parts* in the region construction [AD94], and will be called here as *pattern*.

An *n-pattern* is a surjective function $\varphi : [0 \ldots n] \to [1 \ldots k]$, where $k \leq n+1$. An *n-pattern* abstracts a tuple of $n$ integer values $c = (c_1, \ldots, c_n)$ to the *order* between the components of the tuple and the *sign* of these components. The *zero n-pattern* is the unique function $\varphi_0 : [0 \ldots n] \to \{1\}$. A tuple $R \in \mathcal{R}_\forall^n$ is said *compatible* with a *n-pattern* $\varphi$ iff, for all $i \in [1 \ldots n]$, $\varphi(i) R_i \varphi(0)$. For example, the tuple $c = (-2, 1, -2, 0)$ is compatible with the 4-pattern $\{0, 4 \mapsto 2; 1, 3 \mapsto 1; 2 \mapsto 2\}$.

The transitions between configurations of $\mathcal{A}$ have three effects on the tuples of integers: they may increment or decrement tuples, or reset certain components of these to zero. We need to define appropriate operations on *n-patterns* that abstract these three operations. To this end, we associate to each *n-pattern* $\varphi$ the following three mappings: $v_1(\varphi) : [0 \ldots n] \to [1 \ldots k-1]$, $v_2(\varphi) : [0 \ldots n] \to [1 \ldots k]$ and $v_3(\varphi) : [0 \ldots n] \to [1 \ldots k+1]$, defined as follows:

1. For each $i \in [1 \ldots n]$ with $\varphi(i) < \varphi(0)$, $v_1(\varphi)(i) = v_2(\varphi)(i) = v_3(\varphi)(i) = \varphi(i)$.
2. $v_1(\varphi)(0) = v_2(\varphi)(0) = \varphi(0) - 1$ and $v_3(\varphi)(0) = \varphi(0)$.
3. For each $i \in [1 \ldots n]$ with $\varphi(i) \geq \varphi(0)$, $v_1(\varphi)(i) = \varphi(i) - 1$, $v_2(\varphi)(i) = \varphi(i)$ and $v_3(\varphi)(i) = \varphi(i) + 1$.

The *increment* of $\varphi$ is the set of *n-patterns*

$$\varphi^{++} = \begin{cases} \{\varphi\} \cup \{v_1(\varphi) \mid \varphi(0) \geq 2\} & \text{iff } card(\varphi^{-1}(\varphi(0))) = 1 \\ \{v_3(\varphi)\} \cup \{v_2(\varphi) \mid \varphi(0) \geq 2\} & \text{iff } card(\varphi^{-1}(\varphi(0))) \geq 2 \end{cases}$$

We may very similarly define the *decrement* of an *n-pattern* $\varphi$, denote it $\varphi^{--}$. The formal definition is left to the reader, due to space limitations. Finally, given $X \subseteq [1 \ldots n]$, the *X-reset* of a *n-pattern* $\varphi$ is the *n-pattern* $\varphi[X := 0]$ defined as follows:

1. $\varphi[X := 0] = card\{\varphi(i) \mid \varphi(i) \leq \varphi(0), i \notin X\}$.
2. For each $i \in X$, $\varphi[X := 0](i) = \varphi[X := 0](0)$.
3. For each $i \in X$, $\varphi[X := 0](i) = card\{\varphi(j) \mid \varphi(j) \leq \varphi(i), j \notin X\}$.

We then construct the *finite* automaton $\overline{\mathcal{A}} = (\overline{Q}, \overline{\theta}, \overline{Q}_0, \overline{Q}_f)$ where $\overline{Q} = Q \times$ Pat$_n$, $\overline{Q}_0 = Q_0 \times \{\varphi_0\}$, $\overline{Q}_f = Q_f \times$ Pat$_n$ and

$$\overline{\theta} = \{(q, \varphi) \xrightarrow{\delta^+} (q, \varphi') \mid \varphi' \in \varphi^{++}\} \cup \{(q, \varphi) \xrightarrow{\delta^-} (q, \varphi') \mid \varphi' \in \varphi^{--}\} \cup$$
$$\{(q, \varphi) \xrightarrow{x} (q', \varphi') \mid x \in \Sigma \cup \{1\} \text{ and } \exists (q, x, R, X, q') \in \theta \text{ s.t. } \varphi \text{ is compatible}$$
$$\text{with } R, X = \{i \in [1 \ldots n] \mid R_i \neq \forall\} \text{ and } \varphi' = \varphi[X := 0]\}$$

An upper bound for the number of states in $\overline{\mathcal{A}}$ is $n! \cdot 2^n \cdot card(Q)$.    $\square$

# 5  From Timed Automata to Oscillator Automata and Back

In this section we show that oscillator automata represent exactly all discretizations of languages of timed automata. The proof of the right-to-left inclusion is based upon the Kleene theorem 1, while the proof in the other direction actually constructs a region automaton, viewed as a timed automaton.

**Theorem 3.** *For each timed regular expression $E$ there exists an oscillator automaton whose accepted timed language is $\|E\|$.*

*Proof.* We recursively construct *weakly saturated* oscillator automata for the subexpressions of $E$ and show that union, intersection, concatenation, star and renaming preserves the property of being weakly saturated.

For the expression $\underline{t}z$, the oscillator automaton has two states $q_1$ and $q_2$, with $q_1$ initial and $q_2$ final, and loops in $q_1$ with clock ticks and delay symbols. On $z$, it moves to $q_2$, which has no outgoing transitions. Observe that this automaton is saturated.

For expressions $\langle E \rangle_I$, observe first that $\|\langle E \rangle_I\| = \left\|E \wedge \langle \underline{t} + \sum_{a \in \Sigma} a \rangle_I\right\|$. Hence we only need to construct oscillator automata for expressions like $\langle \underline{t} + \sum_{a \in \Sigma} a \rangle_I$. We consider the case $I = ]k, k+1[$ ($k \in \mathbb{N}$), the case $I = \{k\}$ being similar, and the general case following from these by decomposing $I$ into point or open unit length intervals.

Hence, we construct a saturated 1-oscillator automaton with $2k + 4$ states $Q = \{q_0, \ldots, q_{k+1}, r_1, \ldots, r_{k+2}\}$, with $q_0$ initial and $q_{k+1}, r_{k+2}$ final. It loops in any state with any symbol $x \neq 1$, and moves from $q_{i-1}$ to $q_i$ on symbol 1, where $i \in [2 \ldots k]$. It also moves from $q_0$ to $r_1$ and from $r_{i-1}$ to $r_i$ on symbol 1, with $i \in [2 \ldots k+1]$. An $\varepsilon$-transition moves the automaton from $q_k$ to $q_{k+1}$, and this transition is guarded by the relation $>$, and another $\varepsilon$-transition moves the automaton from $r_{k+1}$ to $r_{k+2}$, and this transition is guarded by the relation $<$. The resulting oscillator automaton is saturated.

Suppose now we have two oscillator automata $\mathcal{A}_1$ and $\mathcal{A}_2$ with weakly saturated languages. We may suppose, without loss of generality, that $\mathcal{A}_i$ ($i = 1, 2$) is obtained from a saturated oscillator automaton $\overline{\mathcal{A}}_i$ by replacing, on all transitions, all labels not in $\Sigma$ with $\varepsilon$. That is, $\mathcal{A}_i$ and $\overline{\mathcal{A}}_i$ have the same state space and

- For all $a \in \Sigma \cup \{1, \delta^+, \delta^-\}$, $(q, a, R, X, q') \in \delta_{\mathcal{A}_i}$ iff $(q, a, R, X, q') \in \delta_{\overline{\mathcal{A}}_i}$
- For all $x \notin \Sigma$, if $(q, x, R, X, q') \in \delta_{\overline{\mathcal{A}}_i}$ then $(q, \varepsilon, R, X, q') \in \delta_{\mathcal{A}_i}$,
- If $(q, \varepsilon, R, q') \in \delta_{\mathcal{A}_i}$ and $q$ is not initial and $q'$ not final, then there exists $x \notin \Sigma$ such that $(q, x, R, X, q') \in \delta_{\overline{\mathcal{A}}_i}$.

We will also suppose that, in both $\mathcal{A}_1$ and $\mathcal{A}_2$, the only transitions leaving initial states are either $\delta^+$-loops, or $\delta^-$-loops, or $\varepsilon$-transitions with constraint $R_\forall^n$. Similarly, we suppose that, in both $\mathcal{A}_1$ and $\mathcal{A}_2$, the only transitions entering final states are either $\delta^+$-loops, or $\delta^-$-loops, or $\varepsilon$-transitions with constraint $R_\forall^n$.

We may then build easily a saturated oscillator automaton $\mathcal{B}$ for $D(\overline{\mathcal{A}}_1) \cdot z \cdot D(\overline{\mathcal{A}}_2)$, by simply connecting each final state $q$ of $\overline{\mathcal{A}}_1$ to each initial state $q'$ of $\overline{\mathcal{A}}_2$ via a $z$-transition with constraint $R_\forall^n$, where $z$ is a fresh symbol, not used in any of the automata. By proposition 2, $D(\mathcal{B})$ is saturated. Then, by proposition 3, if we replace in $\mathcal{B}$ all labels not in $\Sigma$ with $\varepsilon$, the resulting automaton accepts $D(\mathcal{A}_1) \cdot D(\mathcal{A}_2)$. It is also clear that this automaton is weakly saturated, as it's $\delta$-language is $[\Sigma_1 \cup \Sigma_2 \cup \{z\} \mapsto \varepsilon]D(\mathcal{B})$.

A similar construction works for $D(\mathcal{A}_1)^+$: we draw $z$-transitions from each final state of $\mathcal{A}_1$ to each of its initial state, with constraint $R_\forall^n$. Coping with $D(\mathcal{A}_1)^*$ is then trivial: we only need to append a small saturated automaton discretizing exactly the timed language containing only the empty timed word.

A straightforward union construction for $\overline{\mathcal{A}}_1$ and $\overline{\mathcal{A}}_2$ assures also that $D(\mathcal{A}_1) \cup D(\mathcal{A}_2)$ is accepted by a weakly non-saturated oscillator automaton. Also, for some $a \in \Sigma$ and $b \in \Sigma \cup \{\varepsilon\}$, the $\delta$-language $[a \mapsto b]D(\mathcal{A}_1)$ is accepted by the oscillator automaton which results from $\mathcal{A}_1$ by replacing $a$ labels with $b$ labels on all $a$-transitions.

The remaining case is intersection. For this operation, we need to modify $\overline{\mathcal{A}}_1$ by appending, in each non-initial non-final state, loops with all symbols in $\Sigma_2$, with constraint $R_\forall^n$. Denote $\tilde{\mathcal{A}}_1$ the resulting automaton. Similarly, we put loops in each state of $\overline{\mathcal{A}}_2$ with all symbols in $\Sigma_1$ and modify initial and final states accordingly, and denote $\tilde{\mathcal{A}}_2$ the resulting automaton. Hence, $D(\mathcal{A}_i) = [\Sigma_1 \cup \Sigma_2]D(\tilde{\mathcal{A}}_i)$, $i = 1, 2$. Also note that both $\tilde{\mathcal{A}}_1$ and $\tilde{\mathcal{A}}_2$ are saturated. Then the classical intersection construction will provide a saturated oscillator automaton $\tilde{\mathcal{A}}$ for $D(\tilde{\mathcal{A}}_1) \cap D(\tilde{\mathcal{A}}_2)$, from which, by deleting all symbols in $\Sigma_1 \cup \Sigma_2$, we get a weakly saturated oscillator automaton for $D(\mathcal{A}_1) \cap D(\mathcal{A}_2)$. The automaton $\tilde{\mathcal{A}}$ is $\tilde{\mathcal{A}} = (Q_1 \times Q_2, \tilde{\delta}, Q_0^1 \times Q_0^2, Q_f^1 \times Q_f^2)$ where

$$\tilde{\delta} = \big\{(q_1, q_2, x, R, X, q_1', q_2') \mid \exists X = X_1 \cup X_2, R = R_1 \times R_2 \text{ s.t.}$$
$$(q_i, x, R_i, X_i, q_i') \in \tilde{\delta}_i, i = 1, 2\big\}$$

**Theorem 4.** *For each oscillator automaton $\mathcal{A}$, $L(\mathcal{A})$ is accepted by a timed automaton.*

*Proof.* Suppose that $\mathcal{A}$ has $n$ counters. For each counter $c_i$ of $\mathcal{A}$ we will use a clock $y_i$ in the timed automaton. Moreover, we will employ a new clock $y_0$, which will be used for marking clock ticks. Hence, almost each 1-transition in $\mathcal{A}$ will give a transition in which we check $y_0 = 1$ and reset $y_0$. The only transitions which will not be transformed into such clock tick transitions will be the first 1-transitions. On each accepted run, the first 1-transition is either ignored, or transformed into a $(y_0 = 1?, y_0 := 0)$ transition.

Note that, on any accepting run in $\mathcal{B}$ we have the property that the interval separating any pair of actions $a$, $b$, is $]n-1, n+1[$, where $n$ is the number of clock ticks on the same run, but in $\mathcal{A}$. If we have a run in which a counter $c_i$ is reset on $a$ and tested only to be positive on $b$, then the time elapse between $a$ and $b$ would be in the interval $]n, n+1[$, but not in $]n-1, n]$. We therefore have to use the test on $c_i$ on the $b$-transition for generating a constraint on the corresponding clock $y_i$ which eliminates the part $]n-1, n]$ from the interval separating $a$ and $b$.

To this end, for each pair of states $q, q' \in Q$ we put in a set $N(q, q')$ all the numbers of 1s occurring on a sequence of $\mathcal{A}$-transitions between $q$ and $q'$. Note that we do not need to check whether that sequence of transitions is feasible. The computation of $N(q, q')$ is a matter of regular sets. We then separate $N(q, q')$ into two sets: $N^0(q, q')$, containing even numbers, and $N^1(q, q')$, with odd numbers. Then, for each $1 \leq i \leq n$ and on each transition $\tau_1 = (q, x, R, X, r)$ on which counter $c_i$ is reset, we start counting modulo 2 the 1s into a register $h_i$.

Take a subsequent transition $\tau_2 = (q', x', R', X', r')$ on which $R_i$ is $<$. This transition is transformed in $\mathcal{B}$ into a transition with constraint $y_i \in ]N^{h_i}(q, r') - 1, N^{h_i}(q, r')[$. When $R_i$ is $=$ we put the constraint $y_i = N^{h_i}(q, r')$, and similarly for $R_i$ being $>$. To see how this solves our problem, denote $n_i$ the number of 1s encountered since $c_i$ was reset. Obviously $n \in N^{h_i}(q, r')$. Therefore, if no transition leaving $q'$ has $R_i = \,' > \,'$ then the constraints we have designed assure that the interval between passing through $\tau_1$ and $\tau_2$ has an empty intersection with $]n - 1, n[$.

Formally, denote $\mathcal{A} = (Q, \Sigma, \delta, Q_0, Q_f)$ the oscillator automaton, whose set of counters is $\mathcal{Y} = \{y_1, \ldots, y_n\}$. The timed automaton $\mathcal{B}$ has tuples $(q, \bar{q}, p, h) \in Q \times Q^n \times \{-, +\} \times \{0, 1\}^n$ as states, where $p$ tells whether the first 1-transition has been encountered or not, $\bar{q}_i$ remembers the last state where the counter $c_i$ was reset, while $h_i$ are the registers modulo 2 whose use was shown above. $\mathcal{B}$ has the following transitions:

1. For each $(q, x, R, X, q') \in \delta$ with $x \in \Sigma \cup \varepsilon$ we put $(q, \bar{q}, p, h) \xrightarrow{a, C, Y} (q', \bar{q}', p, h')$ where $Y \subseteq \mathcal{Y}$ and $C$ is a conjunction of constraints as follows: for each $1 \leq i \leq n$,
   (a) $y_i \in Y$ iff $c_i \in X$ and then $h'_i = 0$ and $\bar{q}'_i = q$; otherwise $h'_i = h_i$ and $\bar{q}'_i = \bar{q}_i$.
   (b) $C$ contains $y_0 \in ]0, 1[$ as a conjunct;
   (c) If $R_i$ is $<$ then $C$ contains $y_i \in ]N^{h_i}(\bar{q}_i, q') - 1, N^{h_i}(\bar{q}_i, q')[$.
   (d) If $R_i$ is $=$ then $C$ contains $y_i = N^{h_i}(\bar{q}_i, q')$.
   (e) If $R_i$ is $>$ then $C$ contains $y_i \in ]N^{h_i}(\bar{q}_i, q'), N^{h_i}(\bar{q}_i, q') + 1[$.
2. For each $(q, 1, R, X, q') \in \delta$ we have a transition $(q, \bar{q}, p, h) \xrightarrow{\varepsilon, C, Y} (q', \bar{q}', p', h')$ where $Y \subseteq \mathcal{Y}$ and $C$ is a conjunction of constraints, both satisfying the requirements 1a, 1c, 1d and 1e from the previous point, as well as the following:
   (f) If $p$ is $+$ then $p' = p$ and $C$ contains $c_0 = 1 \wedge c_0 \in Y$; otherwise $p'$ is $+$.
   (g) If $c_i \notin X$ then $h'_i = (h_i + 1) \mod 2$.

The $\delta^+$ and $\delta^-$ transitions of $\mathcal{A}$ do not generate transitions in $\mathcal{B}$. The initial states of $\mathcal{B}$ are tuples $(q, \bar{q}, p, h)$ with $q \in Q_0$, $\bar{q}_i = q_0$, $h_i = 0$ and $p$ being $-$. The final states of $\mathcal{B}$ are tuples $(q, \bar{q}, p, h)$ with $q \in Q_f$.    □

# 6  Conclusions

We have presented a new discretization technique for timed automata, which corrects the non-compositionality of the region construction w.r.t. intersection. The discretization we have proposed makes use of two delay symbols with "nonarchimedian" semantics. Our discretizations are words over an alphabet containing the action symbols, clock tick symbols and the delay symbols. We have also investigated a class of automata that accepts discretizations of languages of timed automata.

The significance for model-checking is yet unclear. As noted at the end of section 4, the size of the finite automaton built for checking whether an $n$-oscillator automaton has an empty language is at most $n! \cdot 2^n \cdot card(Q)$. Comparing with the region construction, we observe that the largest constants used within clock constraints are hidden within $card(Q)$. On the other hand, the common basis for comparison should be the emptiness checking for timed regular expressions. But timed regular expressions are a very inefficient specification language, since, by [ACM02] in order to specify the language of a timed automaton, its untimed behavior is specified $n$ times, once for each clock. More research is also needed in order to check whether compositional model checking or test generation may benefit from our approach.

Oscillator automata can be used on their own as a model for timed systems. We draw the attention also to the special form of timed automata which are built from oscillator automata. It is easy to check, for those automata, whether the accepted language is $k$-bounded, and hence whether the automaton is determinizable and complementable.

We believe that our approach may help in importing more results from classical language theory. The first subject that may take advantage of our approach is the adaptation of results concerning monoid recognizability. Our discretization might also give an alternative framework for laying the basis of a timed trace theory [DT99].

# References

[ACM02]  E. Asarin, P. Caspi, and O. Maler. Timed regular expressions. Journal of ACM, 49:172–206, 2002. 168, 169, 171, 180

[AD94]  R. Alur and D. Dill. A theory of timed automata. Theoretical Computer Science, 126:183–235, 1994. 168, 170, 171, 176

[AD03]  E. Asarin and C. Dima. Balanced timed regular expressions. ENTCS, 68, issue 5, 2003. 168

[Bea98]  D. Beauquier. Pumping lemmas for timed automata. In Proceedings of FoSSaCS '98, volume 1378 of LNCS, pages 81–94, 1998. 168

[Bel57]  R. Bellmann. Dynamic Programming. Princeton University Press, 1957.

[BP99]  P. Bouyer and A. Petit. Decomposition and composition of timed automata. In Proceedings of ICALP '99, volume 1644 of LNCS, pages 210–219, 1999. 168

[BP02]  P. Bouyer and A. Petit. A Kleene/Büchi-like theorem for clock languages. Journal of Automata, Languages and Combinatorics, 2002. To appear. 168

[BPT01]  P. Bouyer, A. Petit, and D. Thrien. An algebraic characterization of data and timed languages. In Proceedings of CONCUR 2001, volume 2154 of LNCS, pages 248–261, 2001. 168

[Dim01]   C. Dima. An algebraic theory of real-time formal languages. PhD thesis, Université Joseph Fourier Grenoble, France, 2001. 174

[Dim02]   C. Dima. Computing reachability relations in timed automata. In Proceedings of LICS '02, pages 177–186, 2002. 171, 173

[DT99]    D. D'Souza and P. S. Thiagarajan. Product interval automata: A subclass of timed automata. In Proceedings of FSTTCS '99, volume 1738 of LNCS, pages 60–71, 1999. 180

[Her98]   Ph. Herrmann. Timed automata and recognizability. Information Processing Letters, 65:313–318, 1998. 168

[HU92]    J. E. Hopcroft and J. D. Ullman. Introduction to Automata Theory, Languages and Computation. Addison-Wesley/Narosa Publishing House, 1992.

[LPY97]   K. G. Larsen, Paul Petterson, and Wang Yi. Uppaal: Status & developments. In Proceedings of CAV '97, LNCS, pages 456–459, 1997. 168

[Yov98]   S. Yovine. Model-checking timed automata. In Lectures on Embedded Systems, volume 1494 of LNCS, pages 114–152, 1998. 168

# Folk Theorems on the Determinization and Minimization of Timed Automata

Stavros Tripakis

VERIMAG
Centre Equation
2, avenue de Vignate, 38610 Gières, France
www-verimag.imag.fr

**Abstract.** Timed automata are known not to be complementable or determinizable. Natural questions are, then, could we check whether a given TA enjoys these properties? These problems are not algorithmically solvable, if we require not just a yes/no answer, but also a witness. Minimizing the "resources" of a TA (number of clocks or size of constants) are also unsolvable problems. Proofs are provided as simple reductions from the universality problem. These proofs are not applicable to the corresponding decision problems, the decidability of which remains open.

## 1 Introduction

Timed automata [2] (TA) have been established as a convenient model for describing timed systems. This is despite the fact that the model does not enjoy a number of important properties which hold, for instance, in its untimed counter-part, finite-state automata. In particular, timed automata are not *complementable* in general, meaning that, given a TA $A$, there does not always exist a TA accepting the complement of the language accepted by $A$. This holds even if we interpret timed automata as accepting finite-length words, which is the framework we follow in this paper. Timed automata are also not *determinizable* in general, meaning that, given a (non-deterministic) TA $A$, there does not always exist a deterministic TA accepting the same language.

Complementation is important for capturing the negation of logical specification by automata, in so-called automata-theoretic verification. Determinization is crucial for implementability and essential in problems of observation, fault diagnosis and test generation (e.g., [13]). Often, works in such domains (e.g., [11, 10]) are restricted to determinizable sub-classes of TA, for instance, so-called event-clock automata [3].

Given these facts, it is natural to ask: "can it be *checked* whether a given TA $A$ is complementable/determinizable ?". Unfortunately, as we show in this paper, this cannot be done algorithmically, assuming we require not just a "yes/no" answer but also a *witness*, that is, a TA complementing/determinizing $A$. Interestingly, we do not know if the decision problems (admitting a "yes/no" answer) are decidable. The proofs we provide rely on the construction of a witness and are based on a reduction of the universality problem, known to be undecidable.

K.G. Larsen and P. Niebert (Eds.): FORMATS 2003, LNCS 2791, pp. 182–188, 2004.
© Springer-Verlag Berlin Heidelberg 2004

Another set of questions concerns TA *minimization*, in the sense of reduction of "resources" of timed automata.[1] While the resources of untimed automata can be seen to be states and transitions, in timed automata, the clocks and the constants used in the guards are also important resources. In fact, these are in some sense more "expensive" resources than states and transitions, since most decidable problems concerning timed automata have worst-case complexity polynomial in the states and transitions and exponential in the clocks and constants [2, 7].

Given these facts, it is natural to ask: "can it be checked whether the number of clocks or the size of the constants of a given TA can be reduced ?". Unfortunately, this cannot be done algorithmically (assuming, as previously, that we require not just a "yes/no" answer but also a witness).

These results are probably folk theorems in the timed automata community. However, to the best of our knowledge, they have not yet been published. An exception is a similar result appearing in [15]. There, it is shown that computing the *clock-degree* of a given timed language (represented as a timed automaton) cannot be done algorithmically. The clock-degree is the minimum number of clocks necessary to recognize the language.

Reduction of clocks by removing *inactive* clocks has been considered in [8]. The idea is that the value of some clocks is irrelevant in certain discrete states, because the clock is not tested and will be reset upon leaving that state. However, the static analysis technique used in [8] to remove inactive clocks is not powerful enough to answer the question we are asking. Indeed, active clocks may still be redundant with respect to language equivalence. Also, *minimizable timed automata* (MTA) are introduced in [12]. In an MTA, clocks have bounded time domains. The MTA is also equipped with a set of *relevance formulas* permitting to identify equivalent states modulo inactive clocks. The authors show how a minimal MTA can be algorithmically obtained from a given MTA, where minimality is with respect to states and bisimulation equivalence.

**Preliminaries:** We assume the reader is familiar with timed automata. We consider a basic TA model, namely, automata with a finite set of discrete states and transitions, where each transition is labeled with a letter in a finite alphabet $\Sigma$, has a so-called *diagonal-free* clock guard [5] (i.e., no constraints of the form $x - y \leq c$) and a set of clocks to reset to zero. We use the following notation. $\mathcal{R}$ is the set of positive reals. $\mathcal{U} = (\Sigma \times \mathcal{R})^*$ is the set of all finite-length timed words over $\Sigma$. Given $L \subseteq \mathcal{U}$, $\overline{L}$ is the complement of $L$, that is, $\overline{L} = \mathcal{U} - L$. Given a timed automaton $A$ over $\Sigma$, $L(A) \subseteq \mathcal{U}$ is the set of all finite-length timed words accepted by $A$. The *universality problem* is to check, given a TA $A$, whether $L(A) = \mathcal{U}$. The problem is known to be undecidable [2]. The *untimed language* of $A$, denoted $L_u(A)$, is equal to the set of all finite-length words in $\Sigma^*$

---

[1]  We use the term minimization in the sense of reduction of "resources", and not in the sense of computing the quotient with respect to a bisimulation relation, as in [1, 16, 14].

accepted by $A$ if we interpret $A$ as a finite-state automaton, that is, ignoring its clock constraints.

## 2    Complementability

*Problem 1.* Given a TA $A$, does there exist a TA $B$ such that $L(B) = \overline{L(A)}$ ? If so, construct such a $B$.

**Theorem 1.** *Problem 1 is not Turing computable.*[2]

*Proof.* We can reduce the universality problem to Problem 1, as follows. Given $A$, solve Problem 1. If $B$ exists, $L(A) = \mathcal{U}$ iff $L(B) = \emptyset$. If $B$ does not exist, then $L(A) \neq \mathcal{U}$, because the empty language can be accepted by a timed automaton with no accepting states.

Note that the proof relies on the fact that we have a witness and can check emptiness on it. We do not know whether the decision problem corresponding to Problem 1 (which only asks whether $B$ exists) is decidable. Also notice that knowing the existence of a witness does not help in finding one. Enumerating all possible witnesses does not help, since checking for a given $B$ whether $L(B) = \overline{L(A)}$ is undecidable.

## 3    Determinizability

*Problem 2.* Given a TA $A$, does there exist a deterministic TA $B$ such that $L(B) = L(A)$ ? If so, construct such a $B$.

**Theorem 2.** *Problem 2 is not Turing computable.*

*Proof.* We can reduce the universality problem to Problem 2, as follows. Given $A$, solve Problem 2. If $B$ exists, compute $C$ such that $L(C) = \overline{L(B)}$: since $B$ is deterministic, this can be done simply by turning accepting states into non-accepting states and vice-versa. Then, $L(A) = \mathcal{U}$ iff $L(C) = \emptyset$. If $B$ does not exist, then $L(A) \neq \mathcal{U}$, because the language $\mathcal{U}$ can be accepted by a deterministic timed automaton with a single accepting state, no clocks, and a self-loop for each letter in $\Sigma$.

## 4    Minimization

### 4.1    Reducing the Number of Clocks

*Problem 3.* Given a TA $A$ with $n$ clocks, does there exist a TA $B$ with $n-1$ clocks, such that $L(B) = L(A)$ ? If so, construct such a $B$.

---

[2]    With a slight language abuse, when we say a problem is computable we mean that the implicitly defined function solving the problem is computable. For instance, in the case of Problem 1, this function takes a TA $A$ and returns, either a TA $B$ such that $L(B) = \overline{L(A)}$, or $\perp$, when such a $B$ does not exist.

We should note that the technique of clock reduction by removing inactive clocks, proposed in [8], does not solve Problem 3. Indeed, consider the timed automaton that performs $a$ and resets $x := 0$, then has two transitions with $b$, one with guard $x > 1$ and another with guard $x \leq 1$. In this automaton, clock $x$ is redundant: the two transitions labeled with $b$ can be replaced by a single transition without any guard. However, the method of [8] finds that clock $x$ is active, because it is tested after it is reset.

Solving Problem 3 can be used in minimizing the number of clocks of $A$: just keep trying to remove clocks one by one until no clocks are left or until no more clocks can be removed. In particular, the problem, given $A$, to find, if it exists, an automaton $B$ without any clocks, such that $L(B) = L(A)$, can be reduced to Problem 3. This observation is used in the proof below.

**Theorem 3.** *Problem 3 is not Turing computable.*

*Proof.* We can reduce the universality problem to the minimization problem, as follows. Given $A$, check whether there exists $B$ with no clocks such that $L(B) = L(A)$ and if so construct such a $B$. If $B$ exists, check whether the untimed language of $B$, $L_u(B)$, is equal to $\Sigma^*$. If $L_u(B) = \Sigma^*$, then $L(A) = L(B) = \mathcal{U}$. Indeed, $B$ interpreted as a regular automaton accepts all finite words over $\Sigma$, and since it has no clocks, it cannot put any time constraints on them. Thus, when interpreted as a timed automaton, $B$ accepts all finite timed words over $\Sigma$. If $L_u(B) \neq \Sigma^*$, then there is some $w \in \Sigma^* - L_u(B)$. Again, since $B$ has no timing constraints, $w \notin L(B)$, thus, $L(A) \neq \mathcal{U}$. If $B$ does not exist then $L(A) \neq \mathcal{U}$, since $\mathcal{U}$ can be accepted by an automaton with no clocks.

## 4.2  Reducing the Size of Constants

*Problem 4.* Given a TA $A$ where constants are not greater than $c$, does there exist a TA $B$ where constants are not greater than $c-1$, such that $L(B) = L(A)$ ? If so, construct such a $B$.

Solving Problem 4 is enough for minimizing the size of constants of $A$: just keep trying to reduce the size of constants by one until it becomes zero or until it can be reduced no more. In particular, the problem, given $A$, to find, if it exists, an automaton $B$ with constants at most zero, such that $L(B) = L(A)$, can be reduced to Problem 3.

**Lemma 1.** *Let $A$ be a TA over $\Sigma$ with constants at most 0. There exists a finite-state automaton $A_u$ over $\Gamma = \Sigma \cup \{\tau\}$, where $\tau \notin \Sigma$, such that $L(A) = \mathcal{U}$ iff $L(A_u) = \Gamma^*$.*

*Proof.* There are two types of guards in $A$: $x > 0$ or $x = 0$, where $x$ is a clock. For each clock $x$ of $A$, $A_u$ will have one variable $b_x \in \{0, 1\}$. $A_u$ will have the same set of discrete states as $A$. For every discrete transition of $A$, $A_u$ will have a transition labeled with the same letter. For every reset $x := 0$ of the transition of $A$, we add a reset $b_x := 0$ to the transition of $A_u$. For every guard $x = 0$

(resp. $x > 0$) of the transition of $A$, we add a guard $b_x = 0$ (resp. $b_x = 1$) to the transition of $A_u$. At every discrete state of $A_u$ we add a *self-loop* transition labeled by $\tau$, which sets each variable $b_x$ to 1. Notice that the language of $A_u$ is closed under "stuttering" of $\tau$, for instance, if $\tau a_0 \tau a_1 \cdots \tau a_n \in L(A_u)$ then $\tau^+ a_0 \tau^+ a_1 \cdots \tau^+ a_n \subseteq L(A_u)$ and vice-versa. This is because taking a $\tau$ transition two or more times in a row leaves the state of $A_u$ unchanged.

Define the following equivalence between states of $A$ and states of $A_u$. Given a state $(q, v)$ of $A$ ($q$ is a discrete state and $v$ is a vector of values for each clock $x$) and a state $(q', u)$ of $A_u$ ($q'$ is a discrete state and $u$ is a vector of values for each variable $b_x$), the two states are equivalent, denoted $(q, v) \sim (q', u)$, if $q = q'$ and for all $i$, $v(i) = 0 \Leftrightarrow u(i) = 0$. We claim that if $s_1 \sim s_2$ then:

1. for each $a \in \Sigma$ and state $s_1'$ of $A$ such that $s_1 \overset{a}{\to} s_1'$, there exists state $s_2'$ of $A_u$ such that $s_2 \overset{a}{\to} s_2'$ and $s_1' \sim s_2'$;
2. for each $a \in \Sigma$ and state $s_2'$ of $A_u$ such that $s_2 \overset{a}{\to} s_2'$, there exists state $s_1'$ of $A$ such that $s_1 \overset{a}{\to} s_1'$ and $s_1' \sim s_2'$;
3. for each $t \in \mathcal{R}$ and state $s_1'$ of $A$ such that $s_1 \overset{t}{\to} s_1'$, there exists state $s_2'$ of $A_u$ such that $s_2 \overset{\tau}{\to} s_2'$ and $s_1' \sim s_2'$;
4. for each state $s_2'$ of $A_u$ such that $s_2 \overset{\tau}{\to} s_2'$, for each $t \in \mathcal{R}$, there exists state $s_1'$ of $A$ such that $s_1 \overset{t}{\to} s_1'$ and $s_1' \sim s_2'$.

The above four properties allow us to prove that $L(A) = \mathcal{U}$ iff $L(A_u) = \Gamma^*$.

**Theorem 4.** *Problem 4 is not Turing computable.*

*Proof.* By Lemma 1, checking universality of a TA with constants at most zero is decidable. Since checking universality of a general TA is undecidable, Problem 4 is not computable.

## 5   Similar Problems with "Bounded Resources"

One might think that the above negative results could be remedied if one bounds the resources of the witness automaton. A similar approach is taken in [9, 4], where it actually results in a decidable version of an otherwise undecidable problem. Unfortunately, this is not the case for the problems defined in this paper.

More precisely, given non-negative integers $n$ and $c$, let $TA(n, c)$ be the class of timed automata having at most $n$ clocks and where constants are at most $c$. Then, the bounded-resource versions of Problems 1, 2, 3, and 4 can be stated as follows.

*Problem 5.* Given a TA $A$ and non-negative integers $n, c$, does there exist a TA $B \in TA(n, c)$ such that $L(B) = \overline{L(A)}$ ? If so, construct such a $B$.

*Problem 6.* Given a TA $A$ and non-negative integers $n, c$, does there exist a deterministic TA $B \in TA(n, c)$ such that $L(B) = L(A)$ ? If so, construct such a $B$.

*Problem 7.* Given a TA $A$ with $n$ clocks and non-negative integer $c$, does there exist a TA $B \in TA(n-1, c)$, such that $L(B) = L(A)$ ? If so, construct such a $B$.

*Problem 8.* Given a TA $A$ with constants not greater than $c$ and non-negative integer $n$, does there exist a TA $B \in TA(n, c-1)$, such that $L(B) = L(A)$ ? If so, construct such a $B$.

It turns out that all four problems above are not computable. The proofs are almost identical to the ones for the unbounded-resource versions, with the addition that we set $n$ and/or $c$ to zero when reducing the universality problem to the problem in question. For example, in the case of Problem 5, if there exists no $B$ in $TA(0, 0)$ such that $L(B) = \overline{L(A)}$ then $L(A) \neq \mathcal{U}$, since there is a TA with no clocks accepting the empty language. If $B \in TA(0, 0)$ exists then $L(A) \neq \mathcal{U}$ iff $L(B) = \emptyset$.

# 6  Conclusions and Open Questions

The folk theorems presented in this paper confirm some inherent undecidability properties of the timed automata model. A number of open questions remain. We do not know whether the decision problems corresponding to the problem defined in this paper are decidable.

Another interesting problem is minimization of the number of discrete states of a TA (while possibly increasing the number of clocks or size of constants). The interesting cases are when diagonal guards or resets to constants other than zero are not allowed. Otherwise, a discrete state can be encoded as the ordering $x_1 < x_2 < \cdots < x_m$ of a sufficient number of extra clocks $x_1, ..., x_m$ and moving to this state can be encoded with an appropriate reset, such as $x_1 := 0, x_2 := 1, ..., x_m := m$. Note that, although these features do not add to the expressiveness of the model, removing them can only be done at the expense of adding discrete states [5, 6].

# Acknowledgements

The author would like to thank Kim Larsen for his invitation to BRICS, Aalborg, where part of this work was conducted. Thanks to Yassine Lakhnech for pointing out Wilke's Ph.D. thesis. Thanks to Rajeev Alur, Eugene Asarin and Joel Ouaknine for valuable discussions.

This work has been partially supported by European IST project "Next TTA" under project No IST-2001-32111.

# References

[1] R. Alur, C.Courcoubetis, N. Halbwachs, D.L. Dill, and H. Wong-Toi. Minimization of timed transition systems. In *3rd Conference on Concurrency Theory CONCUR '92*, volume 630 of *Lecture Notes in Computer Science*, pages 340–354. Springer-Verlag, 1992. 183

[2] R. Alur and D. L. Dill. A theory of timed automata. Theoretical Computer Science, 126:183–235, 1994. 182, 183

[3] R. Alur, L. Fix, and T. Henzinger. A determinizable class of timed automata. In CAV '94, volume 818. LNCS, Springer, 1994. 182

[4] P. Bouyer, D. D'Souza, P. Madhusudan, and A. Petit. Timed control with partial observability. In CAV '03, 2003. 186

[5] P. Bouyer, C. Dufourd, E. Fleury, and A. Petit. Are timed automata updatable? In CAV '00. LNCS 1855, 2000. 183, 187

[6] P. Bouyer, C. Dufourd, E. Fleury, and A. Petit. Expressiveness of updatable timed automata. In MFCS '00. LNCS 1893, 2000. 187

[7] C. Courcoubetis and M. Yannakakis. Minimum and maximum delay problems in real-time systems. In Computer-Aided Verification, LNCS 575. Springer-Verlag, 1991. 183

[8] C. Daws and S. Yovine. Reducing the number of clock variables of timed automata. In Proc. 17th IEEE Real-Time Systems Symposium, RTSS '96, 1996. 183, 185

[9] D. D'Souza and P. Madhusudan. Controller synthesis for timed specifications. In Springer LNCS, editor, STACS '02, volume 2285, 2002. 186

[10] B. Nielsen and A. Skou. Automated test generation from timed automata. In TACAS '01. LNCS 2031, Springer, 2001. 182

[11] J. Springintveld, F. Vaandrager, and P. D'Argenio. Testing timed automata. Theoretical Computer Science, 254, 2001. 182

[12] J. G. Springintveld and F. W. Vaandrager. Minimizable timed automata. In FTRTFT '96, pages 130–147. LNCS 1135, 1996. 183

[13] S. Tripakis. Fault diagnosis for timed automata. In FTRTFT '02. LNCS 2469, Springer, 2002. 182

[14] S. Tripakis and S. Yovine. Analysis of timed systems using time-abstracting bisimulations. Formal Methods in System Design, 18(1):25–68, January 2001. 183

[15] Thomas Wilke. Automaten und Logiken zur Beschreibung zeitabhängiger Systeme. PhD thesis, Institut Für Informatik und Praktische Mathematik, Christian-Albrechts Universität, Kiel, 1994. In German. 183

[16] M. Yannakakis and D. Lee. An efficient algorithm for minimizing real-time transition systems. Formal Methods in System Design, 11(2), 1997. 183

# Control Synthesis
# for a Smart Card Personalization System
# Using Symbolic Model Checking*

Biniam Gebremichael and Frits Vaandrager

Nijmegen Institute for Computing and Information Sciences
University of Nijmegen, P.O. Box 9010, 6500 GL Nijmegen, The Netherlands
{biniam,fvaan}@cs.kun.nl

**Abstract.** Using the Cadence SMV symbolic model checker we synthesize, under certain error assumptions, a scheduler for the smart card personalization system, a case study that has been proposed by Cybernetix Recherche in the context of the EU IST project AMETIST. The scheduler that we synthesize, and of which we prove optimality, has been previously patented. Due to the large number of states (which is beyond $10^{13}$), this synthesis problem appears to be out of the scope of existing tools for controller synthesis, which typically use some form of explicit state enumeration. Our result provides new evidence that model checkers can be useful to tackle industrial sized problems in the area of scheduling and control synthesis.

## 1 Introduction

### 1.1 Background

Model checking involves analyzing a given model of a system and verifying that this model satisfies some desired properties. System models are typically described as finite transition systems, while properties are described in terms of temporal logic. Once the definition of the system, $\mathbf{S}$, and its property, $\psi$, are fixed, the model checking problem is easily described as $\mathbf{S} \models \psi$? (does $\mathbf{S}$ satisfy $\psi$?). Thanks to the symbolic representation of transition systems, state-of-the-art model checking tools are now capable of solving such problems for models with more than $10^{20}$ states [4].

Control synthesis, on the contrary, does not assume the existence of a model of the full system. Instead, it considers the uncontrolled plant and tries to synthesize a controller by finding a possible instance of a model that satisfies a desired property. Control synthesis for Discrete Event Systems (DES) has been extensively studied over the past two to three decades, and a well-established theory has been developed by Ramadge and Wonham [16]. The Ramadge and Wonham framework (RW) is based on the formal (regular) language generated by a finite

---

* This work was supported by the European Community Project IST-2001-35304 AMETIST, http://ametist.cs.utwente.nl.

state machine. The RW plant model $P$ (*generator*) is obtained by describing the plant processes in terms of a formal language which is generated by a finite automaton. A *means of control* is adjoined to this *generator* by identifying the events that can be enabled or disabled by the controlling agent. The specifications $S_p$ are described in terms of formal language generated by $P$. The controller is then constructed from a recognizer for the specified language given by $S_p$.

In this paper, we consider a problem which in theory could very well be solved using the Ramadge and Wonham supervisory control theory. However, given the size of the state space involved, existing control synthesis tools are (to the best of our knowledge) unable to actually compute a solution. Therefore, instead, we tackled the problem using the symbolic model checker SMV [14].[1] This approach allows us to benefit from the (BDD-based) symbolic representation technique of SMV and to (partially!) solve a problem which, because of its size (more than $10^{13}$ states), would be intractable otherwise. Our results demonstrate that model checkers can be useful to solve problems in the area of scheduling and control synthesis.

## 1.2  Outline

Using SMV we synthesize a scheduler for a smart card personalization system, which has previously been patented by Cybernetix Recherche. We also show that this scheduler, known as the "super single mode" [2] is optimal in the absence of errors. Finally, we synthesize a set of schedulers for defective card treatment that stabilize the system back to the super single mode. Together, these schedulers constitute a controller for the system under the assumption that a certain amount of time elapses between faults.

The paper is structured as follows: Section 2 provides a formal definition of the uncontrolled plant of the smart card personalization system, and defines the correctness and optimality criteria. Section 3 explains the super single mode, and how it was generated using SMV. Section 4 deals with systems with faulty cards. We list the errors that may occur during the operations of the machine, show how to deal with such errors, and give an overview of the synthesized error treatment methods. We conclude the paper by pointing out some observations and directions for future work in Section 5.

A full version of this paper appeared as [8]. An electronic copy of SMV code and also of the trace simulator that we developed to visualize schedules are available via the URL

    http://www.cs.kun.nl/ita/publications/papers/biniam/cyber.

## 1.3  Related Work

The Ramadge and Wonham framework has been implemented by several research groups and industries. One of the tools developed by Wonham and his research

---

[1] We use the version of SMV developed at Cadence Berkeley Laboratories, see http://www-cad.eecs.berkeley.edu/~kenmcmil/smv/.

team is CTCT (C based Toy Control Theory)[2], a tool that was basically built for research purposes only, and uses an exhaustive list to represent the model. Its capacity, as the name indicates, has never extended beyond toy examples. A new approach, Vector Discrete Event Systems, was studied in [12, 22] to alleviate the shortcoming of CTCT by exploiting the structural properties of DES. Although this approach resulted in better performance, its structural analysis approach cannot be generalized [5].

Other notable developments on this area are: The UMDES-LIB library from University of Michigan [18], Bertil Brandin's tool for DES control synthesis with heuristics [3], a tool for Condition/Event Systems [19], other tool by Martine Fabian and Knut Åkesson [1].

All the above tools lack symbolic representation of state transitions, and suffer from state space explosion problems. A Binary Decision Diagram (BDD) like data structure called Integer Decision Diagram (IDD) has been used to represent sets of states symbolically. For example, Gunnarsson in [9] and Zhang and Wonham in [23] have used IDDs in their implementation. This approach is quite promising for dealing with large systems, but it is still in laboratory stage, and not available to the public.

Our main motivation for using SMV is thus to overcome this deficiency and benefit from symbolic representation of SMV. The smart card personalization system is quite a large system and cannot be handled with a tool that does not use symbolic representation. Our paper shows how the scheduler synthesis can be solved using a model checker and presents new evidence that model checkers can be useful in solving problems in the area of scheduling and synthesis. Our work has been inspired by similar approaches that were employed in [7, 10, 15] to synthesize schedulers for industrial size problems.

We were the first to model the smart card personalization system and to synthesize a scheduler for it. However, the same case study has also been addressed by other members of the AMETIST consortium. T. Krilavicius and Y. Usenko [11] constructed models using UPPAAL and $\mu$CRL, and used these to synthesize controllers. Whereas in our model production of cards is essentially an infinite process, Krilavicius and Usenko only consider scheduling of a finite number of cards. As a consequence, they do not synthesize the super single mode. Inspired by [11], T. Ruys used SPIN to synthesize a controller for the smart card personalization machine [17]. Also this model only considers scheduling of a finite number of cards (the largest parameter values considered are 5 cards and 4 stations). In order to handle the state space explosion, Ruys encodes branch & bound search strategies in SPIN. In addition, he has to instruct SPIN to use a number of heuristics, which in our view are both complex (the code for the heuristics is longer than the code of our entire model!) and debatable (Ruys assumes that cards cannot overtake each other; in the real machine this is possible with the help of the personalization stations). A. Mader in [13] applied decomposition and mixed strategies to model and synthesize a controller for the extended smart card personalization machine that include printers and flippers.

---

[2] See http://odin.control.toronto.edu/people/profs/wonham.

G. Weiss employed Life Sequence Charts (LSC) to synthesize a scheduler with smart play-in/play-out approach [21]. None of the mentioned approaches deals with error handling.

## 2   Smart Card Personalization System

The "smart card personalization system" is a case study that has been proposed by Cybernetix Recherche in the context of the EU IST project AMETIST [2]. The case study concerns a machine for smart card personalization, which takes piles of blank smart cards as raw material, programs them with personalized data, prints them and tests them.

The machine has a throughput of approximately 6000 cards per hour. It is required that the output of cards occurs in a predefined order. Unfortunately, some cards may turn out to be defective and have to be discarded, but without changing the output order of personalized cards. Decisions on how to reorganize the flow of cards must be taken within fractions of a second, as no production time is to be lost.

The goal of the case study is to model the desired production requirements as well as the timing requirements of operations of the machine, and on this basis synthesize the coordination of the tracking of defective cards. More specifically, the goal is to synthesize optimal schedules for the personalization machine in which defective cards are dealt with, i.e., schedules in which

1. cards are produced in the right order (safety). The order of cards is important as no other sorting mechanism should exist in the system,
2. throughput is maximal (liveness).

### 2.1   The Uncontrolled Plant Model

Figure 1 shows a simplified smart card personalization machine. The machine consist of a conveyor belt and personalization stations mounted on top of it. The machine also has an input station and an output station, which are situated on the left and right side of the belt respectively. New cards enter the system through the input station and advance to the right one step at a time. At some point, a card is lifted up to one of the personalization stations, spends some time there (is personalized), and is then dropped back onto the belt. The card then moves towards the output station for testing and delivery. The actual machine is considerably more complicated than the machine in Figure 1, but our aim is to find a scheduler that effectively utilizes the personalization stations and optimizes throughput. The simplified model of the machine appears to be adequate for this purpose.

The SMV model for the uncontrolled machine is a collection of processes running concurrently: forward (moving a belt one step to the right) and, for each personalization station $j$, lift_drop$_j$ (lifting/dropping a card from/to the belt to/from station $j$). We employ a discrete model of time, in which one time

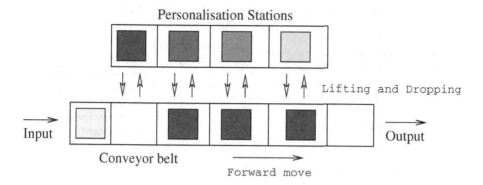

**Fig. 1.** Simplified smart card personalization machine

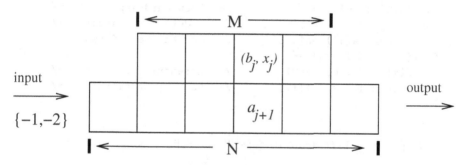

**Fig. 2.** The model of the smart card personalization machine

unit is equivalent to one forward move of the belt. All personalization stations are identical and need S time units to personalize a card. We assume lifting and dropping takes no time.

We assume there are M stations (denoted by $b_j$), and N = M+2 slots in the belt (denoted by $a_j$) as shown in Figure 2. To make model checking possible, the number of different personalizations is assumed to be bounded by some value K, which is a multiple of M. Each slot or station will have a value as shown in Table 1.

An empty slot/station is coded twice (as -3 and -2) in order to distinguish between the initial value (-3) and the slot/station being emptied along the way (-2). This allows us to control intermediate blank slots more efficiently, as will be explained below. We also use an integer variable $x_j$, ($0 \le j <$ M) as a clock to record how long a card has been held in station $j$.

**Table 1.** System parameters and encoding of values

| parameter | represents | slot/station value | meaning |
|-----------|------------|--------------------|---------|
| M | number of stations | -3 | empty (initial value) |
| N | total number of slots | -2 | emptied |
| K | different number of personalizations | -1 | new card |
| | | $j$, $0 \leq j < K$ | personalized with j |
| S | time needed for personalization | K | defective card |

Formally, the process **forward** is defined as follows (for a complete specification of the system we refer the reader to [8]).

```
module forward(a,b,x){

    next(a[0]):={-1,-2};                /* a~new card appears    */
    for(j=1;j<=N-1;j=j+1)               /*non-deterministicaly   */
            next(a[j]):=a[j-1];         /* move the belt forward */
    for(j=0;j<=M-1;j=j+1){
        if(x[j]<S & b[j]>=0)            /* increment clocks of   */
            next(x[j]):= x[j]+1;        /* the busy stations     */
    }
}
```

and the processes **lift_drop**$_j$ $(0 \leq j < M)$ are defined as:

```
module lift_drop(a,b,x,j){

if(b[j] <= -2 & a[j+1] = -1){ /* idle station and new card*/
    next(b[j]):= 0..K;        /* generate a~personalization */
    next(a[j+1]):=b[j];       /* reset the slot             */
    next(x[j]):=0;            /* reset the clock            */
}
else if(b[j] >= 0 &  x[j] = S /* card personalized          */
        & a[j+1] = -2 ){      /* a~blank slot beneath       */
    next(a[j+1]):=b[j];       /* drop the card              */
    next(b[j]):= -2;          /* reset the station          */
  }
}
```

**Correctness.** The desired correctness property is:

*There exists a run that always produces personalized cards in the right order.*

To formalize the concept of "right order", an observer process is introduced that compares the output value with the expected value. Formally, the observer is defined as follows. We introduce a new state variable out, which initially is 0 and assume K is a multiple of M, say 2.M. The behavior of the observer is specified by:

```
if(out = a[N-1])        next(out):= (out+1) mod K;
   else if(a[N-1]>-2)   next(out):= K;
```

If cards are not produced in the right order or if a card is output that has not been personalized, the observer sets the value of out to the "error" value K. The control objective then becomes to ensure that the observer will never detect an error. We can synthesize a scheduler that realizes this (if it exists) by asking SMV whether the following CTL formula holds:

$$\text{AF}\neg(\text{out} < \text{K}). \tag{1}$$

If this formula does *not* hold then there exists an infinite run in which for all states out < K, i.e., the observer never detects an error. In this case SMV will provide a counter example, which essentially is an infinite schedule for the machine that meets the control objective.

**Optimization.** Obviously, there are many runs in which all states satisfy out < K, for instance, a run in which the machine produces no cards at all. The interesting runs are those with high throughput, or more specifically with less number of blank slots in the output.

To minimize the blank slots in the output and in order to guide SMV towards optimal schedules, we introduce the "blank tolerance condition" of the machine, in the form of a new state variable tl, which is initially 0, and is incremented and decremented as follows:

```
if(a[N-1]=-2)                              next(tl):=tl-1;
else if( a[N-1]>=0 & (a[N-1] mod S) = S-1) next(tl):=tl+1;
```

We add 1 to tl each time S cards have been produced ($a_{N-1}$ modulo S = S-1). We decrement tl with 1 whenever a blank slot arrives ($a_{N-1}$ = -2). However, we start decrementing only after the leading blank slots (a[N-1] = -3) have passed. In all other cases we leave the value of tl unchanged.

Now we ask SMV whether the following CTL formula holds:

$$\text{AF}\neg(\text{out} < \text{K} \wedge \text{tl} \geq 0). \tag{2}$$

If this formula does not hold, there exists an infinite scheduler that maintains the invariant tl $\geq$ 0. This means that each time when the system has produced S cards, the observer tolerates a single blank slot.

**Table 2.** The super single mode for 4 personalization stations

For each time the upper line shows the values of the stations (columns 0–3); the lower line shows the values of the slots in the conveyor belt (input, under stations 0–3, output). An empty cell means idle, a box (□) represents a new card, and a number represents the personalization value.

| time | in put | 0 | 1 | 2 | 3 | out put |
|------|--------|---|---|---|---|---------|
| 0 |  |  |  |  |  |  |
|  |  |  |  |  |  |  |
| 1 |  |  |  |  |  |  |
|  | □ |  |  |  |  |  |
| 2 |  | 0 |  |  |  |  |
|  | □ |  |  |  |  |  |
| 3 |  | 0 |  |  |  |  |
|  | □ | □ |  |  |  |  |
| 4 |  | 0 | 1 |  |  |  |
|  | □ | □ |  |  |  |  |
| 5 |  | 0 | 1 |  |  |  |
|  |  | □ | □ |  |  |  |
| 6 |  |  | 1 | 2 |  |  |
|  | □ | 0 | □ |  |  |  |
| 7 |  | 4 | 1 | 2 |  |  |
|  | □ |  | 0 | □ |  |  |
| 8 |  | 4 |  | 2 | 3 |  |
|  | □ | □ | 1 | 0 |  |  |
| 9 |  | 4 | 5 | 2 | 3 |  |
|  | □ | □ |  | 1 | 0 |  |
| 10 |  | 4 | 5 |  | 3 |  |
|  |  | □ | □ | 2 | 1 | 0 |
| 11 |  |  | 5 | 6 | 3 |  |
|  | □ | 4 | □ |  | 2 | 1 |
| 12 |  | 8 | 5 | 6 |  |  |
|  | □ |  | 4 | □ | 3 | 2 |
| 13 |  | 8 |  | 6 | 7 |  |
|  | □ | □ | 5 | 4 |  | 3 |
| 14 |  | 8 | 9 | 6 | 7 |  |
|  | □ | □ |  | 5 | 4 |  |
| 15 |  | 8 | 9 |  | 7 |  |
|  |  | □ | □ | 6 | 5 | 4 |
| 16 |  |  | 9 | 10 | 7 |  |
|  | □ | 8 | □ |  | 6 | 5 |
| 17 |  | 12 | 9 | 10 |  |  |
|  |  | □ | 8 | □ | 7 | 6 |
| 18 |  | 12 |  | 10 | 11 |  |
|  | □ | □ | 9 | 8 |  | 7 |

## 3   The Super Single Mode

Using the approach outlined in the previous section, the example run in Table 2 was generated. With a "normal-speed" PC we were able to generate example runs for M ≤ 5 (in the real machine M could be 8, 16 or 32). The runs exhibit the schedule of the super single mode as patented by Cybernetix. Table 2 shows the first 19 configurations of the the super single mode with M = 4, S = 4, K = 12. Each row represents a single configuration at a given time. The upper part of the row shows the values of the stations, while the lower part shows the values of the slots in the conveyor belt. An empty cell means the slot or the station is idle, a box (□) represents a new card, and a number represents the personalization value of the card contained in the station or in the slot. Table 2 can be read as:

- time 0: the machine is empty.
- time 1: first new card arrives on the conveyor belt.
- time 2: the first card is lifted to station 0.
- time 4: the second card is lifted to station 1 and it continues likewise.
- time 5: there is no card from the input.
- time 6: station 0 finishes personalizing a card with value 0. In super single mode, M (4 in this example) time units are required to personalize a card.
- time 7: station 0 proceeds with personalizing another card with a different value (namely 4). Note that value 3 is not taken yet. This pattern shows that the order of output is exactly the same as the order of the cards when they are fed into the machine, but the production order is different, and there is an overlap between rounds. This overlap is even more clearly visible when a machine with 8 (instead of 4) personalization stations is considered.

If in our model a station is allowed to take more than M time units for personalizing a card, i.e., S > M, then CTL formula (2) holds. In other words: if the conveyor belt is rolling faster than the personalization stations can handle then personalizing M consecutive cards becomes impossible.

Similarly, for a personalization time of M time units, if we have M+1 consecutive new cards followed by empty slots (even with lots of empty slots), then it becomes impossible to personalize all of them. This result implies that the super single mode is optimal in the absence of errors.

## 4   Error Recovery

The control objective for the smart card personalization machine is to personalize cards in the right order even in the presence of errors. The super single mode, as explained above, only works for a perfect machine that makes no errors. In general, it is difficult to prevent errors from occurring (even though errors are rare, approximately 1 in 6000 cards), and so it makes our approach more realistic if we allow for the occurrence of errors in our model, and provide a means of recovering from them.

There are several methods to achieve fault-tolerant behavior. Our approach is inspired by the concept of *self-stabilization* [6, 20], which is well-known from the area of distributed algorithms. An algorithm is called stabilizing if it eventually starts to behave correctly (i.e., according to the specification of the algorithm), regardless of the initial configuration.

Figure 3 shows the production cycle of the personalization machine under the super single mode. In the normal mode of operation the machine loops on the super single mode cycle (the continuous line). This loop is also shown in Table 2 with actual figures. The configurations of the machine at time 9, 10, 11, 12, 13 are equivalent (personalization value modulo M = 4) to the configurations at time 14, 15, 16, 17 and 18 respectively. Thus the super single mode enters the loop at time 9 and loops forever with a period of 5 time units.

However, when an error occurs (dashed line in figure 3), an error recovery treatment (dotted line) should be conducted to stabilize the system and bring it back to the loop. We use SMV to synthesize an error recovery treatment that brings the machine back to the loop. Basically, our approach is as follows:

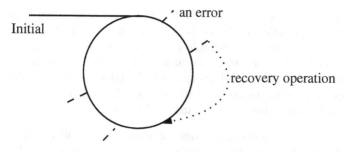

**Fig. 3.** Stabilization of the smart card personalization system

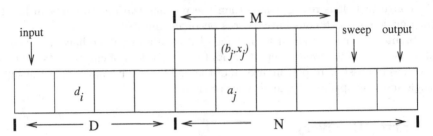

**Fig. 4.** Expanded model of the smart card personalization machine

1. Use SMV to synthesize a regular super single mode run, as described in the previous section.
2. Pick a state on this run and manually introduce an error; the new error state $s$ now becomes the start state of the model.
3. Pick an arbitrary state $t$ on the super single mode cycle, and encode this as an SMV state formula $\varphi$.
4. Ask SMV whether the following formula holds

$$\mathsf{AG}\neg\varphi. \tag{3}$$

If formula (3) does not hold then SMV generates a counterexample; this counterexample is the schedule for a recovery operation that brings the system from state $s$ back into super single mode.

Note that, unlike the theory of self-stabilization, we do not consider arbitrary initial configurations, but only configurations that have been obtained by introducing a single error into a super single mode configuration.

### 4.1   Types of Errors

It is easy to list many scenarios that can make the system behave erratically. In this paper we will only consider errors that may occur in the card. That is:

1. Type 1 errors (E1) are errors in a smart card originating from physical damage or other reasons. This type of error is detected by the personalization stations. In $E1_a$ and $E1_b$ in Table 3 are examples of E1 error.
2. Type 2 error (E2) are errors originating from the personalization station when cards are personalized wrongly, which makes them unusable. This type of error is detected by a tester situated at the end of the personalization stations. $E2_a$ in Table 3 is an example of E2 error.

To make our system recoverable from these errors, we will modify our model in two ways: by adding extra operations and by expanding the belt in both directions.

**Table 3.** The super single mode for 8 personalization stations with error. Only card values in station is shown

| time | input | personalization stations | | | | | | | | output (tester) |
|------|-------|---|---|---|---|---|---|---|---|---|
| | | 0 | 1 | 2 | 3 | 4 | 5 | 6 | 7 | |
| 9 | | | | | | | | | | |
| 10 | □ | | | | | 4 | | | | |
| 11 | □ | 8 | | | | | | | | |
| 12 | □ | | | | | 5 | | | | |
| 13 | □ | | $E1_a$ | | | | | | | |
| 14 | □ | | | | | | | $E1_b$ | | |
| 15 | □ | | | 10? | | | | | | |
| 16 | □ | | | | | | | | 7? | |
| 17 | □ | | | | 11? | | | | | |
| 18 | | | | | | | | | | 0 |
| 19 | □ | | | | 12? | | | | | 1 |
| 20 | □ | 16? | | | | | | | | 2 |
| 21 | □ | | | | | 13? | | | | $E2_a$ |
| 22 | □ | | 17? | | | | | | | 4 |
| 23 | □ | | | | | | | 14? | | 5 |
| 24 | □ | | | 18? | | | | | | $E1_b$ |
| 25 | □ | | | | | | | | 15? | 7? |
| 26 | □ | | | | 19? | | | | | |
| 27 | | | | | | | | | | 8 |

## 4.2 Recovery Operations

If a defective card is detected in the tester then, in order to maintain correctness (i.e., produce personalized cards in the right order), the defective card has to be removed, a replacement card has to be produced, and inserted in the right position. In order to realize this, first the defective card has to be swept off the belt, and then the belt has to go back to one of the personalization stations to retrieve a replacement card and place it in the right position. For these purpose we enrich our model with 'backward' and 'sweep' operations.

The backward move is the same as the forward move except that it moves the belt in the opposite direction. The forward move is the "normal" way of moving the belt, the backward move is used only to handle defective cards [2]. We assume that a backward move takes 1 time unit per step.

When the belt moves backward, the leftmost cards on the belt are also pushed back to the edge. For technical reasons explained in [2], the preferred way of treatment is to expand the belt to the left. As shown in Figure 4, the gap between the input station and the first personalization station, denoted by $d_i$ ($0 \leq i \leq D$, $D = M$), is important for backward movement. Similarly, the belt is also expanded to the right: $N$ ($= M+2$) covers the extended slots in the right side.

Table 4. Safety requirements for belt operations

| Operation | Safety requirements | meaning |
|---|---|---|
| backward | $d_0 < 0$ | no processed card reaches input station, unprocessed (new) cards can return back to the input station |
| forward | $a_{N-1} = $ out $\vee$ $a_{N\ 1} = -2$ | no unexpected card reaches the tester station |
| sweep | $a_M = K$ | only defective cards are swept |

A sweeper is a device that kicks defective cards from the belt. In the physical machine, a sweeper is situated after the personalization station. Formally the sweep operation is defined as:

```
module sweep(a){
  if(a[M]=K)     next(a[M]):=-2;
}
```

### 4.3  Safety Requirements

During the stabilization process, the machine executes operations that are not performed in super single mode. Even if the machine is allowed to perform these special operations, there are some safety requirements that have to be obeyed by the control program. These are shown in Table 4. Complete SMV code for error recovery treatment is given in [8].

### 4.4  Results

For a single error scenario as defined above, there are 2.M possible error configurations in one cycle of the super single mode. Using these error configurations as an initial state and the formula (3) we generated a recovery path that could stabilize the system back to the super single mode. Obviously, each path is different for different initial state, however, they share similar pattern. Thus we groups similar paths together and explain their property below.

1. When the error type is E1 and the faulty card was detected in the first half stations ($b_i$: $0 \leq i \leq \lfloor \frac{M}{2} \rfloor$), then the faulty card remain in the station until a free slot is available. And the personalization value remains unused until next. For example When the faulty card $E1_a$ in Table 3 was detected the personalization value (which is 9) was used in station 2

2. Using the same technique, for E1 errors in the second half stations ($b_i$: $\lfloor \frac{M}{2} \rfloor < i \leq M-1$) will not solve the problem, instead it will introduce another error. The generated recovery path for this scenario is to skip the personalization value for now and let the error evolve to E2 error. The personalization value (6) of $E1_b$ in Table 3 was skipped and $E1_b$ will be again an error of type E2 at time 24.

**Table 5.** Defective card treatment for error type 2

| time | input | 0 | 1 | 2 | 3 | 4 | 5 | 6 | 7 | tester |
|------|-------|---|---|---|---|---|---|---|---|--------|
| 12 | □ |  |  |  |  |  | 5 |  |  |  |
| 21 | □ |  |  |  |  |  | 13 |  |  | 3 |
| 22 | □ |  | 17 |  |  |  |  |  |  | 4 |
| 23 | □ |  |  |  |  |  |  | 14 |  | (E2) |
| 24 | □ |  |  |  | 5 |  |  |  |  | 6 |
| 25* | □ | □ |  | 10 | 9 | 8 | □ | 7 | 6 |  |
| 26* | □ |  | 10 | 9 | 8 | □ | 7 | 6 |  |  |
| 32* | □ | 7 | 6 | 5 |  |  |  |  |  |  |
| 38 | □ |  |  |  |  |  |  | 14 |  | 5 |

3. The recovery path for E2 errors consists:
    - finding a station with a fresh card, this station should be in the first half. Otherwise an error like $E1_b$ will happen again. See also Example 1.
    - rolling the belt backward to this station,
    - personalizing the card with the personalizations value which is missing, and
    - dropping the card to the belt and forward it to the tester.

*Example 1.* In Table 5, at time 23 the $5^{th}$ card is found defective. At the same time station 6 starts with a fresh card. If a replacement card would be produced in this station, then personalization number 14 would be skipped. But this will introduce another error, because the $16^{th}$ and $17^{th}$ cards are already in preparation and they can not be altered. Instead we can produce the card in the next station (station 2) that becomes available.

### 4.5   Cost of Error Recovery

An upper bound on the number of time units spent recovering from an error can be calculated as follows.

1. Once an error is detected by the tester, one step forward may be necessary if it is an error like in Example 1.
2. To reproduce a replacement card we will require $S = M$ time units, during this time the belt rolls back to the station.
3. Once the card is reproduced, it will take another M time units for the new card to reach the tester. In practice the belt can move forward faster than M time units, and the time spent to reach the tester will be smaller.

Thus, based on the above observation, $2.M + 1$ time units are required in the worst case to recover from a single error. It is possible to tighten this upper bound by introducing fast forward and fast backward moves.

# 5  Conclusions

Using SMV, we rediscovered the super single mode that has previously been
patented by Cybernetix. This result gives us a new evidence that model checking
can also be useful as a design aid for new machines. Our approach also allowed
us to generate defective card treatments, that may arise due to damaged cards
and wrong personalization. The present work shows error treatments for single
error, we believe the same technique can be easily extended to multiple error
treatment. In this way, using model checking, we go beyond scheduler synthesis
and actually solve a control synthesis problem.

The input language of Cadence SMV is sufficiently expressive to encode in
a natural and compact way a simplified model of the personalization machine.
However, safety and liveness properties for multiple error treatments (of single
or multiple types) are complicated to express in temporal logic, especially when
dealing with the uncontrolled plant. Nevertheless, by decreasing the degree of
uncontrollability of the plant, we believe multiple errors can be handled and
more complex discrete time models of the actual Cybernetix design (including
the controller) can be described.

A possible disadvantage of our approach is that the SMV descriptions are
difficult to understand for people who are not familiar with formal methods
(unlike say Petri nets). However, a clear advantage is that our description can
serve directly as input for a powerful model checker.

# References

[1] K. Åkesson and M. Fabian. Implementing supervisory control for chemical batch
    processes. In International Conf. on Control Applications, pages 1272–1277. IEEE
    Computer Society Press, 1999. 191
[2] Sarah Albert. Cybernetix case study – informal description. Technical report,
    Cybernétix - LIF, 2002. Available through URL http://ametist.cs.utwente.nl.
    190, 192, 199
[3] B. A. Brandin. The real-time supervisory control of an experimental manufactur-
    ing cell. In IEEE Transactions on Robotics and Automation, volume 12, pages
    1–13, 1996. 191
[4] J. R. Burch, E. M. Clarke, K. L. McMillan, D. L. Dill, and L. J. Hwang. Symbolic
    model checking $10^{20}$ states and beyond. Information and Computation, 98(2):142–
    170, June 1992. 189
[5] C. G. Cassandras and S. Lafortune. Introduction to Discrete Event Systems.
    KLuwer Academic, 1999. 191
[6] E. W. Dijkstra. Self-stabilizing systems in spite of distributed control. Communi-
    cations ACM, 17:643–644, 1974. 197
[7] Ansgar Fehnker. Scheduling a steel plant with timed automata. In Sixth
    International Conference on Real-Time Computing Systems and Applications
    (RTCSA '99), Hong Kong, China, pages 280–287. IEEE Computer Society Press,
    1999. 191

[8] Biniam Gebremichael and Frits Vaandrager. Control synthesis for a smart card personalization system using symbolic model checking. Technical Report NIII-R0312, Nijmegen Institute for Computing and Information Sciences, University of Nijmegen, 2003. 190, 194, 200

[9] J. Gunnarsson. Symbolic Methods and Tools for Discrete Event Dynamic Systems. PhD thesis, Linköping Studies in Science and Technology, 1997. 191

[10] Thomas Hune, Kim G. Larsen, and Paul Pettersson. Guided Synthesis of Control Programs using UPPAAL. Nordic Journal of Computing, 8(1):43–64, 2001. 191

[11] Tomas Krilavicius and Yaroslav Usenko. Smart card personalisation machine in UPPAAL and $\mu$CRL, 2003. In preparation. 191

[12] Y. Li and W. M. Wonham. Control of vector discrete-event systems I - the base model. In IEEE Trans. on Automatic Control, volume 38, pages 1214–1227. IEEE Computer Society Press, 1993. 191

[13] A. Mader. Deriving schedules for the cybernetix case study, 2003. Available through URL http://ametist.cs.utwente.nl. 191

[14] K. L. McMillan. Symbolic Model Checking: An Approach to the State Explosion Problem. Kluwer Academic Publishers, 1993. 190

[15] Peter Niebert and Sergio Yovine. Computing optimal operation schemes for multi batch operation of chemical plants. In Nancy A. Lynch and Bruce H. Krogh, editors, Hybrid Systems: Computation and Control, Third International Workshop, HSCC '00, volume 1790 of Lecture Notes in Computer Science, pages 338–351. Springer, 2000. 191

[16] P. J. G. Ramadge and W. M. Wonham. The control of discrete event systems. Proceedings of the IEEE, 77:81–98, 1989. 189

[17] Theo Ruys. Optimal Scheduling Using Branch and Bound with SPIN 4.0. In Thomas Ball and Sriram K. Rajamani, editors, Model Checking Software – Proceedings of the 10th International SPIN Workshop (SPIN 2003), volume 2648 of Lecture Notes in Computer Science, pages 1–17, Portland, OR, USA, May 2003. Springer-Verlag, Berlin. 191

[18] M. Sampath, R. Sengupta, S. Lafortune, K. Sinnamohideen, and D. Teneketzis. Diagnosability of discrete event systems. IEEE Transactions on Automatic Control, 40(9):1555–1575, September 1995. 191

[19] R. S. Sreenivas and B. H. Krogh. On condition/event systems with discrete state realizations. In Discrete Event Dynamic Systems. Theory and Applications 1, pages 209–236. Flumer Academic, 1991. 191

[20] G. Tel. Introduction to Distributed Algorithms. Cambridge University Press, 1994. 197

[21] Gera Weiss. Modeling smart-card personalization machine with LSCs. Research report, Weizmann, 2003. 192

[22] W. M. Wonham Y. Li. Control of vector discrete-event systems II - controller synthesis. In IEEE Transactions on Autom. Control, volume 39, pages 512–531, 1994. 191

[23] Z. Zhang and W. M. Wonham. STCT: An efficient algorithm for supervisory control design. In Symposium on Supervisory Control of Discrete Event Systems, 2001. 191

# On Timing Analysis of Combinational Circuits[*]

Ramzi Ben Salah, Marius Bozga, and Oded Maler

VERIMAG
2, av. de Vignate, 38610 Gieres, France
{ramzi.salah,marius.bozga,oded.maler}@imag.fr

**Abstract.** In this paper we report some progress in applying timed automata technology to large-scale problems. We focus on the problem of finding maximal stabilization time for combinational circuits whose inputs change only once and hence they can be modeled using acyclic timed automata. We develop a "divide-and-conquer" methodology based on decomposing the circuit into sub-circuits and using timed automata analysis tools to build conservative low-complexity approximations of the sub-circuits to be used as inputs for the rest of the system. Some preliminary results of this methodology are reported.

## 1 Introduction

It is well known that timed automata (TA) [AD94] are well suited for modeling delays in digital circuits [D89, L89, MP95]. Although some applications of TA technology for solving timing-related problems for such circuits have been reported [MY96, BMPY97, TKB97, TKY+98, BMT99, BJMY02], the state- and clock-explosion associated with such models, restricted the applicability of TA to small circuits. In this work we try to treat larger combinational circuits by using the old-fashioned recipe of abstraction and approximation. When viewed from a purely-functional point of view, combinational circuits realize instantaneous Boolean functions. However, when gate delays are taken into account, the computation of that function is not considered anymore as an atomic action but rather as a process where changes in the inputs are gradually propagated to the outputs. The question of finding the worst-case propagation delay of the circuit, that is, the maximal time that may elapse between a change in the inputs and the last change in the outputs, is of extreme practical importance as it determines, for example, the frequency of the clock with which a circuit can operate. Static techniques currently practiced in industry are based on finding the longest (in terms of accumulated delays) path from inputs to outputs in the circuit. While these bounds are easy to compute (polynomial in the size of the circuit), they can be over pessimistic because they abstract from the particular logic of the circuit which may prevent such longest paths from being exercised.[1]

---

[*] This work was partially supported by a grant from Intel and by the European Community Projects IST-2001-35304 AMETIST (Advanced Methods for Timed Systems), http://ametist.cs.utwente.nl

[1] A lot of effort has been invested in the problem of detecting such "false paths".

K.G. Larsen and P. Niebert (Eds.): FORMATS 2003, LNCS 2791, pp. 204–219, 2004.

On the other hand, models based on timed automata do express the interaction between logic and timing and hence can lead to more accurate results. Alas, TA-based techniques are still very far from being applicable to industrial-size circuits.

The present paper attempt to find a better trade-off between accuracy and tractability by using timed automata as an underlying semantic model and by applying abstraction techniques to parts of the circuit in order to build for them small over-approximating timed automata that can be plugged as inputs to other parts of the circuit. Our abstraction technique takes advantage of the acyclic nature of the circuits and their corresponding automata, which implies, among other things, that every variable changes finitely many times before stabilization in every run of the automaton.

The rest of the paper is organized as follows: in Section 2 we give a formal definition of circuits, their "languages" and the maximal stabilization time problem. In section 3 we explain the modeling of such circuits as timed automata. Section 4 is devoted to our abstraction technique, its properties and the way it is implemented using the tools IF/Kronos and Aldebaran. Preliminary experimental results are reported in Section 5 followed by a discussion of related work and future directions.

## 2    Timed Boolean Circuits

Throughout this paper we restrict ourselves to acyclic circuits.

**Definition 1 (Boolean Circuits).** *A Boolean circuit is $C = (V, \rightsquigarrow, F)$ where $V$ is a set of nodes, $\rightsquigarrow$ is an irreflexive and anti-symmetric binary relation and $F$ is a function that assigns to every non-input node $v$ a Boolean function $F_v$.*

Here $v \rightsquigarrow v'$ means that $v$ influences $v'$ directly. The transitive closure of $\rightsquigarrow$, $\stackrel{*}{\rightsquigarrow}$, induces a strict partial order $(V, \stackrel{*}{\rightsquigarrow})$ where the minimal elements are called *input* nodes and are denoted by $V_x$. The rest of the nodes are called non-input nodes and denoted by $V_y$. A subset $V_z$ of $V$ consists of output nodes, those that are observable from the outside. An example appears in Figure 1-(a). The set of immediate *predecessors* of a node is $\pi(v) = \{v' : v' \rightsquigarrow v\}$ and the set of its predecessors (backward cone) is $\pi^*(v) = \{v' : v' \stackrel{*}{\rightsquigarrow} v\}$.

By substitution we define for every node $v$ a function $G_v$ defined on the inputs in its backward cone. We will use $X = \mathbb{B}^{|V_x|}$, $Y = \mathbb{B}^{|V_y|}$ and $Z = \mathbb{B}^{|V_z|}$ to denote the sets of possible assignments to input, non-input and output nodes, respectively. The whole circuit can be viewed as computing a function $G : X \rightarrow Y$. The *stable state* of the circuit associated with an input vector $\mathbf{x} \in X$ is $\mathbf{y} = G(\mathbf{x})$.

This concludes the formalization of Boolean circuits and their functions. These functions are *instantaneous* with no notion of time. The next step is to lift them to functions (operators) on signals, that is, on functions that specify the evolution of a value over (continuous) time.

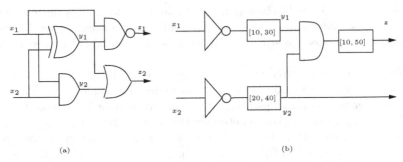

**Fig. 1.** (a) A Boolean circuit; (b) A timed Boolean circuit

**Definition 2 (Signals).** *Let $A$ be a set and let $\mathcal{T} = \mathbb{R}_+$ be a time domain. An $A$-valued signal over $\mathcal{T}$ is a partial function $\alpha : \mathcal{T} \to A$ whose domain of definition is an interval $[0, r)$ for some $r \in \mathcal{T}$.*

We use $\alpha[t]$ to denote the value of $\alpha$ at $t$. When $A$ is finite, signals are piecewise-constant and make discontinuous jumps at certain points in time. This is formalized as follows. The *left limit* of a signal $\alpha$ at time $t$ is defined as $\alpha[t^-] = \lim_{t' \to t} \alpha[t']$. For every piecewise-constant signal $\alpha$ we define:

- The ordered set of *jump points*, $\mathcal{J}(\alpha) = \{t : \alpha[t^-] \neq \alpha[t]\} = \{t_0, t_1, \ldots\}$.
- The set of *maximally-uniform intervals* $\mathcal{I}(\alpha) = \{I_1, I_2, \ldots\}$ where $I_i = [t_{i-1}, t_i)$ for $t_{i-1}, t_i \in \mathcal{J}(\alpha)$.

Clearly, the value of $\alpha$ is uniform over any subset of a maximally-uniform interval. We restrict our attention to well-behaving signals i.e. those for which $\mathcal{J}(\alpha)$ has finitely-many elements in any finite interval. We denote the set of $A$-valued signals by $\mathcal{S}(A)$.

When a gate or any other I/O device gets a signal as an input, it transforms it into an output signal. This is captured mathematically by what is called a *transducer*, or a *signal operator*, a function that maps signals to signals. We restrict such functions to be *causal*, that is, the value of the output at time $t$ can depend only on the value of the input in times $[0, t]$ and not on later values. The simplest type of operators are memoryless (instantaneous) operators defined as follows.

**Definition 3 (Memoryless Operators).** *A memoryless signal operator is a function $f : \mathcal{S}(A) \to \mathcal{S}(B)$ obtained as a pointwise extension of a function $f : A \to B$, that is, $\beta = f(\alpha)$ if $\beta[t] = f(\alpha[t])$ for every $t$ in the domain of $\alpha$.*

In reality, since gates are realized by continuous physical processes, it takes some time to propagate changes from input to output ports. To define this phenomenon mathematically we need the basic operator with memory for discrete-valued signals, the delay, which takes a signal and "shifts" it in time. One can define a variety of delay operators differing from each other in complexity and in

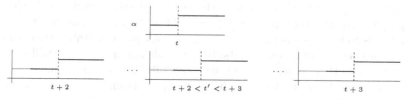

**Fig. 2.**   An input signal $\alpha$ and few of the elements of $D_{[2,3]}(0, \alpha)$

physical faithfulness. The class of models that we consider is called *bi-bounded* inertial delays [BS94] and is characterized by an interval $I = [l, u]$ which gives lower and upper bounds on the propagation delay. For the purpose of this paper we will use the model introduced in [MP95] but since the choice of the delay model is orthogonal to the rest of the methodology we will defer the exact definition of the operator to Section 3 where it will be defined in terms of its corresponding timed automaton and use meanwhile a general semi-formal definition.

**Definition 4 (Delay Operators).** *A delay operator is a non-deterministic function of the form $D_I : A \times \mathcal{S}(A) \to 2^{\mathcal{S}(A)}$ where $I = [l, u]$ is a parameter of the operator with $l > 0$. A signal $\beta$ is in $\Delta_I(b, \alpha)$ if*

1. *The value of $\beta$ is $b$ at the initial interval $[0, t)$;*
2. *Changes in $\alpha$ are not propagated to $\beta$ before $l$ time elapses;*
3. *Changes in $\alpha$ must be propagated to $\beta$ if they persist for $u$ time;*
4. *Changes in $\alpha$ that persist for less then $l$ time are not propagated at all to $\beta$.*

Figure 2 illustrates such an operator which, typically, will have uncountably-many output signals for an input signal. All signal operators can be lifted naturally into operators on sets of signals.

A timed circuit model is obtained from a Boolean circuit by connecting the output of every non-input node to a delay operator which models the delay associated with the computation of that node (see Figure 1-(b). In other words, a gate with a propagation delay is modeled as a composition of a memoryless Boolean operator and a delay operator (see [MP95]).

**Definition 5 (Timed Boolean Circuits).** *A timed Boolean circuit is $C = (V, \leadsto, F, I)$ where $(V, \leadsto, F)$ is a Boolean circuit and $I$ is a function assigning to every non-input node $v$ a delay interval $I_v = [l_v, u_v]$ with $0 < l_v \leq u_v < \infty$.*

The semantics of a timed circuit is given in terms of a non-deterministic transducer $F_C : Y \times \mathcal{S}(X) \to 2^{\mathcal{S}(Y)}$ such that $\beta \in F_C(\mathbf{y}, \alpha)$ if $\alpha$ and $\beta$ satisfy the set of signal inclusions associated naturally with the circuit [MP95] and $\mathbf{y}$ is the initial state of the non-input part of the circuit.

The stabilization time problem is motivated by the use of Boolean circuits in synchronous sequential machines (the hardware name for automata). At the beginning of every clock cycle new input values together with the values of memory elements (computed in the previous cycle) are fed into the circuit and the changes are propagated until the circuit stabilizes and the clock falls. The "width" of the clock needs to be large enough to cover the longest possible stabilization time of the circuit over all admissible inputs. In our modeling approach we will consider primary inputs that change at most once and within a bounded amount of time and hence, due to acyclicity and the finite upper-bounds associated with the delays, they induce finitely many changes throughout the circuit.

**Definition 6 (Ultimately-Constant Signals).** *A signal $\alpha$ is ultimately-constant (u.c.) if it has a finite number of jump points (i.e. there is some time $t$ such that the signal remains constant after $t$). The minimal such $t$ for $\alpha$ is called its stabilization time and is denoted by $\theta(\alpha)$. This definition extends to sets of signals by letting $\theta(L) = \max\{\theta(\alpha) : \alpha \in L\}$.*

The following properties hold for every *u.c.* signal $\alpha$:

1. The signal $f(\alpha)$ is also u.c. for every Boolean function $f$.
2. For a delay operator $D_I$ with $I = [l, u]$ and for every $\beta \in D_I(\alpha)$, $\theta(\beta) \leq \theta(\alpha) + u$.

Consequently, u.c. inputs to acyclic timed circuits produce u.c. outputs. Constant signals constitute a special class of u.c. signals and we will use $\alpha_{\mathbf{x}}$ to denote a signal whose value is constantly $\mathbf{x}$.

We can now define the problem of maximal stabilization time of a circuit with respect to a pair of input vectors $\mathbf{x}$ and $\mathbf{x}'$ where $\mathbf{x}$ is the input presented in the preceding cycle, and which determines the initial (stable) state, and $\mathbf{x}'$ is the value of a new constant signal. We denote by $L(C, \mathbf{x}, \mathbf{x}')$ the set of $Y$-signals $\beta \in F_C(\mathbf{y}, \alpha_{\mathbf{x}'})$ when the circuit is initialized with the stable state $\mathbf{y} = G(\mathbf{x})$.

**Definition 7 (Stabilization Time of a Circuit).** *Given a timed Boolean circuit $C = (V, \rightsquigarrow, F, I)$ and two input vectors $\mathbf{x}, \mathbf{x}' \in X$ the stabilization time associated with $(\mathbf{x}, \mathbf{x}')$ is $\theta(C, \mathbf{x}, \mathbf{x}') = \max\{\theta(\beta) : \beta \in L(C, \mathbf{x}, \mathbf{x}')\}$ and the maximal stabilization time of the circuit is $\theta(C) = \max\{\theta(C, \mathbf{x}, \mathbf{x}') : \mathbf{x}, \mathbf{x}' \in X\}$.*

## 3  Modeling with Timed Automata

Timed automata are automata augmented with continuous clock variables whose values grow uniformly at every state. Clocks can be reset to zero at certain transitions and tests on their values can be used in conditions for enabling transitions.

**Definition 8 (Timed Automaton).** *A timed automaton is $\mathcal{A} = (Q, C, I, \Delta)$ where $Q$ is a finite set of states, $C$ is a finite set of clocks, $I$ is the staying condition (invariant), assigning to every $q \in Q$ a conjunction $I_q$ of inequalities of the form $c \leq u$, for some clock $c$ and integer $u$, and $\Delta$ is a transition relation consisting of elements of the form $(q, \phi, \rho, q')$ where $q$ and $q'$ are states, $\rho \subseteq C$ and $\phi$ (the transition guard) is a conjunction of formulae of the form $(c \geq l)$ for some clock $c$ and integer $l$.*

A *clock valuation* is a function $\mathbf{v} : C \to \mathbb{R}_+ \cup \{0\}$ and a *configuration* of the automaton is a pair $(q, \mathbf{v})$ consisting of a discrete state (location) and a clock valuation. Every subset $\rho \subseteq C$ induces a reset function $\mathrm{Reset}_\rho$ on valuations which resets to zero all the clocks in $\rho$ and leaves the other clocks unchanged. We use $\mathbf{1}$ to denote the unit vector $(1, \ldots, 1)$ and $\mathbf{0}$ for the zero vector. We will use the term *constraints* to refer to both guards and staying conditions. A *step* of the automaton is one of the following:

- A discrete step: $(q, \mathbf{v}) \xrightarrow{\delta} (q', \mathbf{v}')$, for some transition $\delta = (q, \phi, \rho, q') \in \Delta$, such that $\mathbf{v}$ satisfies $\phi$ and $\mathbf{v}' = \mathrm{Reset}_\rho(\mathbf{v})$.
- A time step: $(q, \mathbf{v}) \xrightarrow{t} (q, \mathbf{v} + t\mathbf{1})$, $t \in \mathbb{R}_+$ such that $\mathbf{v} + t\mathbf{1}$ satisfies $I_q$.

A *run* of the automaton starting from a configuration $(q_0, \mathbf{v}_0)$ is a finite sequence of steps

$$\xi : \quad (q_0, \mathbf{v}_0) \xrightarrow{t_1} (q_1, \mathbf{v}_1) \xrightarrow{t_2} \cdots \xrightarrow{t_n} (q_n, \mathbf{v}_n).$$

We model timed circuits as a composition of timed automata such that each automaton may observe the states of other automata and refer to them in its transition guards and staying conditions.[2] The automaton for a Boolean gate of the form $y = f(x_1, x_2)$ is just a trivial one-state automaton that has self-looping transitions for all tuples $(x_1, x_2, y)$ that satisfy the equation. In fact, this is not really an automaton but an instantaneous logical constraint that must always be satisfied. The automaton for the delay operator $D_{[l,u]}$ (Figure 3) has four states, $0, 0', 1, 1'$. The 0 and 1 states are stable, that is, the values of the output of the delay is consistent with its input $x$. When at state 0, if the input changes to 1, the automaton moves to an unstable state $0'$ and resets a clock $C$ to zero. It can stay at $0'$ as long as $C < u$ and can switch to stable state 1 as soon as $C \geq l$. If the input changes back to 0 before the transition to 1 the automaton returns to 0. We call these three types of transitions *excite*, *stabilize* and *regret*, respectively. Note that states 0 and $0'$ are indistinguishable from the outside and another automaton will see a change from 0 to 1 only after the "stabilize" transition.

Composing all the automata, together with the model of their inputs we obtain a closed automaton as in Definition 8 whose semantics is identical to that of the timed circuit [MP95]. To be more precise, an automaton whose semantics is $L(C, \mathbf{x}, \mathbf{x}')$ is obtained by letting the initial state be the stable state corresponding to $G(\mathbf{x})$ and composing it with a static automaton for the input $\mathbf{x}'$. The obtained automaton is acyclic and all paths converge in finite time to the only stable state that corresponds to $G(\mathbf{x}')$. The maximal stabilization time is hence the maximal time that the automaton can stay in any unstable state. Note that in such a state at least one of the components is in a $0'$ or $1'$ state and hence its staying condition forces it to leave the state.

We recall some definitions commonly-used in the verification of timed automata [HNSY94, Y97, LPY97, BDM+98, A99]. A *zone* is a set of clock valuations consisting of points satisfying a conjunction of inequalities of the form $c_i -$

---

[2] To avoid over-formalization we do not define "open" interacting automata. Such definitions can be found in [MP95].

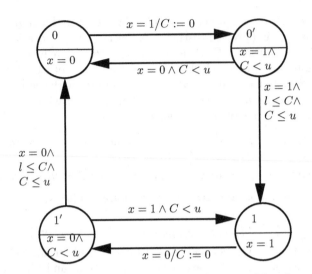

**Fig. 3.** The timed automaton for a delay element. The $x$ variable refers to the observable state of the input automaton which is 0 at $\{0, 0'\}$ and 1 at $\{1, 1'\}$

$c_j \geq d$ or $c_i \geq d$. A *symbolic state* is a pair $(q, Z)$ where $q$ is a discrete state and $Z$ is a zone. It denotes the set of configurations $\{(q, \mathbf{z}) : \mathbf{z} \in Z\}$. Symbolic states are closed under the following operations:

- The *time successor* of $(q, Z)$ is the set of configurations which are reachable from $(q, Z)$ by letting time progress without violating the staying condition of $q$:
$$Post^t(q, Z) = \{(q, \mathbf{z} + r\mathbf{1}) : \mathbf{z} \in Z, r \geq 0, \mathbf{z} + r\mathbf{1} \in I_q\}.$$
We say that $(q, Z)$ is *time-closed* if $(q, Z) = Post^t(q, Z)$.
- The *$\delta$-transition successor* of $(q, Z)$ is the set of configurations reachable from $(q, Z)$ by taking the transition $\delta = (q, \phi, \rho, q') \in \Delta$:
$$Post^\delta(q, Z) = \{(q', \text{Reset}_\rho(\mathbf{z})) : \mathbf{z} \in Z \cap \phi\}.$$

- The *$\delta$-successor* of a time-closed symbolic state $(q, Z)$ is the set of configurations reachable by a $\delta$-transition followed by passage of time:
$$Succ^\delta(q, Z) = Post^t(Post^\delta(q, Z)).$$

The forward reachability algorithm for TA starts with an initial zone and generates all successors until termination, while doing so it generates the *reachability graph* (also known as the simulation graph).

**Definition 9 (Reachability Graph).** *The reachability graph associated with a timed automaton starting from a state $s$ is a directed graph $S = (N, \rightarrow)$ such that $N$ is the smallest set of symbolic states containing $Post^t(s, \{\mathbf{0}\})$ and closed under $Succ^\delta$. The edges are all pairs of symbolic states related by $Succ^\delta$.*

The fundamental property of the reachability graph is that it admits a path from $(q, Z)$ to $(q', Z')$ if and only if for every $\mathbf{v}' \in Z'$ there exists $\mathbf{v} \in Z$ and a run of the automaton from $(q, \mathbf{v})$ to $(q', \mathbf{v}')$. Hence the union of all reachable symbolic states gives exactly the reachable configurations.

To compute the maximal stabilization time we add an auxiliary clock $T$ which is never reset to zero and hence in every reachable configuration its value represents the total time elapsed since the beginning of the run. The maximal value of $T$ over all reachable symbolic states $(q, Z)$ with $q$ unstable is the maximal stabilization time (note that due to acyclicity the value of $T$ is bounded in all unstable states). Hence, the problem of maximal stabilization time can, in principle, be solved using standard TA verification tools.

## 4    The Abstraction Technique

Given the complexity of TA verification we move to an abstraction methodology based on the following simple idea. We decompose the circuit into sub-circuits small enough to be handled completely by TA verification tools. We take the automaton $\mathcal{A}$ which corresponds to such a sub-circuit and use its reachability graph to construct an automaton $\hat{\mathcal{A}}$ having two important properties:

1. The set $L(\hat{\mathcal{A}})$ of signals that it generates is a reasonable over-approximation of the projection of $L(\mathcal{A})$ on the output variables of the sub-circuit.
2. It is much smaller than $\mathcal{A}$ in terms of states and clocks.

Hence if we replace $\mathcal{A}$ by $\hat{\mathcal{A}}$ as a model of the sub-circuit we are guaranteed to over-approximate the semantics of the circuit and hence to over-approximate the stabilization time.

To better understand the technique it is worth looking at the reachability graph from a different angle. In timed automata, as in any other automata augmented with auxiliary variables, the transition graph is misleading because a discrete state stands for many possible clock valuation which may differ in the constraints they satisfy and hence in the behaviors that can be generated from them. It might be the case that a state $q$ will never be reached with a clock valuation satisfying some transition guard and hence the corresponding transition will never be taken. By performing the reachability algorithm for $\mathcal{A}$ starting from an initial state we obtain a graph which represents the "feasible part" of $\mathcal{A}$, excluding behaviors that violate timing constraints. Figure 5-(a) shows the

(a)                                    (b)

**Fig. 4.** Projection on the absolute time introduces spurious runs

reachability graph for the circuit of Figure 1-(b) where the inputs change from $(0, 1)$ to $(1, 0)$. In fact the reachability graph can serve as a skeleton of another timed automaton $\mathcal{A}'$ whose semantics in terms of runs is equivalent to that of $\mathcal{A}$. To see that, one just has to associate with each symbolic state $(q, Z)$ the staying condition $Z$ and label each transition $(q, Z) \xrightarrow{\delta} (q', Z')$ by the guard and reset of $\delta$. The resulting automaton $\mathcal{A}'$ differs from $\mathcal{A}$ in two aspects: certain states of $\mathcal{A}$ are split into several copies according to clock values, and all transitions that are not possible in $\mathcal{A}$ due to timing constraints do not appear in $\mathcal{A}'$ at all.

Now if we relax some timing constraints in $\mathcal{A}'$ we may introduce spurious behaviors that violate these constraints, however we will *not* add any new *qualitative* behavior (sequence of events) that was not possible in $\mathcal{A}$ because such behaviors have already been eliminated while computing the reachability graph. The most straightforward way to relax timing constraints is to project the constraints on a subset of the clocks and discard the rest. In particular if we throw away all clocks except $T$ which measures the absolute time, the relaxed guard for any transition will be of the form $T \in [t_1, t_2]$. Clearly, a transition can be taken in the new automaton iff there is a run of the original automaton in which the corresponding transition could be taken at some time $t \in [t_1, t_2]$. However, this abstraction can add additional runs which are impossible in the original automaton as the following example shows. Consider the automaton of Figure 4-(a) where the first transition could take place in $[l_1, u_1]$ while the second can take place between $l_2$ and $u_2$ *after* the occurrence of the first. Applying the above procedure we obtain the automaton of Figure 4-(b) where the second transition could be taken anywhere in $[l_1 + l_2, u_1 + u_2]$ regardless of the time of the first.

The next step is to hide transitions which are not observable from the outside, i.e. all transitions of non-output variable and all non-visible transitions ("excite" and "regret") of the output variables $y_2$ and $z$. The one-clock automaton thus obtained for our example appears in Figure 5-(b). We then apply a minimization algorithm which merges states that are indistinguishable with respect to the remaining visible transitions. More formally we consider the congruence relation $\sim$ on the nodes of the labeled reachability graph defined as the largest relation satisfying:

$$q_1 \sim q_2 \text{ iff } \forall \delta, I \quad q_1 \xrightarrow{\tau^* \cdot (\delta, I)} q_1' \Rightarrow (\exists q_2' \text{ s.t. } q_2 \xrightarrow{\tau^* \cdot (\delta, I)} q_2' \wedge q_1' \sim q_2'). \quad (1)$$

Here $(\delta, I)$ stands for a transition-interval pair and $\tau^*$ to an arbitrary sequence of unobservable transition. This relation is the "safety bisimulation" of [BFG+91]. The minimized automaton, whose states are congruence classes of $\sim$, can be seen in Figure 6-(a).

Relation (1) looks at transition labels in a purely-syntactic manner, that is, the label $-y_2[20, 30]$ in Figure 6-(a) is considered distinct from $-y_2[20, 40]$ and hence the transitions are not merged. To obtain a more aggressive abstraction we define a weaker equivalence $\sim'$ that ignores differences in intervals:

$$q_1 \sim' q_2 \text{ iff } \forall \delta, I \quad q_1 \xrightarrow{\tau^* \cdot (\delta, I)} q_1' \Rightarrow (\exists q_2', I' \text{ s.t. } q_2 \xrightarrow{\tau^* \cdot (\delta, I')} q_2' \wedge q_1' \sim' q_2'). \quad (2)$$

**Table 1.** Maximal stabilization time for all input pairs for the circuit of Figure 8

| x | 00 | | | 01 | | | 10 | | | 11 | | |
|---|---|---|---|---|---|---|---|---|---|---|---|---|
| x′ | 10 | 01 | 11 | 11 | 00 | 10 | 00 | 11 | 01 | 01 | 10 | 00 |
| stab-time | 510 | 340 | 340 | 170 | 510 | 425 | 510 | 0 | 255 | 255 | 0 | 510 |

The states of the minimized automaton are equivalence classes of $\sim'$ and the transitions between these classes are labeled by $(\delta, \bar{I})$ where $\bar{I}$ is the join (convex hull) of all the intervals $I_i$ such that there are transition labeled by $(\delta, I_i)$ between elements of the corresponding classes (see Figure 7).[3] The result of minimization with respect to $\sim'$ appears in Figure 6-(b) and one can see that it gives a succinct over-approximation of the behavior of $y_2$ and $z$.

We have implemented the above mentioned technique. Our tool chain starts with a circuit description as Boolean equations with delays and generates from it automatically a network of interacting timed automata written in the IF format [BGM02]. After generating the reachability graph with the interval labels we apply the Aldebaran tool set ([BFKM97]), slightly modified to implement minimization with respect to $\sim'$ to obtain the abstract model.

## 5   Experimental Results

We have conducted some preliminary experiments with our approach on some sample circuits that we have constructed. First, to demonstrate the semantic advantage of timed automata we analyzed the circuit of Figure 8 which has a false path. We use delays of $[83, 85]$ for all gates (except the inverters that have zero delay) and compare our results with static timing analysis which gives stabilization time of $7 \times 85 = 594$. Since our method works is currently restricted to one pair of input vectors, we repeat the analysis for all 12 pairs and obtain the results of Table 1. As one can see, the TA-based analysis discovers that the maximal stabilization time is only $6 \times 85 = 510$.

The major set of experiments was conducted on circuits consisting of a sequential concatenation of an increasing number of copies of the circuit of Figure 1-(a) (the $y_3$ and $y_4$ of stage $n$ are the $x_1$ and $x_2$ of stage $n+1$). We assume that input $x_1$ may rise anywhere in $[10, 35]$ and $x_2$ in $[15, 63]$. In general, the complexity of the reachability graph is sensitive to the choice of delay bounds: for an interval $[l, u]$, the larger is the ratio $(u - l)/l$, more "scenarios" are possible and transitions at "deep" gates can precede transitions in gates closer to the input.[4] Table 2 shows the performance of our technique (computation time and size of the reachability graph) as a function of the number of stages for three choices of gate delay intervals $[1, 2]$, $[10, 12]$ and $[100, 102]$. All the experiments

---

[3] Another choice might be to join only intervals that have a non-empty intersection.
[4] In fact, if we assume no lower-bound on the delay (the "up-bounded" model of [BS94]), events can happen in any order.

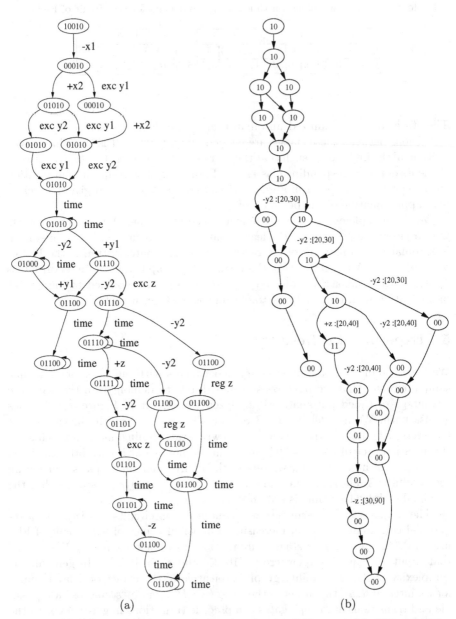

**Fig. 5.** (a) The reachability graph for the circuit of Figure 1-(b). The transition labels exc z, reg z, +z and -z correspond, respectively, to excitation, regret, rising and falling of the variable z. (b) The corresponding one-clock automaton after hiding internal transitions. The label +z[20,30] means that z may change from 0 to 1 anytime inside the interval [20, 30]

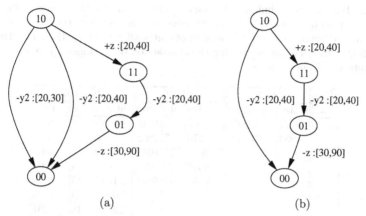

**Fig. 6.** (a) The results of applying standard minimization. (b) The result of minimization with interval fusion

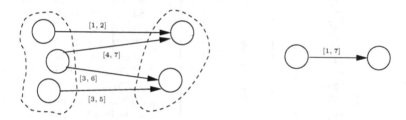

**Fig. 7.** Minimization by joining intervals

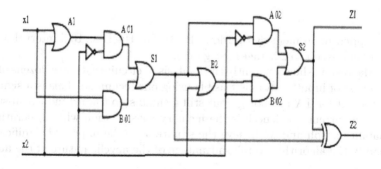

**Fig. 8.** A circuit with a false path

**Table 2.** Testing our technique with varying delay bounds. The 'states' column indicates the number of symbolic states in the model of stage $n$ before the last minimization and the 'min' columns indicate the number of states after minimization. The 'time' column indicates the time for computing the abstraction of all stages up to $n-1$ and the reachability graph for stage $n$

| $[l, u]$ | [1, 2] | | | [10, 12] | | | [100, 102] | | |
|---|---|---|---|---|---|---|---|---|---|
| no. | states | min | time | states | min | time | states | min | time |
| 1 | 71 | 4 | 0:01 | 65 | 3 | 0:01 | 65 | 3 | 0:01 |
| 2 | 934 | 12 | 0:02 | 270 | 7 | 0:02 | 270 | 7 | 0:02 |
| 3 | – | – | – | 2690 | 11 | 0:03 | 2690 | 11 | 0:04 |
| 4 | | | | 5397 | 23 | 0:05 | 4080 | 16 | 0:06 |
| 5 | | | | 217951 | 144 | 9:44 | 21498 | 30 | 0:12 |
| 6 | | | | – | – | – | 50543 | 39 | 0:30 |
| 7 | | | | | | | 73502 | 48 | 1:01 |
| 8 | | | | | | | 95619 | 57 | 1:54 |
| 9 | | | | | | | 117736 | 66 | 3:12 |
| 10 | | | | | | | 139853 | 75 | 5:08 |
| 11 | | | | | | | 161970 | 84 | 7:32 |
| 12 | | | | | | | 184087 | 93 | 10:05 |
| 13 | | | | | | | 206204 | 102 | 14:42 |
| 14 | | | | | | | 228321 | 111 | 20:39 |
| 15 | | | | | | | 250438 | 120 | 28:15 |
| 16 | | | | | | | 272555 | 129 | 36:46 |
| 17 | | | | | | | 117736 | 138 | 49:36 |
| 18 | | | | | | | 316789 | 147 | 1:04:04 |
| 19 | | | | | | | 338906 | 156 | 1:21:48 |
| 20 | | | | | | | 361023 | 165 | 1:42:59 |
| 21 | | | | | | | 383140 | 174 | 1:58:56 |
| 22 | | | | | | | 405257 | 183 | 2:30:31 |
| 23 | | | | | | | – | – | – |

were stopped upon memory overflow (1GB). For the $[100, 102]$ interval we were able to analyze up to 22 stages (88 gates).

As the results show, currently the analysis of circuits with few dozens of gates for one pair of input vectors is feasible using our technique. This is a significant improvement for TA technology but still a small step toward industrial-size circuits. The current bottleneck is the memory consumption while generating the reachability graph and we believe the situation can be improved significantly if we modify the algorithm to take advantage of the acyclic nature of the automaton.

## 6    Discussion

There have been numerous publications on abstraction in general and abstraction of timed systems in particular, e.g. [AIKY95, WD94, B96, PCKP00], some based on relaxing the timing constraints and refining them successively if the abstract system cannot be verified. In [TAKB96] an assume-guarantee framework is defined for timed automata, which is used later to verify a multi-stage asynchronous circuit [TB97] by using small abstractions for each stage. These abstractions are generated manually. The closest work to ours is [ZMM03] which uses timed Petri nets for describing circuits and their desired properties. To abstract a circuit they apply "safe transformations" that consist of hiding of internal actions and clocks, and possibly over-approximating the set of behaviors. This work does is not specialized to acyclic circuits and the formal properties of the abstraction (defined in terms of *trace theory*) seem to be more complicated. Other attempts to solve the maximal stabilization time using TA are reported in [TKB97, TKY+98].

Due to space limitation we do not discuss here possible variation of the techniques such as different abstraction styles, nor other important ingredients of the methodology such as the partitioning strategy. The adaptation of the technique to cyclic circuits and to open systems in general is a very challenging goal whose achievement can have a big impact on the design of timed systems.

## Acknowledgement

Many colleagues have contributed throughout the years to the work described in this paper. In particular, Sergio Yovine and Stavros Tripakis helped us a lot in understanding various aspects of the verification of timed automata. They also participated in previous efforts to apply it to circuit analysis. The development of the abstraction methodology was carried out at certain points in time by Nishant Sinha, Fadhel Graiet, Olfa Ben Sik Ali and Bara Diop. Comments and criticism made by Avi Efrati, Ken Stevens and anonymous referees improved the quality of the paper.

## References

[A99]      R. Alur, Timed Automata, Proc. CAV '99 LNCS 1633, 8-22, Springer, 1999. 209

[AD94]     R. Alur and D. L. Dill, A Theory of Timed Automata, Theoretical Computer Science 126, 183–235, 1994. 204

[AIKY95]   R. Alur, A. Itai, R. P. Kurshan and M. Yanakakis, Timing Verification by Successive Approximation, Information and Computation 118, 142-157, 1995. 216

[B96]      F. Balarin, Approximate Reachability Analysis of Timed Automata, Proc. RTSS '96, 52-61, IEEE, 1996. 216

[BGM02]    M. Bozga, S. Graf and L. Mounier, IF-2.0: A Validation Environment for Component-Based Real-Time Systems, In Proc. of CAV '02, LNCS 2404, Springer, 2002. 213

[D89]      D. Dill, Timing Assumptions and Verification of Finite-State Concurrent Systems, in Automatic Verification Methods for Finite State Systems, LNCS 407, Springer, 1989. 204

[BFKM97]   M. Bozga, J.-C. Fernandez, A. Kerbrat and L. Mounier, Protocol Verification with the Aldebaran Toolset, Software Tools for Technology Transfer 1, 166-183, 1997. 213

[BDM+98]   M. Bozga, C. Daws, O. Maler, A. Olivero, S. Tripakis, and S. Yovine, Kronos: a Model-Checking Tool for Real-Time Systems, Proc. CAV '98, LNCS 1427, Springer, 1998. 209

[BJMY02]   M. Bozga, H. Jianmin, O. Maler and S. Yovine, Verification of Asynchronous Circuits using Timed Automata, ENTCS 65, 2002. 204

[BFG+91]   A. Bouajjani, and J.-C. Fernandez, S. Graf, C. Rodriguez and J. Sifakis, Safety for Branching Time Semantics, Proc. ICALP '91, LNCS 510, Springer, 1991. 212

[BMPY97]   M. Bozga, O. Maler, A. Pnueli, S. Yovine, Some Progress in the Symbolic Verification of Timed Automata, in Proc. CAV '97, 179-190, LNCS 1254, Springer, 1997. 204

[BMT99]    M. Bozga, O. Maler and S. Tripakis, Efficient Verification of Timed Automata using Dense and Discrete Time Semantics, in Proc. CHARME'99 , 125-141, LNCS 1703, Springer, 1999. 204

[BS94]     J. A. Brzozowski and C-J. H. Seger, Asynchronous Circuits, Springer, 1994. 207, 213

[HNSY94]   T. Henzinger, X. Nicollin, J. Sifakis, and S. Yovine, Symbolic Model-checking for Real-time Systems, Information and Computation 111, 193–244, 1994. 209

[L89]      H. R. Lewis, Finite-state Analysis of Asynchronous Circuits with Bounded Temporal Uncertainty, TR15-89, Harvard University, 1989. 204

[LPY97]    K. G. Larsen, P. Pettersson and W. Yi, UPPAAL in a Nutshell, Software Tools for Technology Transfer 1/2, 1997. 209

[MP95]     O. Maler and A. Pnueli, Timing Analysis of Asynchronous Circuits using Timed Automata, in Proc. CHARME '95, LNCS 987, 189-205, Springer, 1995. 204, 207, 209

[MY96]     O. Maler and S. Yovine, Hardware Timing Verification using KRONOS, In Proc. 7th Israeli Conference on Computer Systems and Software Engineering, 1996. 204

[PCKP00]   M. A. Pena, J. Cortadella, A. Kondratyev and E. Pastor, Formal Verification of Safety Properties in Timed Circuits, Proc. Async'00, 2-11, IEEE Press, 2000. 216

[TAKB96]   S. Tasiran R. Alur, R. P. Kurshan and R. Brayton, Verifying Abstractions of Timed Systems, in Proc. CONCUR '96, 546-562, Springer, 1996. 217

[TB97]     S. Tasiran and R. K. Brayton, STARI: A Case Study in Compositional and Hierarchical Timing Verification, in Proc. CAV '97, 191-201, LNCS 1254, Springer, 1997. 217

[TKB97]    S. Tasiran, Y. Kukimoto and R. K. Brayton, Computing Delay with Coupling using Timed Automata, Proc. TAU '97, 1997. 204, 217

[TKY+98]   S. Tasiran, S. P. Khatri, S. Yovine, R. K. Brayton and A. Sangiovanni-Vincentelli, A Timed Automaton-Based Method for Accurate Computation of Circuit Delay in the Presence of Cross-Talk, FMCAD '98, 1998. 204, 217

[WD94]     H. Wong-Toi and D. L. Dill, Approximations for Verifying Timing Properties, in Theories and Experiences for Real-Time System Development, World Scientific Publishing, 1994. 216

[Y97]      S. Yovine, Kronos: A verification tool for real-time systems, International Journal of Software Tools for Technology Transfer 1, 1997. 209

[ZMM03]    H. Zheng, E. Mercer and C. Myers, Modular Verification of Timed Circuits using Automatic Abstraction, IEEE Trans. on CAD, to appear, 2003.   217

# Analysis of Real Time Operating System Based Applications

Libor Waszniowski and Zdenek Hanzalek

Czech Technical University
Centre for Applied Cybernetics, Department of Control Engineering
Karlovo nám. 13, 121 35 Prague 2, Czech Republic
{hanzalek,xwasznio}@fel.cvut.cz

**Abstract.** This text is dedicated to modelling of real-time applications running under multitasking operating system. Theoretical background is based on timed automata by Alur and Dill. As this approach is not suited for modelling pre-emption we focus on cooperative scheduling. In the addition, interrupt service routines are considered, and their enabling/disabling is controlled by interrupt server considering the specified server capacity. The server capacity has influence on the margins of the computation times in the application processes. Such systems, used in practical real-time applications, can be modelled by timed automata and further verified since their reachability problem and model checking of TCTL problem is decidable. Use of this methodology is demonstrated on the case study.

## 1 Introduction

The aim of this article is to show, how timed automata [1] can be applied to modelling of real time software applications running under operating system with cooperative scheduling. The application under consideration consists of several process, it includes mechanisms for interrupt handling, and it uses inter-process communication primitives like semaphores, queues etc. Model checking theory based on timed automata and implemented in model checking tools (e.g. UPPAAL[2]) can be used for verifying time parameters or safety and liveness properties of proposed models.

Timing analysis of software (especially with concurrency and synchronisation) is not trivial problem and it requires sophisticated methods and analysis tools. Several special purpose methods have been developed in the area of real time scheduling [3],[7]. These methods, e.g. rate monotonic analysis (RMA) [4], are very successful for analysis of systems with periodic processes. To deal with non-periodic processes, the standard method is to consider the non-periodic process as the periodic one using the minimal inter-arrival time as process period. The analysis based on such model is too pessimistic in some cases since inter-arrival times can vary over time [13]. Incorporation of inter-process communication primitives leads to pessimistic results as well since it does not model any internal process structure and therefore worst-case blocking time must be considered, even though it can never occur (see section 7).

To achieve more precise analysis, process models allowing more precise and complex timing constraints are needed. In [13] the timed automata are extended by

K.G. Larsen and P. Niebert (Eds.): FORMATS 2003, LNCS 2791, pp. 219-233, 2004.

asynchronous processes (i.e. processes triggered by events) to provide model for event-driven systems, which is further used for schedulability analysis. Processes (in [13] called tasks) associated to locations of timed automaton are executable programs characterised by its worst-case execution time, deadline and other parameters for scheduling (e.g. priority). Transition leading to a location in such automaton denotes an event triggering the process and the guard on transition specifies the possible arrival times of the event. Released processes are stored in a process queue and they are assumed to be executed according to a given scheduling strategy. Both non-preemptive and preemptive scheduling strategies are allowed. Such modell can deal with non-periodic processes in more accurate manner than RMA. Moreover there is a possibility to model internal process structure as it is shown in section 2, but the computation time of modelled blocks of code cannot vary.

This drawback is overcome by more detailed process model proposed in [9] providing a method for constructing models of real time Ada tasking programs. Time, safety or liveness properties of produced model based on constant slope linear hybrid automata can be automatically analysed by HyTech verifier. The state of the hybrid automaton consists of various state variables representing an abstraction of program's state and it contains also continuous variables used to measure the amount of processor time allocated to each process. A transition of the hybrid automaton represents execution of the sequential code segment. The timing constraints of the transition are derived from the time bounds of the corresponding code. Even thought the author reports that the analysing algorithm does usually terminate in practice, the reachability problem for hybrid automata is undecidable in general.

Hybrid automaton (or some of its subclass e.g. stopwatch automaton [10]) is needed to model premption since it is necessary to accumulate computing time of each process separately. The continuous variable used to measure the amount of CPU time allocated to each process must progress when the corresponding process is executed and must be stopped when the corresponding process is preempted. Such behaviour cannot be modelled by timed automaton that does not allow stopping of the clock variable (see [1]).

Based on these observations we provide the model of real time system consisting of several concurrent processes scheduled by cooperative scheduler. Since the internal structure of the processes and the scheduler are modelled by timed automata, the model of the system is more accurate than the models used for schedulability analysis (RMA and timed automata extended by processes). Opposite to the model of the system with preemption based on hybrid automata, this approach has guarantied termination of verification algorithm due to decidability of reachability problem and model checking of timed computation tree logic (TCTL) problem. Moreover timed automata are one of the most studied models for real time systems and several model checkers are available (e.g. Kronos[1] and UPPAAL[2] [2]).

Preemptive schedulers are known to provide higher utilisation of processor than the cooperative ones [3]. On the other hand the processor utilisation is less important criterion when the schedulability can be proven for a given set of processes under cooperative policy. Moreover the cooperative scheduling has some advantages

---

[1] http://www-verimag.imag.fr/TEMPORISE/kronos/
[2] http://www.uppaal.com

important especially for hard real time applications where the highest reliability is required. In cooperative scheduling, process specifies when it is willing to release CPU to another process. Then it is easy to make sure that all data structures are in a defined state. Applications using cooperative scheduling are therefore easier to program and to debug. In this paper we present another important advantage of cooperative scheduling that is possibility to create mathematical model of the application based on timed automata and to verify its time, safety and liveness properties.

The rest of this paper is organised as follows: section 2 illustrate on an example of scheduling anomaly that when one wants to make use of the internal process structure, then it is needed to consider also lower margins of computation times. Sections 3 and 4 represents a marginal part of this article. They deal with modelling of applications running under operating system based on cooperative scheduling. Since interrupt handling can play important role in such systems, they are taken into consideration in section 5. Section 0 illustrates an extension of proposed model by inter-process communication. Presented methodology and its comparison with RMA approach is demonstrated on case study in section 7.

## 2    On Scheduling Anomaly in Multitasking Operating System

Several multiprocessor time anomalies are known in the scheduling theory [3],[5],[7]. Similar non-linear behaviour (a shortening of the computation time leading to the prolongation of the completion time) can be found on one processor regardless the scheduling policy (preemptive or cooperative), when the processes contain computations, resource sharing and idle waiting (notice that the idle waiting is processed in parallel with computation of another process).

Example depicted in Fig. 2.1 shows a high priority processes *P-high* and a low priority process *P-low* sharing one resource represented by a semaphore *Sem*. The processes consist of computations with specified deterministic computation time, of idle waiting with specified deterministic delay and of shared resource guarded by semaphore. The computation times and delays given behind slash are assumed to be constants. The computation time of *CompC* is $C=2$ in the instance a) or $C=1$ in the instance b).

The semaphore is taken by *P-high* first in the instance a) regardless the scheduling policy (priority based preemptive or priority based cooperative). Consequently the process *P-high* is completed in 7 time units and the process *P-low* is completed in 9 time units, see Fig. 2.1 a). In the instance b), the semaphore is taken by process *P-low* first and consequently the process *P-high* is completed in 9 time units and the process *P-low* is completed in 10 time units, see Fig. 2.1 b).

The shortening of the computation time in the process P-low (*C* shorted from 2 to 1) leads to the prolongation of the completion time of both processes. As a consequence this example illustrates a necessity to consider also lower margins of computation times when process internal structure is modelled.

This result is important to modelling process internal structure by timed automata extended by tasks [13]. Timed automata extended by tasks allow to model precedence constraints over tasks by boolean variables shared between tasks and automaton.

Therefore it is possible to model each process as timed automaton and to associate to its locations tasks representing corresponding computation. Precedence constraints are used to prevent starting of next computation before the previous one is finished. Since tasks associated to locations is characterised only by its worst-case computation time, some mechanisms must be used to prevent occurrence of anomalies described in this section. One solution can be to leave processor idling when some computation is finished sooner than it was supposed.

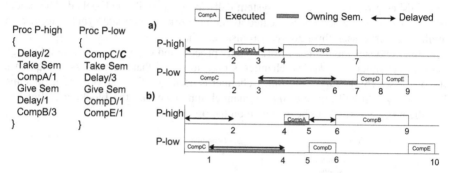

**Fig. 2.1.** Example of the monoprocessor scheduling anomaly

# 3    Cooperative Scheduling Model

Cooperative scheduling enables to deschedule currently executed process only in explicitly specified points, where the system call *yield()* is called or where the process is waiting.

An example of the application process model is depicted in Fig. 3.1. There are four types of locations. *Computation* locations (*Comp1, Comp2, Comp3* for short) corresponding to non-preemptible blocks of code (the *Computations* do not contain any blocking operation). Each two successive *Computation* locations are separated by one *Yield* location corresponding to yield instruction where the process can be descheduled and then it waits until it is scheduled again. On *WaitTimer* location the process does not require the processor. *WaitTimer* location is followed by *WaitProc* location where the signalled process waits until it is scheduled. The double circle used for *WaitTimer* location specifies that this is initial location.

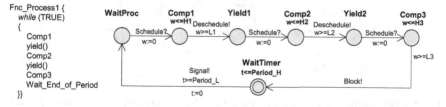

**Fig. 3.1.** Model of the application process executed under cooperative scheduling policy

As each part of the program modelled by *Computation* location cannot be affected by the preemption, its finishing time is equal to the computation time which is supposed to be known a priory and bounded by interval $\langle L,H \rangle$ (lover and upper margins allow to involve uncertainty of execution time due to non-modelled code branching inside the computations, bus errors, cache faults, page faults, cycle stealing by DMA device, etc.).

The following behaviour of the cooperative scheduler is assumed: if the processor is free, the process with the highest priority among all processes in a queue of ready processes is scheduled. The currently executed process will run until it voluntarily relinquishes processor by calling system call *yield()* or until it is blocked. The model of the cooperative scheduler is created as the network of automata synchronised with application processes through synchronisation channels as depicted in Fig. 3.2. The scheduler chooses the highest priority ready process and enables its execution through *Schedule* channel. *Deschedule* channel is used to signal that the process relinquishes the processor (by *yield()*). The *Block* channel is used to relinquish processor on some blocking system call and the *Signal* channel announce that the blocking is finished and the process is ready to be executed on the processor.

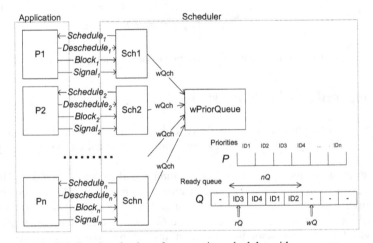

**Fig. 3.2.** Synchronization of cooperative scheduler with processes

One automaton of the cooperative scheduler model (*Sch$_i$*) is depicted in Fig. 3.3.

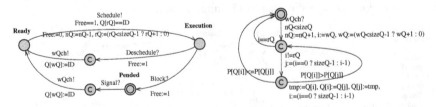

**Fig. 3.3.** One automaton Sch$_i$ of the coopera-  **Fig. 3.4.** Automaton *wPriorQueue* providing
tive scheduler in Fig. 3.2                         reordering of queue $Q$

Each process is identified by unique integer *ID* (0,1,2,...). Priority of the process is stored in global array *P*, indexed by *ID*. IDs of all processes, which are in *Ready* state, are stored in the queue modelled as global array $Q$ representing a circular buffer. The integer $nQ$ is the number of elements in the queue. The integer $rQ$ is the position for reading of the first element in $Q$ and the integer $wQ$ is position of the first empty element in $Q$ as is depicted in Fig. 3.2. Processes are ordered in descending order according to their priorities in $Q$ ($rQ$ points to the ready process with highest priority).

As shown in Fig. 3.3 mutual exclusive access to *Execution* location is guarded by two-state variable *Free*. Moreover, only the highest priority process scheduler automaton (its *ID* is at the top of ready queue) can take transition from *Ready* to *Execution* location. To prevent processor idling, the transition from *Ready* to *Execution* location must be taken as soon as it is enabled. This is provided by declaring the channel *Schedule* as *urgent channel* (no time progress is enabled when there are some enabled transitions synchronised through urgent channel). The two unnamed locations with the letter c inside the circle are so called *committed* locations providing atomicity of traversing of in-coming and out-coming transitions (committed location must be left immediately without any interference of other automaton in the model). These locations are in Fig. 3.3 necessary only due to impossibility to use two synchronizations on one transition in UPPAAL.

*ID* of the process leaving the *Ready* state is deleted from the ready queue by decrementing number of elements in the queue $nQ$ and by moving reading pointer $rQ$ to the next element in the queue. *ID* of the process entering the *Ready* state is written to the end of ready queue. The ready queue must be reordered after this operation. Ordering according priorities is provided by automaton *wPriorQueue* depicted in Fig. 3.4. Reordering mechanism is started by synchronisation channel *wQch*.

**Note on the Modelling of the Context Switch Time:** *Notice that the model of the scheduler automaton proposed in Fig. 3.3 is simplified by assumption that the context switch does not take any time. But for proper exploration of time properties of real-time system the context switch time should be considered. Since the context switch in cooperative scheduling occurs once per Computation location, context switch time can be simply involved in the computation time of each Computation.*

## 4  Modelling Deterministic Behaviour of the Scheduler

Notice that proposed model created as synchronised product of application process automata and corresponding scheduler automata (Fig. 3.3) contain non-deterministic behaviour, which does not correspond to real behaviour of the scheduler. This non-determinism occurs when the transition from *Ready* to *Execution* location of one scheduler automaton $Sch_i$ is enabled and simultaneously the transition from *Pended* to *Ready* location of another scheduler automaton $Sch_j$ is enabled. In such case the transition from *Pended* to *Ready* should be taken first since the scheduler updates states of processes first. Then the highest priority ready process should be chosen and the scheduler automaton of this process should take the transition to *Execution* location. Please realise that the model adopted in previous paragraph allows also other behaviour: the transition from *Ready* to *Execution* location of the first scheduler automaton is taken first and the transition from *Pended* to *Ready* location of the second scheduler automaton is taken afterwards. In such case the second process looses the chance to compete the processor that is undesirable since the lower priority process can take the processor even though there is some higher priority ready process at the same time.

The objective of this paragraph is to eliminate such undesirable behaviour, which does not correspond to reality. The transition priorities will be used to determine the order of transitions. High priority 2 will be assigned to the transitions from *Pended* to *Ready* locations in all scheduler automata. Lower priority 1 will be assigned to all remaining transitions. Since the transition priority is not concerned in timed automata, it is incorporated by modifying guards on transitions.

This approach is demonstrated on simple example of two periodic application processes modelled by time automata depicted in Fig. 4.1. Process P1 is the low priority one and process P2 is the high priority one. Both processes are scheduled by cooperative scheduler modelled by two scheduler automata Sch1 and Sch2 depicted in Fig. 3.3.

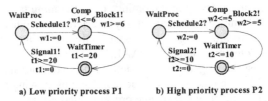

**a) Low priority process P1**          **b) High priority process P2**

**Fig. 4.1.** Automata of application processes

Resulting model of whole application is a synchronised product of all concerned automata (Sch1, Sch2, P1, P2, wPriorQueue) and it is depicted in Fig. 4.2. The location names consist of the first letters of the location names of the original automata (in the order Sch1, Sch2, P1, and P2). Priorities are assigned to the transitions were non-deterministic choice can occur (high priority 2 to the transitions from *Pended* to *Ready* and low priority 1 to other transitions). Notice that urgent locations (symbol ∪ inside the location) are used to prevent processor idling (the time progress is disabled when some automaton resides in urgent location). This function was provided by urgent channel *Schedule* in automaton *Sch$_i$* in previous section.

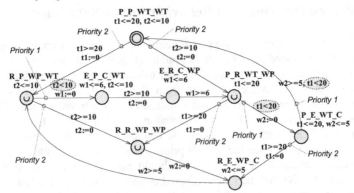

**Fig. 4.2.** Resulting model concerning transition priorities (synchronised product of Sch1, Sch2, P1, P2, wPriorQueue)

Our approach to transition priority is the following. Suppose the transition from *Pended* to *Ready* to be taken non-deterministically between lower and upper margin of the signalling time (within interval $<L, H>$). Since any process can be scheduled prior to another process becoming ready infinitely short time after scheduling decision, the transition priority has no sense in interval $<L, H)$. In other words it is desired to preserve the non-determinism in interval $<L, H)$. On the other hand, the priority of transition from *Pended* to *Ready* must be high at time $H$ since the scheduler updates states of processes prior to scheduling decision (as explained above). Our approach to give priority to the transitions is to restrict the lower priority transition guard $g_1$ to $g_1'=g_1 \wedge (t<H)$, where $t$ is a clock and $t \leq H$ is invariant of the location where the higher priority transition begin. Restricted guards of lower priority transitions are in doted grey filled ellipsis in Fig. 4.2.

# 5   Interrupts

Interrupts are usually used for fast handling of asynchronous external events. Interrupt is particularly important in cooperative scheduling since low priority process cannot be preempted and therefore high priority process cannot be used to handle asynchronous event when short response time is required. When the interrupt request (IRQ) arrives from the environment and corresponding interrupt is enabled, currently executed process is interrupted and interrupt service routine (ISR) is executed. The *relative finishing time F* of currently executed *Computation* is therefore prolonged by computation time of ISR ($C_{ISR}$) and it is no more equal to the known *computation time*. Therefore it is needed to change upper margin $H$ of each computation location in the timed automata process model. Each $H$ is prolonged by *MaxSC* (maximum server capacity), the value corresponding to the processor time reserved for all interrupt service routines. Since the number of interrupt requests depends on the environment, the total computation time of all ISR ($\Sigma C_{ISR}$) is not known a priory and moreover the existence of its upper bound is not guaranteed.

The *interrupt server* limiting amount of processor time spent for interrupts is used to guarantee that $\Sigma C_{ISR}$ does not exceed *MaxSC* value. Contrary to servers used for handling aperiodic tasks in scheduling theory (pooling, deferrable, sporadic servers [3], [6]), the prevention of servicing interrupt must be done at the hardware level (by disabling IRQ) and before the IRQ occur. The architecture of the system with *interrupt server* is depicted in Fig. 5.1. Interrupt service routines are not called directly when some interrupt is requested, but they are wrapped by the code of *ISR_Server()* function (see Fig. 5.3). The *interrupt server* has specified *server capacity SC*, which is filled by the value *MaxSC* at the beginning of each computation. The function *Fill_Server(MaxSC)* listed in Fig. 5.3 is used for it. When an interrupt occurs the *server capacity SC* is decreased by the value of corresponding $C_{ISR}$ and *interrupt server* checks if the remaining capacity *SC* is sufficient for handling next *ISR*. If not the corresponding *IRQ* is disabled. This check is provided when *SC* changes, once by *Fill_Server()* and repeatedly on each interrupt by *ISR_Server()*. Notice that $C_S$, the computation time of *ISR_Server()*, is considered. Further $H$ has to be prolonged by $C_{FS}$, the computation time of the function *Fill_Server()* (see Fig. 5.2). The lower margin $L$ of any computation location is affected only by $C_{FS}$.

Notice that the function *ISR_Server()* supposes that the hardware does not support nested interrupts (*ISR_Server()* cannot be interrupted by another interrupt).

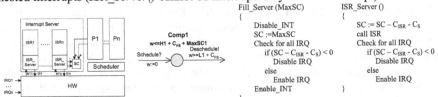

Fill_Server (MaxSC)
{
Disable_INT
SC :=MaxSC
Check for all IRQ
    if (SC – $C_{ISR}$ - $C_S$) < 0
        Disable IRQ
    else
        Enable IRQ
Enable_INT
}

ISR_Server ()
{
SC := SC – $C_{ISR}$ - $C_S$
call ISR
Check for all IRQ
    if (SC – $C_{ISR}$ - $C_S$) < 0
        Disable IRQ
    else
        Enable IRQ
}

**Fig. 5.1.** System architecture with *interrupt server*

**Fig. 5.2.** *Computation location considering interrupts*

**Fig. 5.3.** *Interrupt server* routines

Choice of MaxCS value for different locations depends on application requirements and it is specified at the design stage.

# 6    Inter Process Communication Primitives

Very important part of each multitasking application (and source of many possible errors) is a communication between processes and their synchronisation. Operating system usually provides many facilities to manage inter process communication. It is not intention of this paper to introduce models of all possible kinds of inter process communication.

On example of *semaphore* we show, how to extend the proposed model of the scheduler and application. The semaphore is the primitive used mostly for synchronization and mutual access to resources. It can be taken or given by the process using the system calls *Take()* or *Give()*. When the semaphore is given, its value is increased. When the semaphore is taken, its value is decreased. When the value of the semaphore is zero, it cannot be taken and the process attempting to take it is blocked until the semaphore is given by another process. This blocking time can be bounded by timeout. When more than one processes are blocked on one semaphore, they are waiting in priority queue or FIFO (First In First Out) queue. This basic behaviour of semaphore can be modified according to the purpose it is dedicated to. We suppose the semaphore being of counting type with value ranging from zero to *MaxCount*.

Example of an application process model with semaphore is depicted in Fig. 6.1. The process attempts to take the semaphore by synchronisation *Take!*. Then it waits in location *WaitSem* until the semaphore is taken (synchronisation *Taken?*) or until timeout expires (synchronisation *TOut!*). The synchronisation *Give!* is used to give the semaphore. Notice that giving the semaphore is not blocking operation and therefore the semaphore is given on the transition entering the *Comp3* location. On the other hand taking semaphore is blocking operation and therefore transitions with *Taken?* and *TOut!* lead to the locations *WaitProc2* or *WaitProc3* resp. where the process waits for the processor.

```
Fnc_Process1 {
    while (TRUE) {
        Comp1
        Result := Take (Sem, TimeOut)
        if (Result == TOut) {
            Comp4
            Yield() }
        else {
            Comp2
            Give (Sem) }
        Comp3
        Wait_End_of_Period
    } }
```

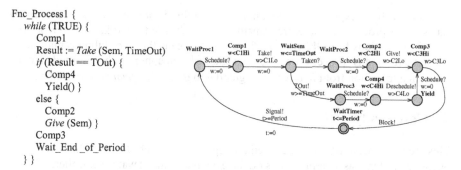

**Fig. 6.1.** Model of process containing *Take* and *Give* one semaphore

The scheduler model for application with two semaphores is depicted in Fig. 6.2. The scheduler of executed process is asked for taking the semaphore by synchronisation *Take?*. If the semaphore is empty (*Sem==0*), the processor is relinquished (*Free:=1*), *ID* of the process is written to the queue of the semaphore (*SemQ*) and the queue (FIFO or priority) is reordered by synchronisation *wSemQch!*. The scheduler and the process then wait in the location *WaitSem* until the semaphore is given by another process or until its time-out expires. If the semaphore is not empty (*Sem>0*) its value is decreased and the synchronisation *Taken!* is immediately followed by synchronisation *Schedule!* to continue in execution. The processor is not relinquished in this case.

The queue of the processes waiting for the semaphore (*SemQ*) can be FIFO queue or priority queue. In the case of priority queue, its elements (*IDs* of processes) must be reordered according to priorities when the next process issues *Take* on the empty semaphore. This is managed by the automaton similar to the one depicted in Fig. 3.4. The only difference is the name of the queue (*SemQ*, *wSemQch*, *nSemQ*, *rSemQ*, *wSemQ*). Reordering is not necessary when FIFO queue is used.

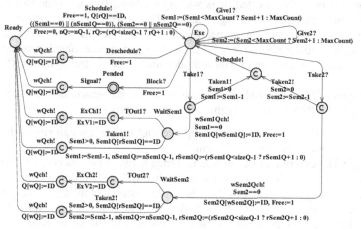

**Fig. 6.2.** Scheduler model containing two semaphores (extension of Fig. 3.3)

As it has been explained in section 4, rescheduling is not possible before updating states of all processes. Therefore transitions from *Ready* to *Exe* have lower priority than the transitions from *WaitSem* to *Ready*. That means that scheduling cannot occur when there is any process waiting on signalled semaphore. This is modelled by restricting transition from *Ready* to *Exe* guard $g$ to $g'=g \wedge \forall_i (Sem_i=0 \vee nSem_iQ=0)$.

## 7 Case Study

This section demonstrates the methodological approach to modeling real-time operating system based applications on the example of the elevator controller.

The elevator cabin either resides in a floor, or it moves between floors, or it goes through a floor (Fig. 7.2 b). The cabin movement is controlled by a three-state variable *Go* having value UP, or DOWN, or STOP. The value of the cabin possition sensor is stored in variable *In* which is equal to 1 when the cabin resides in or goes through any floor. The motor overheating is detected by value Hi of variable *Temper* (see Fig. 7.2 c). In such case the cabin must stop in the forthcoming floor and further movement is disabled by resetting variable *Enable*.

All sensors and actuators are connected to the control system by the buss guaranteeing the message delivering time. The control system software consists of three processes scheduled by cooperative scheduler and one interrupt service routine.

The buss controller generates interrupt request (*IRQ*), when new data are received from the buss or all prepared data were transmitted to the buss (see Fig. 7.2 d). If the interrupt is enabled (*EN=1*), the hardware interrupt controller (see Fig. 7.2 e) interrupts the CPU, the *ISR_Server* is invoked (see Fig. 7.2 f) and semaphore *Sem1* is signalled. The highest priority process *ComProc*, providing communication services, takes semaphore *Sem1*, then it recognises the data receiver and it signals semaphore *Sem3* or *Sem4* (see the code in Fig. 7.1 and corresponding automaton in Fig. 7.2 g). The middle priority process *DiagProc* provides diagnostic and emergency shut-down when the motor is over-heated. It is waiting on *Sem3* (see the code in Fig. 7.1 and corresponding automaton in Fig. 7.2 h). The lowest priority process *CtrlProc* providing the cabin control is waiting on *Sem4* (see the code in Fig. 7.1 and corresponding automaton in Fig. 7.2 i). The semaphore *Sem2* provides mutual exclusive access to all shared data. Pseudocode of the control system software is in Fig. 7.1.

The goal of this case study is to create model of the system and to use the model-checking tool UPPAAL to verify the following properties:

- Prop1: No *IRQ* is lost i.e. a) handling interrupt is short enough and b) the interrupt server capacity is sufficient.
- Prop2: The execution of *ComProc* is started between two successive interrupts.
- Prop3: The execution of *ComProc* is finished within 24 ms after taking *Sem1*.
- Prop4: The execution of *DiagProc* is finished within 24 ms after taking *Sem3*.
- Prop5: The execution of *CtrlProc* is finished within 34 ms after taking *Sem4*.
- Prop6: Usage of elevator is disabled 166 ms after the motor overheating.
- Prop7: The cabin will stop in any floor 5.2 s after the motor overheating.

The automata of the proposed model are depicted in Fig. 7.2. The interconnection of all automata is depicted in Fig. 7.2 a). The scheduler automaton is similar to the one in Fig. 6.2 but it is extended for four semaphores.

```
ComProc ()                  DiagProc ()                   CtrlProc ()
{                           {                             {
  while (true) {              while (true) {                while (true) {
    Take (Sem1)                Take (Sem3)                   Take (Sem4)
    Fill_Server (S1)           Fill_Server (S1)              Take (Sem2)
    Computation1  /12          Computation1  /12             Fill_Server (S1)
    Take (Sem2)                Take (Sem2)                   Computation1  /22
    Fill_Server (S2)           Fill_Server (S2)              Yield
    Computation2 /12           Computation2  /12             Fill_Server (S2)
    if (Data==DIAG)            if (Temper==Hi)       {       Computation2  /12
      Give (Sem3)                Enable:=0                   Give (Sem2)
    if (Data==CTRL)              if (Go!=STOP and In==1)  }
      Give (Sem4)                  Go:=STOP           }
    Give (Sem2)              }
} }                           Give (Sem2)
                            } }
```

**Fig. 7.1.** Control system software pseudocode

The specified properties are formalized in CTL as follow:

- Prop1 a)  $\forall\Box \neg$ IntCtrl.NestedIRQ,      b)  $\forall\Box \neg$IntCtrl.DisabledIRQ
- Prop2:   $\forall\Box$ Sem1<2
- Prop3:   $\forall\Box$ ((ComProc.WaitProc2 $\wedge$ ComProc.t=0) $\Rightarrow$
           $\forall\Diamond$ (ComProc.EndComp $\wedge$ ComProc.t<24))
- Prop4:   $\forall\Box$ ((DiagProc.WaitProc2 $\wedge$ DiagProc.t=0) $\Rightarrow$
           $\forall\Diamond$ (DiagProc.EndComp $\wedge$ DiagProc.t<24))
- Prop5:   $\forall\Box$ ((CtrlProc.WaitProc2 $\wedge$ CtrlProc.t=0) $\Rightarrow$
           $\forall\Diamond$ (CtrlProc.EndComp $\wedge$ CtrlProc.t<34))
- Prop6:   $\forall\Box$ ((Temper=Hi $\wedge$ tTemper=0) $\Rightarrow$
           $\forall\Diamond$ (Enable=0 $\wedge$ tTemper<166))
- Prop7:   $\forall\Box$ ((Temper=Hi $\wedge$ tTemper=0) $\Rightarrow$
           $\forall\Diamond$ (Enable=0 $\wedge$ In=1 $\wedge$ Go=STOP $\wedge$ tTemper<5200))

Result of verification: all properties except Prop6 are satisfied.

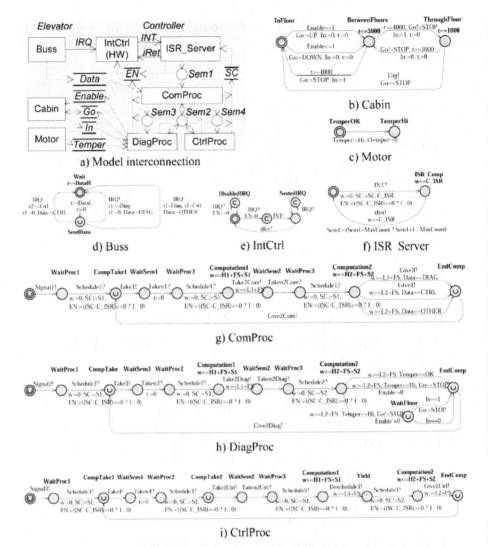

**Fig. 7.2.** Elevator and control system model

**Note:** *Please notice that under the worst-case conditions the cabin will not stop on any floor 5,2 s after increasing of the motor temperature (Prop7 is not satisfied) even though the DiagProc will react on this situation within 166 ms (Prop6 is satisfied) and the maximal time that the cabin spends between two floors is 5 s (time invariant of the state BetweenFloors in cabin model in* Fig. 7.2 b) *is t<=5000). This result would be hart to find by separate analysis of time and logical properties of the system. In fact the property Prop7 is satisfied for the value of tTemper<5332 since 5332<=166+5000+166.*

## 7.1  Comparison with RMA Approach

Notice that properties *Prop3*, *Prop4* and *Prop5* represent exploration of the worst-case completion time for processes *ComProc*, *DiagProc* and *CtrlProc*. Let's compare approach adopted in this article to RMA approach.

Let's suppose that the processes are scheduled by preemptive rate monotonic scheduling:

- the minimal interarrival times are $T_{ComProc}=50$, $T_{DiagProc}=100$ and $T_{CtrlProc}=200$,
- the worst-case computation times are $C_{ComProc}=24$, $C_{DiagProc}=24$ and $C_{CtrlProc}=34$
- critical section (*Sem2*) is locked for durations $D_{ComProc}=12$, $D_{DiagProc}=12$ and $D_{CtrlProc}=34$.

It is obvious that without internal structure knowledge, the worst-case blocking time on *Sem2* must be considered: $B_{ComProc}=34$, $B_{DiagProc}=34$.

Based on these very abstracted assumptions on system behaviour, the RMA evaluates processes *ComProc* and *DiagProc* non-schedulable.

## 8  Conclusion and Future Work

The cooperative scheduling approach given in this article avoids preemption modelling by hybrid automata. The model of the application processes and cooperative scheduler is based on timed automata, for which model checking of TCTL property problem is decidable (opposite to hybrid automata). Interrupts and inter-process communications – the most important aspect of real time embedded applications – are taken into consideration in the proposed model. With respect to the processor utilisation and reaction time the cooperative scheduling conceived in this article is not the most efficient one, but due to simplicity reasons many embedded applications are often based on similar cooperative scheduling mechanisms handling interrupts separately, therefore this approach is not just an academic idea.

Existing approaches for design and analysis of real-time applications, like Rate Monotonic Analysis (using preemptive scheduling based on priority assignment respecting the rate of periodic processes), use very elegant way of deciding whether the application is schedulable or not. But it is needed to mention, that the model checking approach provides a room for verifying more complex properties (e.g. detection of deadlocks in communication, specification of buffer size,...). Model checking provides also room for modelling of more complex time behaviour of the controlled system, running truly in parallel with the control system (modelled as separate automaton).

As the complexity of the model checking remains very huge in a general case it is motivating to set up the rules applied at a design phase, that would lead into the state spaces of reasonable size. Specification of such rules linked to the identification of the controlled systems represents a possible direction of our future work.

## Acknowledgement

This work was supported by the Ministry of Education of the Czech Republic under Project LN00B096.

## References

[1]    Alur, R., Dill, D.L.: A theory of timed automata. Theoretical Computer Science 126:183-235, 1994.

[2]    David, A.: Uppaal2k: Small Tutorial. Documentation to the verification tool Uppaal2k. http://www.docs.uu.se/docs/rtmv/uppaal/

[3]    Buttazzo, G., C.: Hard Real-Time Computing Systems: Predictable Scheduling Algorithms and Applications. Kluwer Academic Publishers, Boston (1997)

[4]    Sha, L., Klein, M., Goodenough, J.: Rate Monotonic Analysis for Real-Time Systems. 129-155. Foundations of Real-Time Computing: Scheduling and Resource Management. Boston, MA: Kluwer Academic Publishers (1991)

[5]    Graham, R.L.: Bounds on multiprocessing timing anomalies. SIAM Journal on Applied Mathematics, Vol. 17, pp. 416-429, 1969

[6]    Larsen, K.G., Pettersson, P., Yi, W.: Model-Checking for Real-Time Systems. In Proceedings of the 10th International Conference on Fundamentals of Computation Theory, Dresden, Germany, 22-25 August, 1995. LNCS 965, pages 62-88, Horst Reichel (Ed.)

[7]    Liu, J.W.S.: Real-time systems. Prentice-Hall, Inc., Upper Saddle River, New Jersey 2000. ISBN 0-13-099651-3

[8]    Shaw, A.: Reasoning about time in higher-level language software. IEEE Transactions on Software Engineering, vol. 15, July 1989

[9]    Corbett, J. C.: Timing analysis of Ada tasking programs. IEEE Transactions on Software Engineering. 22(7), pp. 461-483, July 1996

[10]    Cassez F., Larsen K.: The Impressive Power of Stopwatches. In Proceedings of CONCUR 2000 - Concurrency Theory, 11th International Conference, University Park, PA, USA, August 2000 CONCUR'2000. LNCS 1877, p. 138 ff., 2000

[11]    Henzinger, T.A., Kopke, P.W., Puri, A., Varaiya, P.: What's decidable about hybrid automata? *Journal of Computer and System Sciences* 57:94--124, 1998

[12]    Bouyer, P., Dufourd, C., Fleury, E., Petit, A.: Are Timed Automata Updatable?. In Proc. 12th Int. Conf. Computer Aided Verification (CAV'00), LNCS, Vol.1855, pp. 464-479, Springer (2000)

[13]    Fersman, E., Pettersson, P., Yi, W.: Timed Automata with Asynchronous Processes: Schedulability and Decidability. In Proceedings of 8th International Conference on Tools and Algorithms for the Construction and Analysis of Systems, TACAS 2002, Grenoble, France, April 8-12, 2002, pp.67-82, Springer-Verlag, 2002. Lecture Notes in Computer Science, Vol.2280

K.G. Larsen and P. Niebert (Eds.): FORMATS 2003, LNCS 2791, pp. 219-233, 2004.
© Springer-Verlag Berlin Heidelberg 2004

# Using Zone Graph Method
# for Computing the State Space
# of a Time Petri Net

Guillaume Gardey, Olivier H. Roux, and Olivier F. Roux

IRCCyN (Institut de Recherche en Communication et Cybernétique de Nantes)
1, rue de la Noë B.P. 92101, 44321 NANTES cedex 3, France
{guillaume.gardey,olivier-h.roux,olivier.roux}@irccyn.ec-nantes.fr
http://www.irccyn.ec-nantes.fr

**Abstract.** Presently, the method to verify quantitative time properties on Time Petri Nets is the use of observers. The state space is then computed to test the reachability of a given marking. The main method to compute the state space of a Time Petri Net has been introduced by BERTHOMIEU and DIAZ [BD91]. It is known as the "state class method". We present in this paper a new efficient method to compute the state space of a bounded Time Petri Net as a marking graph, based on the region graph method used for Timed Automaton [AD94]. The algorithm is proved to be exact with respect to the reachability of a marking and it computes a graph which nodes are exactly the reachable markings of the Time Petri Net. The tool implemented computes faster than TINA, a tool for constructing the state space using classes, and allows to test on-the-fly the reachability of a given marking.

**Keywords:** Time Petri Nets, Zone, State Space, Reachability Analysis, Verification

## 1 Introduction

### Frameworks

The theory of Petri Nets provides a general framework to specify the behavior of reactive systems and time extensions have been introduced to take also temporal specifications into account. The two main time extensions of Petri Nets are Time Petri Nets (TPN) [Mer74] and Timed Petri Nets [Ram74]. While a transition can be fired within a given interval for TPN, in Timed Petri Nets, transitions are fired as soon as possible. There are also numerous way of representing time. It could be relative to places, transitions, arcs or tokens. TPN are mainly divided in P-TPN, A-TPN and T-TPN where a time interval is relative to places (P-TPN), arcs (A-TPN) or transitions (T-TPN). Finally, Time Stream Petri Nets [DS94] were introduced to model multimedia applications.

Concerning the timing analysis of these three models in order to verify properties, few studies have been realized.

K.G. Larsen and P. Niebert (Eds.): FORMATS 2003, LNCS 2791, pp. 246–259, 2004.

Recent works [AN01, dFRA00] consider Timed Arc Petri Nets where each token has a clock representing its "age". Using a backward exploration algorithm [AJ01, FS98], they proved that the coverability and boundedness are decidable for this class of Petri Nets. However, they assume a lazy (non-urgent) behavior of the net: the firing of a transition may be delayed even if its time becomes greater than its latest firing time, disabling the transition.

In [Rok93, RM94], ROKICKI considers an extension of labeled Petri Nets called Orbitals Nets: each transition of the TPN (safe P-TPN) is labeled with a set of events (actions). The state-space is constructed using a forward algorithm very similar to ALUR and DILL region based method. ROKICKI finally uses partial order method to reduce time and space requirements for verification purpose. The semantics used is not formally defined and seems to differ from another commonly adopted proposed by KHANSA [KDC96].

Others approaches aim at translating a TPN into a Timed Automaton (TA) in order to use efficient existent tools on TA. In [CEP00], CORTÈS et al. propose to transform an extension of T-TPN into the composition of several TA. Each transition is translated into an automaton (not necessarily identical due to conflict problems) and it is claimed that the composition capture the behaviour of the TPN. In [CR03], CASSEZ and ROUX propose another structural approach: each transition is translated into a TA using the same pattern. The authors prove the two models are timed-bisimilar. In [SA01], SAVA and ALLA compute the graph of reachable markings of a T-TPN. The result is a TA. Nevertheless, they assume the TPN is bounded and does not include $\infty$ as latest firing time, no proof is given of the timed-bisimilarity between the two models. In [LR03], LIME and ROUX propose a method for building the state class graph of a bounded T-TPN as a TA. The resulting TA is timed-bisimilar and has much lower clocks than previous methods which is of importance for TA model-checking.

Such translations show that TCTL and CTL are decidable for T-TPN and that developed algorithms on TA may be extended to T-TPN.

In this paper, we consider T-TPN in which a transition can be fired within a time interval. For this model, boundedness is undecidable and works report undecidability results, or decidability under the assumption of boundedness of the TPN (as for reachability, decidability [Pop91]). Boundedness and other results are obtained by computing the state-space.

## Related Work

*State Space Computation of a T-TPN.* The main method to compute the state-space of a TPN is the state class graph [Men82, BD91]. A state class $C$ of a TPN is a pair $(M, D)$ where $M$ is a marking and $D$ a set of inequalities called the firing domain. A variable $x_i$ of the firing domain represents the firing time of the enabled transition $t_i$ relatively to the time when the class $C$ was entered in and truncated to nonnegative times. The state class graph preserves markings [BV03] as well as traces and complete traces but can only be used to check untimed reachability properties and is not accurate enough for checking *quantitative* real-time properties. An alternative approach has been proposed by YONEDA *et al.* [YR98] in the form of an extension of equivalence classes (atomic classes) which allow CTL model-checking. LILIUS [Lil99] refined this approach so that it becomes possible to apply partial order reduction techniques that have been developed for untimed systems. BERTHOMIEU and VERNADAT [BV03] propose an alternative construction of the graph of atomic classes of YONEDA applicable to a larger class of nets. In [OY97], OKAWA and YONEDA propose another method to perform CTL model-checking on T-TPN, they use a region based algorithm on safe TPN without $\infty$ as latest firing time. Their algorithm is based on the one of [AD94] and aim at computing a graph conserving branching properties. Nevertheless, the algorithm used to construct the graph seems inefficient (their algorithm do code regions) and no result can be exploited to compare with others methods.

*Zone Based Algorithm.* Another model used to represent timed systems are Timed Automaton (TA) introduced by ALUR and DILL [AD94]. They introduce the construction of the state space based on regions, *i.e.* a representation of clocks values using equivalence classes. The state space is built by analyzing successors of the initial region (forward analysis). Actually efficient forward (and backward) algorithms using regions do not code regions but zones, a finite convex union of regions because regions suffer of a combinatorial explosions and are quite uneasy to manipulate. This method (forward analysis + zone) is implemented in tools like UPPAAL [LPY97] or KRONOS [Yov97] and is efficiently used to model-check CTL or TCTL properties on timed systems.

Nevertheless, recent works proved the limitations of the data structure used to represent zones: Difference Bounded Matrices (DBM). BOUYER [Bou02, Bou03] proved that the use of DBM made in the forward algorithm leads to an over-approximation of the state-space: some states are said to be reachable while, indeed, they are not.

## Contributions

Our aim is to compute efficiently the state space of a bounded T-TPN in order to verify quantitative timing properties.

The paper is devoted to present a different approach to compute the state space of an unsafe bounded T-TPN based on the TA region graph method. Although regions encoding is based on the use of zones implemented with DBM, the algorithm is proved to be exact with respect to the reachability of a marking.

In section 2, we first recall the semantics of T-TPN and present the state class method and its limitations for model-checking. We propose in section 3 a forward algorithm to compute the state space of a bounded T-TPN and prove it is exact with respect to the set of reachable markings. We then present in section 4 some details on the implemented tool and we give an example of the use of observers to check properties on T-TPN. We also compare our tool with a tool using the state class method (TINA). Our experimental tests give encouraging results.

## 2  Definitions

### 2.1  Time Petri Nets

**Definition 1 (T-TPN).** *A Time Petri Net is a tuple* $(P, T, {}^\bullet(.), (.)^\bullet, \alpha, \beta, M_0)$ *defined by:*

- $P = \{p_1, p_2, \ldots, p_m\}$ *is a non-empty set of* places,
- $T = \{t_1, t_2, \ldots, t_n\}$ *is a non-empty set of* transitions,
- ${}^\bullet(.) : T \to \mathbb{N}^P$ *is the backward incidence function,*
- $(.)^\bullet : T \to \mathbb{N}^P$ *is the forward incidence function,*
- $M_0 \in \mathbb{N}^P$ *is the initial marking of the Petri Net,*
- $\alpha : T \to \mathbb{Q}^+$ *is the function giving* the earliest firing time *for a transition,*
- $\beta : T \to \mathbb{Q}^+ \cup \{\infty\}$ *is the function giving* the latest firing time *for a transition.*

A Petri Net marking $M$ is an element of $\mathbb{N}^P$ such that for all $p \in P$, $M(p)$ is the number of tokens in the place $p$.

A marking $M$ enables a transition $t$ if the number of tokens in the corresponding places is greater or equal to the valuation of incoming arcs: $M \geq {}^\bullet t_i$. The set of transitions enabled by a marking $M$ is $enabled(M)$.

A transition $t_k$ is said to be *newly* enabled by the firing of a transition $t_i$ if $M - {}^\bullet t_i + t_i^\bullet$ enables $t_k$ and $M - {}^\bullet t_i$ does not enabled $t_k$. If $t_i$ remains enabled after its firing then $t_i$ is newly enabled. The set of transitions newly enabled by a transition $t_i$ for a marking $M$ is noted $\uparrow enabled(M, t_i)$.

$v \in (\mathbb{R}_{\geq 0})^T$ is a vector of clocks valuations. $v_i$ is the time elapsing since the transition $t_i$ has been newly enabled.

The semantics of T-TPN is defined as a Timed Transition Systems (TTS). Firing a transition is a discrete transition of the TTS, waiting in a marking, the continuous transition.

**Definition 2 (Semantics of a T-TPN).** *The semantics of a T-TPN is defined by the Timed Transition System* $\mathcal{S} = (Q, q_0, \to)$:

- $Q = \mathbb{N}^P \times (\mathbb{R}_{\geq 0})^T$
- $q_0 = (M_0, \bar{0})$
- $\to \in Q \times (T \cup \mathbb{R}) \times Q$ *is the transition relation including a discrete transition and a continuous transition.*
  - *The continuous transition is defined* $\forall d \in \mathbb{R}_{\geq 0}$ *by:*

$$(M, v) \xrightarrow{e(d)} (M, v') \text{ iff } \begin{cases} v' = v + d \\ \forall k \in [1, n]\ M \geq {}^\bullet t_k \Rightarrow v'_k \leq \beta(t_k) \end{cases}$$

  - *The discrete transition is defined* $\forall t_i \in T$ *by:*

$$(M, v) \xrightarrow{t_i} (M', v') \text{ iff } \begin{cases} M \geq {}^\bullet t_i \\ M' = M - {}^\bullet t_i + t_i^\bullet \\ \alpha(t_i) \leq v_i \leq \beta(t_i) \\ \forall k \in [1, n]\ v'_k = \begin{cases} 0 \text{ if } t_k \in\ \uparrow enabled(M, t_i) \\ v_k \text{ otherwise} \end{cases} \end{cases}$$

## 2.2   The State Class Method

The main method to compute the state space of a Time Petri Net is the state class method introduced by BERTHOMIEU and DIAZ in [BD91].

**Definition 3 (State class).** *A State Class $C$ of a TPN is a pair $(M, D)$ where $M$ is a marking and $D$ a set of inequalities called the firing domain. A variable $x_i$ of the firing domain represents the firing time of the enabled transition $t_i$ relatively to the time when the class $C$ was entered in.*

The state class graph is computed iteratively as follows.

**Definition 4 (State Class Method).** *Given a class $C = (M, D)$ and a firable transition $t_j$, the successor class $C' = (M', D')$ by the firing of $t_j$ is obtained by:*

1. *Computing the new marking $M' = M - {}^\bullet t_j + t_j^\bullet$.*
2. *Making variable substitution in the domain: $\forall i \neq j,\ x_i \leftarrow x_i' + x_j$.*
3. *Eliminating $x_j$ from the domain using for instance the Fourier-Motzkin method.*
4. *Computing a canonical form of $D'$ using for instance the Floyd-Warshall algorithm.*

In the state class method, the domain associated to a class is relative to the time when the class was entered and as the transformation (time origin switching) is irreversible, absolute value of clocks cannot be obtained easily. The graph produced is an abstraction of the state space for which temporal information has been lost and generally, the graph has more states than the number of markings of the TPN. Transitions between classes are no longer labeled with a firing constraint but only with the name of the fired transition: the graph is a representation of the untimed language of the TPN.

## 2.3   Limitations of the State Class Method

As a consequence of the graph construction, sophisticated temporal properties are not easy to check. Indeed, the domain associated to a marking is made of relative values of clocks and the function to compute domains is not bijective. Consequently, domains can not be easily used to verify properties involving constraints on clocks.

In order to get rid of these limitations, several works aim to construct a different state class graph by modifying the equivalence relation between classes. To our knowledge, proposed methods [BV03] depend on the property to check. Checking LTL or CTL properties will lead to construct different state class graphs.

Another limitation of methods and proposed tools to check properties is the need to compute the whole state graph while only the reachability of a given marking is needed (safety properties). The graph is then analyzed by a model checker. Using observers is even more costly: actually, for each property to be checked, a new state class graph has to be built and the observer can dramatically increase the size of the state space.

In the next section we will present another method to compute the state space of a bounded T-TPN. The resulting graph keeps in memory temporal information and will allow to test on-the-fly temporal properties. The graph is also more compact: it has exactly as many nodes as the number of reachable markings of the TPN.

# 3  A Forward Algorithm to Compute the State Space of a Bounded T-TPN

The method we propose in this paper is an adaptation, proved to be exact, of the region based method for Timed Automaton [AD94, Rok93].

First, we define a *zone* as a convex union of regions as defined by ALUR and DILL [AD94]. For short, considering $n$ clocks, a zone is a convex subset of $(\mathbb{R}_{\geq 0})^n$. A zone could be represented by a conjunction of constraints on clocks pairs: $x_i - x_j \sim c$ where $\sim \in \{<, \leq, =, \geq, >\}$ and $c \in \mathbb{N}$.

## 3.1  Our Algorithm: One Iteration

Given the initial marking and initial values of clocks, timing successors are iteratively computed by letting time pass or by firing transitions.

Let $M_0$ be a marking and $Z_0$ a zone. The computation of reachable markings from $M_0$ according to the zone $Z_0$ is made as follows:

- Compute the possible evolution of time (future): $\overrightarrow{Z_0}$. This is obtained by setting all upper bounds of clocks to infinity.
- Select only the possible valuations of clocks for which $M_0$ could exist, *i.e.* valuations of clocks must not be greater than the latest firing time of enabled transitions :

$$Z_0' = \overrightarrow{Z_0} \cap \left\{ \bigwedge_i \{x_i \leq \beta_i \mid t_i \in enabled\,(M_0)\} \right\}$$

So, $Z_0'$ is the maximal zone starting from $Z_0$ for which the marking $M_0$ exists.
- Determine the firable transitions: $t_i$ is firable if $Z_0' \cap \{x_i \geq \alpha_i\}$ is a non empty zone.
- For each firable transition $t_i$ leading to a marking $M_{0i}$, compute the zone entering the new marking:

$$Z_i = (Z_0' \cap \{x_i \geq \alpha_i\})\,[X_e := 0]\,, \text{ where } X_e \text{ is the set of newly enabled clocks.}$$

This means that each transition which is newly enabled has its clock reset. Then, $Z_i$ is a zone for which the new marking $M_{0i}$ is reachable.

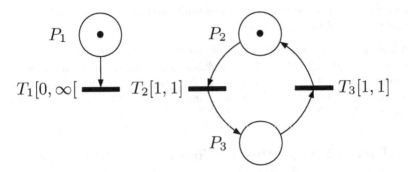

**Fig. 1.** Time Petri Net with an unbounded number of zones

### 3.2 Convergence Criterion

To ensure termination, a list of zones is associated to each reachable marking. It will keep track of zones for which the marking was already analyzed or will be analyzed. At each step, we compare the zone currently being analyzed to the ones previously computed. If the zone is included in one of the list there is no need to go further because it has already been analyzed or it will lead to compute a subgraph.

### 3.3 Overapproximation on Zones

An algorithm to enumerate reachable markings for a bounded TPN could be based on the described algorithm but, generally, it will lead to a non-terminating computation. Though the number of reachable markings is finite for a bounded TPN, the number of zones in which a marking is reachable is not necessarily finite (see figure 1).

Let's consider the infinite firing sequence: $(T_2, T_3)^*$. The initial zone is $\{x_1 = 0 \wedge x_2 = 0 \wedge x_3 = 0\}$ (where $x_i$ is the clock associated to $T_i$), the initial marking $M_0 = (P_1, P_2, P_3) = (1, 1, 0)$. By letting time pass, $M_0$ is reachable until $x_2 = 1$. When $x_2 = x_1 = 1$ the transition $T_2$ has to be fired. The zone corresponding to clock values is : $Z_0 = \{0 \leq x_1 \leq 1 \wedge x_1 - x_2 = 0\}$. By firing $T_2$ and then $T_3$, the net returns to its initial marking. Entering it, values of clocks are: $x_1 = 2$, $x_2 = 0$ and $x_1 - x_2 = 2$. Indeed, $T_1$ remains enabled while $T_2$ and $T_3$ are fired and $x_2$ is reset when $T_3$ is fired because $T_2$ became newly enabled. Given these new values, the initial marking can exists while $x_2 \leq 1$ *i.e.* for the zone: $Z_1 = \{2 \leq x_1 \leq 3 \wedge x_1 - x_2 = 2\}$. By applying infinitely the sequence $(T_2, T_3)$, there exists an infinite number of zones for which the initial marking is reachable.

Actually, the number of zones is not bounded because infinity is used as latest firing time ($T_1$). If infinity is not used as latest firing time, all clocks are bounded and so, the number of different zones is bounded [AD94]. The "naive" algorithm is then exact and can be used to compute the state space of a bounded T-TPN.

We will propose a more general algorithm which computes the state space of a T-TPN as defined in section 2, *i.e.* with infinity as latest firing time allowed. It will be based on the use of an operator on zones which construct an equivalence class. The resulting equivalence class will have a finite number of classes.

### 3.4   Approximation

A common operator on zones is the *k-approx* operator. For a given $k$ value, the use of this operator allows to create a finite set of distinct zones as presented in [AD94]. The algorithm proposed is an extension of the one presented in section 3.1. It consists in applying the *k-approx* operator on the zone resulting from the last step.

This approximation is based on the fact that once the clock associated to an unbounded transition ($[\alpha, \infty[$) has reached the value $\alpha$, its precise value does not matter. Using *k-approx* (with $k = \alpha$) allows to regroup all zones $[x, \infty[, x \geq \alpha$ in one equivalence class.

Unfortunately recent works on Timed Automaton [Bou02, Bou03] have proved that this operator generally leads to an overapproximation of the reachable localities of TA. Nevertheless, for a given class of TA (diagonal-free), there is no overapproximation of the reachable localities.

Results of BOUYER are directly extensible for T-TPN and we could assert the following theorem:

**Theorem 1.** *A forward analysis algorithm using k-approx on zone is exact with respect to TPN marking reachability for bounded TPN.*

A detailed presentation of the result of BOUYER and the demonstration of this theorem is presented in appendix A.

As the approximation is only needed for T-TPN with infinity as latest firing time, the following theorem can be asserted:

**Corollary 1.** *For a bounded T-TPN without infinity as latest firing time, a forward analysis algorithm using zones computes the exact state-space of the T-TPN.* Proof is given in appendix A.

## 4   The Tool: Mercutio

### 4.1   Presentation

We have implemented the algorithm to compute all the reachable markings of a bounded T-TPN using DBM to encode zones. The tool implemented (MERCUTIO) is integrated into ROMEO [Rom00], a software for TPN edition and analysis.

As boundedness of T-TPN is undecidable, MERCUTIO offers stopping criteria: number of reached markings, computation time, bound on the number of tokens in a place.

**Table 1.** TINA 2.5.1 – MERCUTIO (Pentium II, 400 MHz, 256Mb)

| Time Petri Net | TPN (places / trans.) | TINA | MERCUTIO |
|---|---|---|---|
| Example 1 (oex15) | 16 / 16 | 10.5 s | 1.4 s |
| Example 2 (oex7) | 22 / 20 | 30.5 s | 2.4 s |
| Example 3 (oex8) | 31 / 21 | 29 s | 2.4 s |
| Example 4 (P6C7) | 21 / 20 | 31.6 s | 8.5 s |
| Example 5 (P10C10) | 32 / 31 | 4.2 s | 1.8 s |
| Example 6 (landing gear) | 107 / 101 | 18 min 27 s | 27.8 s |
| Example 7 (Gate Controller - 3 trains) | 20 / 23 | 2 s | ¡1 s |
| Example 8 (Gate Controller - 4 trains) | 24 / 29 | 3 min 8 s | 9 s |

## 4.2 Performances

We have compared our tool with TINA [BV]. TINA is developed by BERTHOMIEU and it is the most efficient tool we know for the state class construction of a TPN. The results are given in table 1. We used the last stable version TINA (2.5.1). Though the method is not the same, it can give a time reference to compute all reachable markings of a T-TPN.

Examples 1 to 5 come from real-time systems (parallel tasks [1], periodic tasks[2–3], producer-consumer [4–5]). Example 6 is a larger system representing a simplified landing gear, it counts 107 places and 101 transitions. It is a T-TPN representation of a landing gear case study published in [BC02]. Examples 7 and 8 are the classical level crossing example (3 and 4 trains).

For this set of examples and for all nets we have tested, our tool performs better than TINA.

## 4.3 Reachability Analysis – Observers

TPN observers are a method to model check TPN. It consists in adding to the Petri Net places and transitions to model the property to check. The property is

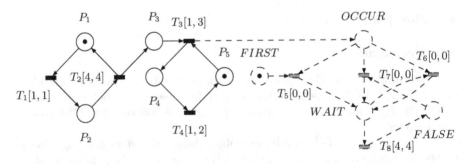

**Fig. 2.** Example of a TPN and an observer (dash point)

transformed in testing for the reachability of a given marking. Then, as for the construction of the state class graph, it is possible to check properties on TPN with observers.

In [TST97], the authors present generic observers to model properties like absolute or relative time between the firing of transitions, causality or simultaneity.

Let's consider the net of figure 2. It represents a simple TPN and an observer in dash point. The observer allows to check the property: "2 successive occurrences of $T_3$ always append in less than 4 time units". The property is false if a run such that the place $FALSE$ as a token exists.

By constructing the state space and look for a run with a token in $FALSE$ allows to conclude that the property is false.

Generally there is no need to compute the whole state space: the algorithm can be stopped at the first marking verifying a property. In its current release, MERCUTIO can perform an on the fly analysis of the TPN. Providing a set of constraints on reachable markings, MERCUTIO will stop at the first marking verifying constraints provided.

## 5   Conclusions

In this paper we proposed an efficient method based on the region graph approach to compute the state space of a bounded T-TPN. The implemented algorithm computes the graph of reachable markings by the use of zones coded with DBM, and we proved it is exact with respect to reachability. Tests on several examples show that our implementation is faster than the most efficient tool we know to compute the state class graph (TINA).

Nevertheless, observers are still a not easy way to model-check a Petri Net. For each new property an observer has to be built and then the state space has to be computed. As the exact state space is computed by our method for a bounded TPN without infinity as latest firing time, we are involved in realizing an on-the-fly TCTL model-checker. Despite the overapproximation issue for TPN with infinity as latest firing time, we think possible, by choosing an appropriate parameter for the approximation, to perform model-checking.

It could also be possible to improve MERCUTIO performances (memory needs) by using compacter data structures. Though DBM is an efficient way to represent zones, several works offers data structures based on Binary Decision Diagram (CDD, RED, CRD) to minimize memory needs. We are currently working on such improvements.

## References

[AD94]    Rajeev Alur and David L. Dill. A theory of timed automata. Theoretical Computer Science, 126(2):183–235, 1994.  246, 248, 251, 252, 253, 258, 259

[AJ01]    Parosh Aziz Abdulla and Bengt Jonsson. Ensuring completeness of symbolic verification methods for infinite-state systems. Theoretical Computer Science, 256:145–167, 2001.  247

[AN01]      Parosh Aziz Abdulla and Aletta Nylén. Timed petri nets and bqos. In
            22nd International Conference on Application and Theory of Petri Nets
            (ICATPN '01), volume 2075 of Lecture Notes in Computer Science, pages
            53–72, Newcastle upon Tyne, United Kingdom, june 2001. Springer-Verlag.
            247

[BC02]      F. Boniol and F. Carcenac. Une étude de cas pour la vérification formelle
            de propriétés temporelles. FAC'2002, Mar 2002. http://www.laas.fr/ fran-
            cois/SVF/FAC02. 254

[BD91]      Bernard Berthomieu and Michel Diaz. Modeling and verification of time
            dependent systems using time petri nets. IEEE transactions on software
            engineering, 17(3):259–273, 1991. 246, 247, 250

[BDFP00]    Patricia Bouyer, Catherine Dufourd, Emmanuel Fleury, and Antoine Petit.
            Are timed automata updatable? In Proc. 12th International Conference
            on Computer Aided Verication (CAV '2000), volume 1855 of LNCS, pages
            464–479. Springer Verlag, July 2000. 258

[Bou02]     Patricia Bouyer. Timed automata may cause some troubles. Technical
            report, LSV, July 2002. 248, 253, 258, 259

[Bou03]     Patricia Bouyer. Unteamable timed automata! In Proc. 20th Annual Sym-
            posium on Theoretical Aspects of Computer Science (STACS '2003), volume
            2607 of LNCS, pages 620–631, Berlin, Germany, February 2003. Springer
            Verlag. 248, 253, 258, 259

[BV]        B. Berthomieu and F. Vernadat. TINA. http://www.laas.fr/tina. 254

[BV03]      Bernard Berthomieu and François Vernadat. State class constructions for
            branching analysis of time petri nets. In 9th International Conference
            on Tools and Algorithms for the Construction and Analysis of Systems
            (TACAS 2003), pages 442–457. Springer–Verlag, Apr 2003. 247, 248, 250

[CEP00]     Luis Alejandro Cortes, Petru Eles, and Zebo Peng. Verification of embed-
            ded systems using a petri net based representation. In 13th International
            Symposium on System Synthesis (ISSS 2000), pages 149–155, Madrid,
            Spain, September 2000. 247

[CR03]      Franck Cassez and Olivier (H.) Roux. Traduction structurelle des réseaux
            de petri temporels vers les automates temporisés. In 4ieme Colloque
            Francophone sur la Modélisation des Systèmes Réactifs, (MSR '03), Metz,
            France, oct 2003. To appear. 247

[dFRA00]    David de Frutos Escrig, Valentín Valero Ruiz, and Olga Marroquín Alonso.
            Decidability of properties of timed-arc petri nets. In 21st International
            Conference on Application and Theory of Petri Nets (ICATPN '00), vol-
            ume 1825 of Lecture Notes in Computer Science, pages 187–206, Aarhus,
            Denmark, june 2000. Springer-Verlag. 247

[DS94]      M. Diaz and P. Senac. Time stream petri nets: a model for timed multime-
            dia information. Lecture Notes in Computer Science, 815:219–238, 1994.
            246

[FS98]      Alain Finkel and Philippe Schnoebelen. Fundamental structures in well-
            structured infinite transitions systems. In 3rd Latin American Theoreti-
            cal Informatics Symposium (LATIN '98), volume 1380 of Lecture Notes in
            Computer Science, pages 102–118, Campinas, Brazil, april 1998. Springer-
            Verlag. 247

[KDC96]     Wael Khasa, J.-P. Denat, and S. Collart-Dutilleul. P-Time Petri Nets
            for manufacturing systems. In International Workshop on Discrete Event
            Systems, WODES'96, pages 94–102, Edinburgh (U.K.), august 1996. 247

[Lil99]      Johan Lilius. Efficient state space search for time petri nets. In M FC S
             W orkshop on C oncurrency '98, volume 18 of ENTC S. Elsevier, 1999. 248
[LPY97]      Kim G. Larsen, Paul Pettersson, and Wang Yi. UPPAAL in a nutshell. In-
             ternationalJournalon Softw are Tools for Technology Transfer, 1(1–2):134–
             152, Oct 1997. http://www.uppaal.com/. 248
[LR03]       Didier Lime and Olivier (H.) Roux. State class timed automaton of a time
             petri net. In The 10th InternationalW orkshop on Petri N ets and Per-
             form ance M odels, (PN PM '03). IEEE Computer Society, Sept. 2003. To
             appear. 247
[Men82]      Miguel Menasche. A nalyse des réseaux de Petri tem porisés et application
             aux systèm es distribués. PhD thesis, Université Paul Sabatier, Toulouse,
             France, 1982. 247
[Mer74]      P. M. Merlin. A study of the recoverability of com puting system s. PhD
             thesis, Department of Information and Computer Science, University of
             California, Irvine, CA, 1974. 246
[OY97]       Yasukichi Okawa and Tomohiro Yoneda. Symbolic ctl model checking of
             time petri nets. In Scripta Technica, editor, Electronics and C om m unica-
             tions in Japan, volume 80, pages 11–20, 1997. 248
[Pop91]      Louchka Popova. On time petri nets. JournalInform ation Processing and
             C ybernetics, EIK , 27(4):227–244, 1991. 247
[Ram74]      C. Ramchandani. A nalysis of asynchronous concurrent system s by tim ed
             Petri nets. PhD thesis, Massachusetts Institute of Technology, Cambridge,
             MA, 1974. Project MAC Report MAC-TR-120. 246
[RM94]       Tomas G. Rokicki and Chris J. Myers. Automatic verification of timed
             circuits.  In International C onference on C om puter-A ided V erification
             (C AV '94), pages 468–480. Springer–Verlag, 1994. 247
[Rok93]      Tomas G. Rokicki. R epresenting an M odeling C ircuits. PhD thesis, Stan-
             ford University, 1993. 247, 251
[Rom00]      Romeo.
             http://www.irccyn.ec-nantes.fr/irccyn/d/fr/equipes/tempsreel/logs.
             A toolfor T im e PetriN ets A nalysis, 2000. 253
[SA01]       A. T. Sava and H. Alla.   Commande par supervision des systemes
             à évènements discrets temporisées. M odélisation des systèm es réactifs
             (M SR 2001), pages 71–86, 2001. 247
[TST97]      Joël Toussaint, Françoise Simonot-Lion, and Jean-Pierre Thomesse. Time
             constraint verifications methods based time petri nets. In 6th W orkshop
             on Future Trends in D istributed C om puting System s (FTD C S '97), pages
             262–267, Tunis, Tunisia, 1997. 255
[Yov97]      Sergio Yovine. Kronos: A verification tool for real-time systems. Interna-
             tionalJournalof Softw are Tools for Technology Transfer, 1(1–2):123–133,
             Oct 1997. http://www-verimag.imag.fr/TEMPORISE/kronos/. 248
[YR98]       Tomohiro Yoneda and Hikaru Ryuba. Ctl model checking of time petri nets
             using geometric regions. IE IC E T ransactions on Inform ation and System s,
             E99-D(3):297–396, march 1998. 248

# A   Proof of Theorem 1 and 2

The proof of the theorem is a consequence of works of BOUYER on Timed Automaton.

## A.1   Timed Automaton and Overapproximation.

In [Bou02, Bou03] the author presents an exact algorithm with respect to reachability for Timed Automaton using an operator called $Closure_k$. Then, it is proved that the operator $k$-approx, commonly used on DBM, leads to a zone included in $Closure_k$. The operator $Closure_k$ has the same aim that $k$-approx, it divides the clock space into a finite number of regions so that the computed number of zones is finite.

Let $\mathcal{A}$ be an updatable timed automaton and $k$ the greatest constant appearing in constraints on clocks of $\mathcal{A}$. $\mathcal{R}_k$ is a finite set of regions as defined in [BDFP00] (similar to region's definition of ALUR and DILL in [AD94]).

The operator $Closure_k$ is defined by:

$$Closure_k(Z) = \cup \{R \in \mathcal{R}_k \mid R \cap Z \neq \emptyset\}$$

Let:

$$Post_{i_k}(Z) = up_{i_k}\left(g_{i_k} \cap \overrightarrow{Z}\right)$$
$$Post'_{i_k}(Z) = Post_{i_k}(Closure_k(Z))$$

where $g_{i_k}$ is a conjunction of diagonal-free constraints on clocks and $up_{i_k}$ a function assigning values to some clocks. Diagonal-free constraints are constraints which do not involved comparisons between clocks, constraints are only of the form: $x_i \sim n, n \in \mathbb{N}, \sim \in \{<, \leq, =\geq, >\}$.

Let:

$$M_1 = Post_{i_n} \circ Post_{i_{n-1}} \circ \cdots \circ Post_{i_1}(Z)$$
$$M_2 = Post'_{i_n} \circ Post'_{i_{n-1}} \circ \cdots \circ Post'_{i_1}(Z)$$
$$M_3 = Closure_k(M1)$$

$M_1$ is the exact zone computed which results from a sequence of transition $i_1, \ldots, i_n$. $M_2$ is the zone computed by a forward algorithm using $Closure_k$.

Using properties on $Closure_k$, it is proved that $M_1 \subseteq M_2 \subseteq M_3$. Precisely, if $M_1$ is empty then $Closure_k(M_1)$ is also empty and then, $M_2$ is also empty. An algorithm using $Closure_k$ as operator is exact with respect to reachability.

Nevertheless, it is not easy to compute $Closure_k$ and DBM is known to be an efficient data structure to perform operations on zones.

It is then proved that for any zone $Z$,

$$Z \subseteq k\text{-}approx(Z) \subseteq Closure_k(Z)$$

$k$-*approx* is computed by replacing in the DBM all values greater than $k$ by $\infty$ and all values lower than $-k$ by $-k$.

Consequently, any algorithm using $k$-*approx* is also exact with respect to reachability for Updatable Timed Automata with diagonal-free constraints. The proof of this theorem is mainly based on the fact that constraints appearing in automaton are diagonal-free.

## A.2    Reachability of a Marking for a TPN Using DBM.

**Theorem 1.** *A forward analysis algorithm using k-approx on DBM is exact with respect to TPN marking reachability for bounded T-TPN.*

*Proof.* We will prove that operations on zones needed are a subset of the ones described in [Bou02, Bou03]. Indeed, we choose $k$ as:

$$k = Max \left( max \left( \alpha \left( t_i \right) \right)_{t_i \in T}, max \left( \beta \left( t_i \right) \right)_{t_i \in T} \right)$$

The reset function is a particular case of the update function $up_{i_k}$ and the intersection we made on zones are intersection with diagonal-free constraints. Actually, we intersect each clock with the value of the latest firing time of the associated transition. Thus, operations on zones for TPN verify the hypothesis of the demonstration presented in [Bou02, Bou03].                    □

**Corollary 1.** *For a bounded T-TPN without infinity as latest firing time, a forward analysis algorithm using zones computes the exact state-space of the T-TPN.*

*Proof.* As infinity is not used as latest firing time, all clocks are bounded and so, the number of different zones is bounded [AD94]. The state space computed is then exact.

# Causal Time Calculus

Franck Pommereau

LACL, Université Paris 12
61, avenue du général de Gaulle, 94010 Créteil, France
pommereau@univ-paris12.fr

**Abstract.** We present a process algebra suitable to the modelling of timed concurrent systems and to their efficient verification through model checking. The algebra is provided with two consistent semantics: a structural operational semantics (as usual for process algebras) and a denotational semantics in terms of Petri nets in which time is introduced through counters of explicit clock ticks. This way of modelling time has been called causal time so the process algebra is itself called the Causal Time Calculus (CTC). It was shown in a separate paper [3] that the causal time approach allowed for efficient verification but suffered from a sensitivity to the constants to which counts of ticks are compared. We show in this paper how this weakness can be removed.

## 1  Introduction

This paper presents a process algebra in which the representation of timing constraints can be explicitly included. With respect to the many such models already defined (a short comparison is given in the conclusion), our contribution is: first, to provide a concurrent semantics instead of the interleaving generally used; second, to propose a multiway communication scheme; and third, to give a way through which efficient model checking can be performed. We thus define a structural operational semantics (SOS), explicitly including concurrency, through SOS rules in Plotkin's style [17]; then a consistent denotational semantics is given by a transformation from process terms to Petri nets on which dedicated verification techniques may be applied [8]. The involved class of Petri nets consists in composable, labelled and coloured nets in which time is introduced by explicitly modelling clocks and counters of clock ticks. This is sometimes called the *causal time* approach and thus, our algebra is called the *Causal Time Calculus (CTC)*. It is worth noting that CTC is actually a descendant of the *Petri Box Calculus* [2] and inherits, in particular, a large part of its syntax, the multiway communication scheme and the concurrent semantics.

A case study [3] made a comparison between the causal time approach and timed automata [1]; it turned out that the verification of Petri nets with causal time using a general model checker for high-level Petri nets (MARIA [13]) was more efficient than the verification of timed automata using well known tools (Kronos [20] and UPPAAL [12]). The approach in [3] was to translate timed automata into the closest possible Petri nets, without any special optimisation.

K.G. Larsen and P. Niebert (Eds.): FORMATS 2003, LNCS 2791, pp. 260–272, 2004.
© Springer-Verlag Berlin Heidelberg 2004

Indeed, an important concern was to avoid a biased comparison. Thus, even if one case study does not allow for any conclusion, this result is very encouraging. However, the causal time approach suffers from a sensitivity to the constants to which ticks counters are compared. The size of the state space actually depends on the product of the largest constants compared to each counter. If one uses $k$ counters, each compared to a value $n$, one gets $n^k$ states only to represent the timing information. We show at the end of this paper that this problem can be removed by identifying states which differ by the values of the ticks counters but are otherwise identical, *i.e.*, lead to the same evolutions. This is very similar to the notion of regions developed for timed automata [1] and allows to use verification techniques based on the concurrent semantics of Petri nets [8] which are generally much more efficient than those based on the interleaving semantics (as in MARIA). The benefits is thus twofold: first, to remove the sensitivity to constants, and second, to improve the good performances obtained in [3].

The next section defines the algebra of terms and its operational semantics. The section 3 presents the Petri nets, called *boxes*, used to define the denotational semantics and gives the transformation from process terms to boxes. These two sections form an extended abstract of the technical report [18] which provides the full definitions, properties and proofs. The section 4 addresses the question of the verification of boxes and is a completely new contribution. We conclude in the section 5 and briefly compare CTC to other timed process algebras.

# 2    CTC Terms: Syntax and Operational Semantics

**Communication.** We assume that there is a set $\mathbb{A}$ of *actions* used to model handshake communication. We also assume that $\tau \notin \mathbb{A}$ and that, for every $a \in \mathbb{A}$, $\widehat{a}$ is also an action in $\mathbb{A}$ such that $\widehat{\widehat{a}} = a$. A *multiaction* is a finite multiset of actions and we denote by $\{\}$ the empty multiaction.

Communication in CTC generalises the synchronisation of CCS [16] (allowed by the parallel composition) followed by the restriction (which forbids the independent execution of synchronised actions). This is formalised through the partial functions $\varphi_{\mathsf{sc}\,a}$, for $a \in \mathbb{A}$, which map the multisets of multiactions allowed to handshake to the multiactions resulting from the communication. For instance, $\varphi_{\mathsf{sc}\,a_1}$ is such that its domain contains $\Gamma \stackrel{\mathrm{df}}{=} \{\{a_1, a_1, a_2\}, \{\widehat{a_1}, a_3\}, \{\widehat{a_1}\}\}$, which denotes that the multiactions of $\Gamma$ may perform a three-way synchronisation. The multiaction corresponding to this communication is given by $\varphi_{\mathsf{sc}\,a_1}(\Gamma) \stackrel{\mathrm{df}}{=} \{a_2, a_3\}$. On the contrary, $\{\{a_1, a_2\}, \{a_2\}\}$ is not in the domain of $\varphi_{\mathsf{sc}\,a_1}$ because the multiactions $\{a_1, a_2\}$ and $\{a_2\}$ cannot handshake.

**Clocks.** The progression of time will be reflected on *clocks* which are nonnegative integer variables that can be tested or updated by the processes. We denote by $\mathbb{N}$ the set of natural numbers. The set $\mathbb{C}$ of clocks is finite and we assume that there exists a function $\mathsf{max} : \mathbb{C} \to \mathbb{N} \setminus \{0\}$ which gives the maximum value allowed for each clock. This allows to specify *deadlines*, *i.e.*, time boundaries within which a process completes [7]. Time progresses when a *tick* occurs incrementing simultaneously all the clocks, which is forbidden when at least one clock has reached its maximum. Notice that we require $\max(c) > 0$ for all $c \in \mathbb{C}$, otherwise no tick could ever occur, resulting in an untimed model.

A *clock vector* is a partial function $\theta : \mathbb{C} \to \mathbb{N}$ such that for all $c \in \mathrm{dom}(\theta)$, $\theta(c) \leq \max(c)$. Such a mapping associates its current value to each clock $c$ in its domain, *i.e.*, the number of ticks which occurred since the last reset of $c$. We denote by $\mathbb{V}$ the set of all clock vectors. For $\theta_1$ and $\theta_2$ in $\mathbb{V}$ such that $\mathrm{dom}(\theta_1) \cap \mathrm{dom}(\theta_2) = \varnothing$, we denote by $\theta_1 + \theta_2$ the clock vector whose domain is $\mathrm{dom}(\theta_1) \cup \mathrm{dom}(\theta_2)$ and which is equal to $\theta_1$ on $\mathrm{dom}(\theta_1)$ and to $\theta_2$ on $\mathrm{dom}(\theta_2)$. By extension, writing $\theta_1 + \theta_2$ will implicitly imply that the domains of $\theta_1$ and $\theta_2$ are disjoint. In the following we denote by $\theta(e)$ the evaluation of the expression $e$ in which the clocks have been replaced by their values as specified by $\theta$. For instance, if $\theta(c) = 3$ then we have $\theta(c + 1) = 4$.

Clocks vectors will be handled through *clock expressions*, attached to the atomic process terms, which are sets of expressions of two kinds: *comparisons* (for instance $c_1 + c_2 > 3$), used to specify a condition under which an atomic process may be executed; and *assignments* (for instance $c_1 := c_2 + 1$) which allow to change the value of a clock. It is required that a clock expression $\delta$ contains at most one assignment for each clock in $\mathbb{C}$. A particular clock expression will be used in the following to represent the occurrence of a tick: $\delta_\tau \stackrel{\mathrm{df}}{=} \{c := c + 1 \mid c \in \mathbb{C}\}$. We say that a clock vector $\theta$ *enables* a clock expression $\delta$ if (1) all the clocks involved in $\delta$ belong to the domain of $\theta$, (2) all the comparisons in $\delta$ evaluate to true through $\theta$, and (3) all the assignments $c := e$ in $\delta$ are such that $\theta(e) \leq max(c)$. In such a case, applying to $\theta$ all the assignments specified in $\delta$ leads to a new vector which is denoted $\delta(\theta)$.

**Syntax.** The syntax of CTC is given in the figure 1. We distinguish *static* terms, denoted by $E$, which cannot evolve, from *dynamic* ones, denoted by $D$, where the current state of the execution is represented by overbars (initial state) and underbars (final state) which may flow through the terms during their execution. We denote by $F$ a static or dynamic term.

The atomic terms are of the form $\alpha\delta$ where $\alpha$ is a multiaction and $\delta$ is a clock expression. Consider for instance the two atomic terms $\{\widehat{a_1}, a_2\}\{\}$ and $\{a_1\}\{c_2 > 0, c_2 := 0\}$. The first one denotes the simultaneous receiving of the signal $a_1$ and sending of the signal $a_2$, which is untimed; the second one can send the signal $a_1$ and reset the clock $c_2$ if the value associated to $c_2$ is greater than zero. Various operators allow to combine terms:

$$E ::= \quad \alpha\delta \quad | \; E \operatorname{sc} a \; | \; E\|E \; | \; E\mathbin{\raisebox{0.2ex}{$\mathrm{\,;\,}$}}E \; | \; E \,\square\, E \; | \; E \circledast E \; | \; E @ \theta$$

$$D ::= \quad \overline{E} \quad | \quad \underline{E} \quad | \; D \operatorname{sc} a \; | \; D\|D \; | \; D\mathbin{\raisebox{0.2ex}{$\mathrm{\,;\,}$}}E \; | \; E\mathbin{\raisebox{0.2ex}{$\mathrm{\,;\,}$}}D$$
$$\qquad\quad | \; D \,\square\, E \; | \; E \,\square\, D \; | \; D \circledast E \; | \; E \circledast D \; | \; D @ \theta$$

**Fig. 1.** The syntax of CTC terms, where $\alpha\delta$ is an atomic term, $a \in \mathbb{A}$ and $\theta \in \mathbb{V}$

- the *sequential composition* $F_1 \, ; F_2$ ($F_1$ is executed first and followed by $F_2$) may be seen as a generalisation of the prefixing operator used in CCS;
- the *choice* $F_1 \, \square \, F_2$ (either $F_1$ or $F_2$ may be executed) corresponds to the choice of CCS;
- the *parallel composition* $F_1 \| F_2$ ($F_1$ and $F_2$ may evolve concurrently) differs from that used in CCS since it does not allow for synchronisation;
- the *iteration* $F_1 \circledast F_2$ ($F_1$ is executed an arbitrary number of times and is followed by $F_2$) allows to represent repetitions while in CCS the recursion would be used;
- the *scoping* $F$ sc $a$ (all the handshakes involving $a$ and $\widehat{a}$ are enforced) was discussed above.

In order to model the clocks, terms are decorated with clock vectors. For instance, we may form $\{a\}\{c := 0\} \, @ \, \{c \mapsto 5\}$ denoting that the clock $c$ has value the 5 for the atomic term $\{a\}\{c := 0\}$.

**Operational Semantics.** An important part of the operational semantics relies on equivalence rules allowing to identify distinct terms which actually correspond to the same state. Formally, we define $\equiv$ as the least equivalence relation on terms such that all the rules in the figure 2 are satisfied. Consider for instance the rule IS2: it states that having the first component of a sequence in its final state is equivalent to have the second component in its initial state, which is indeed the expected semantics of a sequential composition.

Contrasting with CCS where evolutions are expressed by removing prefixes of terms, like in $a.P \xrightarrow{a} P$, the structure of CTC terms never evolves; instead, the overbars may be changed to underbars as in

$$\overline{\{a\}\{c := 0\}} \, @ \, \{c \mapsto 5\} \xrightarrow{(\{a\}, \{c \mapsto 5\})} \underline{\{a\}\{c := 0\}} \, @ \, \{c \mapsto 0\}$$

which produces the *timed multiaction* $(\{a\}, \{c \mapsto 5\})$ denoting the occurrence of $\{a\}$ when the clock $c$ had the value 5, while the reset $c := 0$ has been reflected on the new clock vector. In order to have a concurrent semantics, several timed multiactions may be combined, denoting their concurrent occurrence. Given $\Gamma_1$ and $\Gamma_2$ two multisets of multiactions and $\theta_1, \theta_2 \in \mathbb{V}$ having disjoint domains, $A_1 \stackrel{df}{=} (\Gamma_1, \theta_1)$ and $A_2 \stackrel{df}{=} (\Gamma_2, \theta_2)$ are *timed multisets of multiactions* and $A_1 + A_2 \stackrel{df}{=} (\Gamma_1 + \Gamma_2, \theta_1 + \theta_2)$.

Then, we define a ternary relation $\longrightarrow$ as the least relation comprising all $(F, A, F')$ where $F$ and $F'$ are terms and $A$ is a timed multiset of multiactions, such that the rules in the figure 3 hold. Notice that we use $F \xrightarrow{A} F'$ to denote $(F, A, F') \in \longrightarrow$. When used with $\diamond = \|$, the rule EOP is the way through which true concurrency is introduced. When used with another operator, the syntax ensures that at most one of $A_1$ or $A_2$ has a nonempty multiset of multiactions since at least one of the operands must be a static term. In these cases, the rule EOP shall be used in conjunction with EQ1 in order to compose a static term with a dynamic one. Concerning the rule ETICK, it should be noted that the side condition "$\theta$ enables $\delta_\tau$" implies that $\mathrm{dom}(\theta) = \mathbb{C}$; thus the occurrence of a tick always simultaneously increments all the clocks.

| | | | |
|---|---|---|---|
| EX | $\dfrac{E \equiv E'}{E \equiv E'}$ | ENT | $\dfrac{E \equiv E'}{\overline{E} \equiv \overline{E'}}$ |
| CON1 | $\dfrac{F \equiv F'}{F \text{ sc } a \equiv F' \text{ sc } a}$ | CON2 | $\dfrac{F_1 \equiv F_1',\ F_2 \equiv F_2'}{F_1 \diamond F_2 \equiv F_1' \diamond F_2'}$ |
| ISC1 | $\overline{E} \text{ sc } a \equiv \overline{E \text{ sc } a}$ | ISC1 | $\underline{E} \text{ sc } a \equiv \underline{E \text{ sc } a}$ |
| IPAR1 | $\overline{E_1 \| E_2} \equiv \overline{E_1} \| E_2$ | IPAR2 | $\underline{E_1 \| E_2} \equiv \underline{E_1} \| E_2$ |
| IC1L | $\overline{E_1 \square E_2} \equiv \overline{E_1} \square E_2$ | IC2L | $\underline{E_1 \square E_2} \equiv \underline{E_1} \square E_2$ |
| IC1R | $\overline{E_1 \square E_2} \equiv E_1 \square \overline{E_2}$ | IC2R | $\underline{E_1 \square E_2} \equiv E_1 \square \underline{E_2}$ |
| IS1 | $\overline{E_1 \mathbin{\fatsemi} E_2} \equiv \overline{E_1} \mathbin{\fatsemi} E_2$ | IS2 | $\underline{E_1 \mathbin{\fatsemi} E_2} \equiv E_1 \mathbin{\fatsemi} \overline{E_2}$ |
| IS3 | $E_1 \mathbin{\fatsemi} \underline{E_2} \equiv \underline{E_1 \mathbin{\fatsemi} E_2}$ | IIT1 | $\overline{E_1 \circledast E_2} \equiv \overline{E_1} \circledast E_2$ |
| IIT2 | $\underline{E_1} \circledast E_2 \equiv \overline{E_1 \circledast E_2}$ | IIT3 | $\underline{E_1} \circledast E_2 \equiv E_1 \circledast \overline{E_2}$ |
| IIT4 | $\overline{E_1} \circledast E_2 \equiv E_1 \circledast \overline{E_2}$ | IIT5 | $E_1 \circledast \underline{E_1} \equiv E_1 \circledast \underline{E_2}$ |
| IAT1 | $\dfrac{F \equiv F'}{F @ \theta \equiv F' @ \theta}$ | IAT2 | $\overline{E @ \theta} \equiv \overline{E} @ \theta$ |
| IAT3 | $\underline{E @ \theta} \equiv \underline{E} @ \theta$ | IAT4 | $(F \text{ sc } a) @ \theta \equiv (F @ \theta) \text{ sc } a$ |
| IAT5 | $(F_1 \diamond F_2) @ (\theta_1 + \theta_2) \equiv (F_1 @ \theta_1) \diamond (F_2 @ \theta_2)$ | | |

**Fig. 2.** Similarity relation, where $a \in \mathbb{A}$, $\diamond \in \{\|, \mathbin{\fatsemi}, \square, \circledast\}$ and $\{\theta, \theta_1, \theta_2\} \subset \mathbb{V}$

| | | |
|---|---|---|
| EQ1 | $F \xrightarrow{(\{\}, \theta_\varnothing)} F$ | where $\mathsf{dom}(\theta_\varnothing) = \varnothing$ |
| EQ2 | $\dfrac{F \equiv F',\ F' \xrightarrow{A} F'',\ F'' \equiv F'''}{F \xrightarrow{A} F'''}$ | |
| EAT | $\dfrac{F \xrightarrow{(\Gamma, \theta_1)} F'}{F @ \theta_2 \xrightarrow{(\Gamma, \theta_1 + \theta_2)} F' @ \theta_2}$ | |
| EA | $\overline{\alpha\delta} @ \theta \xrightarrow{\{(\alpha, \theta)\}} \underline{\alpha\delta} @ \delta(\theta)$ | if $\theta$ enables $\delta$ |
| ETICK | $\dfrac{F @ \theta \xrightarrow{A} F' @ \theta}{F @ \theta \xrightarrow{A + (\{\tau\}, \theta)} F' @ \delta_\tau(\theta)}$ | if $\theta$ enables $\delta_\tau$ |
| ESC | $\dfrac{F \xrightarrow{(\Gamma_1, \theta_1) + \cdots + (\Gamma_k, \theta_k)} F'}{F \text{ sc } a \xrightarrow{(\varphi_{\text{sc } a}(\{\Gamma_1, \ldots, \Gamma_k\}), \theta_1 + \cdots + \theta_k)} F' \text{ sc } a}$ | if $\tau$ does not appear in any $\Gamma_i$ and $\{\Gamma_1, \ldots, \Gamma_k\} \in \mathsf{dom}(\varphi_{\text{sc } a})$ |
| EOP | $\dfrac{F_1 \xrightarrow{A_1} F_1',\ F_2 \xrightarrow{A_2} F_2'}{F_1 \diamond F_2 \xrightarrow{A_1 + A_2} F_1' \diamond F_2'}$ | if $\tau$ does not appear in $A_1$ neither in $A_2$ |

**Fig. 3.** Evolution rules, where $\alpha\delta$ is an atomic term, $\{\theta, \theta_1, \theta_2, \theta_\varnothing\} \subset \mathbb{V}$, $a \in \mathbb{A}$, $\diamond \in \{\|, \mathbin{\fatsemi}, \square, \circledast\}$ and assuming that all the applications of $+$ are well defined

# 3   Denotational Semantics

**The Algebra of Boxes.** We start by introducing the labelled coloured Petri nets called *boxes* and the operations used to compose them. These operations exactly correspond to those defined on terms: for each operator on terms, there exists a similar operator defined on boxes.

The labelling of boxes allows to distinguish the *entry*, *internal* and *exit* places; all together, they are called the *control* places since their role is to represent the current state of the control flow. The marking of the entry places corresponds to the initial marking of a box and thus we define $\overline{N}$ as the box $N$ in which one token is added to each entry place. Similarly, the exit places correspond to the final marking and $\underline{N}$ is defined as expected. The internal places correspond to intermediate states during the execution of a box. Except for the scoping, the operators of CTC are also based on the labels of places. For instance, the sequential composition $N_1 \mathbin{\fatsemi} N_2$ is defined by combining the exit places of $N_1$ with the entry places of $N_2$, resulting in internal places whose marking represent both the final marking of $N_1$ and the initial one of $N_2$.

Another class of places is distinguished thanks to their labels, these are the *clock* places in which clock values are modelled. A box has exactly one clock place labelled by $c$ for each $c \in \mathbb{C}$. When several nets are combined, for instance using the parallel composition, clock places with the same label are automatically merged (with their markings) ensuring a unique representation of each clock. While the control places are only allowed to carry the ordinary black token $\bullet$, each clock place may carry any integer from $\mathbb{N}$. Thus, the clock places are the only coloured ones.

The labelling of an arc consists in a multiset of values, variables or expressions which represents the tokens flowing on the arc when the attached transition is executed.

The labelling of a transition contains a multiaction as in atomic terms. This allows to define the scoping w.r.t. $a \in \mathbb{A}$ whose role is to merge sets of transitions whose labels $\alpha_1, \ldots, \alpha_k$ belong to the domain of $\varphi_{\mathsf{sc}\,a}$, the newly created transition being labelled by $\varphi_{\mathsf{sc}\,a}(\{\alpha_1, \ldots, \alpha_k\})$. Transitions are also labelled by guards (boolean conditions involving the variables used in adjacent arcs) which must evaluate to true in order to allow the execution of the transitions. When this occurs, the variables in the guard and the adjacent arcs are associated to values through a *binding* $\sigma$ and we denote by $t_\sigma$ the occurrence of $t$ under the binding $\sigma$.

**From Terms to Boxes.** Let $\alpha\delta$ be an atomic term, its denotational semantics is given by the box $N_{\alpha\delta}$ defined as follows. Its places are: one entry place $s_e$, one exit place $s_x$, and one clock place $s_c$ labelled by $c$ for each $c \in \mathbb{C}$. The marking is empty for the control places and $\{0\}$ for the clock places. The box has one transition $t$ labelled by $\{\tau\}(\bigwedge_{c \in \mathbb{C}} c < \mathsf{max}(c))$ which models the tick and, for each $c \in \mathbb{C}$, there is one arc labelled by $\{c\}$ from $s_c$ to $t$ and one arc labelled by $\{c + 1\}$ from $t$ to $s_c$. There is also one transition $u$ labelled by $\alpha\gamma$, where $\gamma$

is the disjunction of all the comparisons in $\delta$ (or true if there is no comparison in $\delta$), which models the atomic action $\alpha\delta$. This transition $u$ has one incoming arc from $s_e$ labelled by $\{\bullet\}$ and one outgoing arc to $s_x$ with the same label. The other arcs on $u$ correspond to the clocks involved in $\delta$; for all $c \in \mathbb{C}$:

- if $c$ appears in $\delta$ in comparisons only, then there is one arc from $s_c$ to $u$ labelled by $\{c\}$ and one arc from $u$ to $s_c$ with the same label;
- if $c$ appears in $\delta$ in an assignment $c := e$, where $e$ is an expression, then there is one arc from $s_c$ to $u$ labelled by $\{c\}$ and one arc from $u$ to $s_c$ with the label $\{e\}$;
- if $c$ does not appear in $\delta$ then there is no arc between $s_c$ and $u$.

The denotational semantics is then defined by induction:

$$\mathsf{box}(\alpha\delta) \stackrel{\mathrm{df}}{=} N_{\alpha\delta} \qquad \mathsf{box}(\overline{E}) \stackrel{\mathrm{df}}{=} \overline{\mathsf{box}(E)} \qquad \mathsf{box}(\underline{E}) \stackrel{\mathrm{df}}{=} \underline{\mathsf{box}(E)}$$

$$\mathsf{box}(F \text{ sc } a) \stackrel{\mathrm{df}}{=} \mathsf{box}(F) \text{ sc } a \qquad \mathsf{box}(F_1 \diamond F_2) \stackrel{\mathrm{df}}{=} \mathsf{box}(F_1) \diamond \mathsf{box}(F_2)$$

where $\alpha\delta$ is an atomic term, $a \in \mathbb{A}$ and $\diamond \in \{\|, \talloblong, \square, \circledast\}$. Moreover, for $\theta \in \mathbb{V}$, $\mathsf{box}(F \text{ @ } \theta)$ is $\mathsf{box}(F)$ in which the marking of each clock place labelled by $c \in \mathsf{dom}(\theta)$ is set to $\{\theta(c)\}$. For example, assuming $\mathbb{C} \stackrel{\mathrm{df}}{=} \{c\}$ and $\mathsf{max}(c) \stackrel{\mathrm{df}}{=} 4$, the box on the left of the figure 4 is

$$\mathsf{box}\left(\overline{\{a_1\}\{c := 0\} \talloblong \{a_2\}\{c \geq 2\}}\right) \quad .$$

The operational and denotational semantics are closely related: they are actually consistent in arguably the strongest sense since a term and the corresponding box generate isomorphic transitions systems.

## 4   Verification through Unfoldings

A well known technique to perform efficient model checking on Petri nets is to use their concurrent semantics expressed by prefixes of their *unfolding* which are also Petri nets, see [8]. The traditional definition of unfoldings is based on low-level Petri nets, but it was shown in [10] that coloured Petri nets like boxes may be unfolded as well (producing a low-level net). An example of a box and a prefix of its unfolding is given in the figure 4.

In the unfolding, places are called *conditions* and are labelled by the name and the marking of the place to which they correspond in the original net; similarly, transitions are called *events* and correspond to the transition occurrence which labels them. The labelling function is an *homomorphism* and will be denoted by $h$ in the following. In the figure 4, this labelling is indicated inside the nodes and is simplified: a condition labelled by an integer $n$ denotes the presence of $n$ in the place $s_c$ and conditions labelled by $s_1$, $s_2$ or $s_3$ denote the token $\bullet$ in the corresponding place.

An unfolding may be executed by putting one token in each condition with no predecessor. One can check on the example that this allows to perfectly mimic the

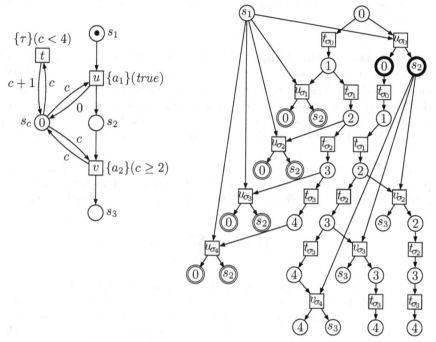

**Fig. 4.** The box of $\overline{\{a_1\}\{c := 0\}} \,\mathring{,}\, \{a_2\}\{c \geq 2\}$ (on the left) and a prefix of its unfolding (on the right), where $\sigma_i \stackrel{\mathrm{df}}{=} \{c \mapsto i\}$ for $0 \leq i \leq 4$; assuming $\mathbb{C} \stackrel{\mathrm{df}}{=} \{c\}$ and $\mathsf{max}(c) \stackrel{\mathrm{df}}{=} 4$

behaviour of the original net. Notice that, when the pairs of conditions depicted with double lines become marked, the execution may be continued from the conditions depicted with thick lines. Indeed, these double-lined pairs are *cuts* where the unfolding have been truncated since the corresponding markings were already represented by the thick-lined pair of conditions. This allows to consider only *prefixes* of the full unfolding which may be itself infinite (if the net has an infinite run). Such a prefix is *complete* (w.r.t. reachability properties) if every reachable marking of the original net is represented in the prefix. This guarantees that reachability properties can be verified on the prefix rather than on the original net.

The notion of completeness actually depends on the properties that should be verified. Usually, those related to reachability are considered, but different ones may be envisaged like in [9]. In our case, if only control flow properties have to be verified on a box, the occurrences of ticks and the markings of clock places could be removed from the unfolding of this box. In the following, we present an intermediate simplification which keeps some timing information but without its full precision: it will not be possible to exactly know the values of the clocks when an event occurs; instead, we will obtain a range of possible values. Moreover, in the simplified unfolding, an event labelled by an occurrence of the tick transition will denote that "time is passing" instead of the more accurate "one tick occurs".

**Simplification of Unfoldings.** We now show how to collapse chains of ticks (as, *e.g.*, at the bottom-right of the figure 4) thus removing the sensitivity of model checking to the constants used in clock expressions. It should be stressed that, for practical applications, the transformation described below should be applied on-the-fly during the computation of the unfolding; but, the principle being independent of the algorithm actually used, we prefer the current presentation.

Let $x$ be an event or a condition, we denote by ${}^\bullet x$ the set of nodes immediately preceding $x$ and by $x^\bullet$ those immediately succeeding $x$. This notation naturally extends to sets of nodes. For a set $E$ of events, we denote by $\mathsf{trans}(E)$ the multiset of transitions involved in $E$, *i.e.*,

$$\mathsf{trans}(E) \stackrel{\mathrm{df}}{=} \sum_{e \in E \wedge h(e) = w_\sigma} \{w\} \quad .$$

To start with, we change the labelling of conditions to triples $(s, p, q)$ where $s$ is a place of the original net and $p$, $q$ are integers such that $0 \le p \le q$. If $s$ is a control place, this label indicates that the condition corresponds to $s$ marked by $\bullet$; but if $s$ is a clock place, the condition corresponds to the place $s$ whose marking is any integer in $\{p, \dots, q\}$. So, the labelling is changed as follows: for each condition which corresponds to the marking of the control place $s$, the label becomes $(s, 0, 0)$; for each condition which corresponds to the marking of the clock place $s'$ by the integer $n$, the label is changed to $(s', n, n)$.

Then, we consider an event $e$ labelled by an occurrence $t_\sigma$ of the tick transition. We call $e$ a *tick event*. One can show that, if ${}^\bullet e = \{c_1, \dots, c_k\}$ with $h(c_i) = (s_i, p_i, q_i)$ for $1 \le i \le k$, then, because the tick transition is connected to clock places through side loops (and not connected to any other place), we must have $e^\bullet = \{c'_1, \dots, c'_k\}$ and $h(c'_i) = (s_i, p'_i, q'_i)$ for $1 \le i \le k$. We distinguish two sets of events: $E \stackrel{\mathrm{df}}{=} ({}^\bullet e)^\bullet$ which contains all the events in conflict with $e$ (including $e$) and $E' \stackrel{\mathrm{df}}{=} (e^\bullet)^\bullet$ which contains all the events enabled by the occurrence of $e$. Then, if $\mathsf{trans}(E) = \mathsf{trans}(E')$, it means that the tick do not change the enabling in the net (it may change the bindings but not the transitions which are enabled). So, $e$ is removed and the conditions in $e^\bullet$ are merged to those in ${}^\bullet e$. Each condition $c'_i$ (whose label is $(s_i, p'_i, q'_i)$) is merged to the corresponding $c_i$ (labelled by $(s_i, p_i, q_i)$) as follows:

- the condition $c'_i$ is removed and the label of $c_i$ is changed to $(s_i, p_i, q'_i)$;
- each tick event $e' \in E'$ becomes a successor of $c_i$;
- each non tick event $e' \in E'$ is removed as well as all its successors nodes. This allows to remove branches which were already possible before the occurrence of the tick.

This simplification step has to be repeated iteratively for all the tick events. We already remarked that, during each step, for $1 \le i \le k$, $c_i \in E$ and $c'_i \in E'$

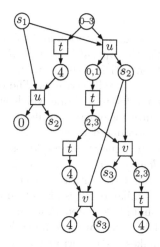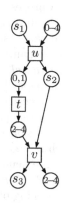

**Fig. 5.** On the left, the prefix generated using the first method, and on the right, using the second one. The bindings are no more relevant and thus not indicated

are such that $h(c_i) = (s_i, p_i, q_i)$ and $h(c_i') = (s_i, p_i', q_i')$. One can now show that we also have $p_i' = q_i + 1$ and that this remains true after some tick events have been removed. It may also be shown that the order in which tick events are considered has no influence on the final result.

By applying this transformation, we obtain the prefix given on the left of the figure 5, notice that some conditions are now labelled by lists or ranges of integers when they correspond to several possible markings of $s_c$. One can see that the left part of the original prefix have been simplified and that the only remaining visible tick is the one which leads to have 4 in $s_c$ thus disabling any further tick. Similarly, the right branch was also simplified and the two remaining occurrences of $v$ correspond to the two following situations: both $v$ and $t$ are enabled; or, only $v$ is enabled.

It may be considered that too much information is still present in the prefix. In particular, one can distinguish between states from which tick can or cannot occur, which is an information only related to our particular modelling of time. In order to simplify again, we can use the same transformation scheme but, instead of removing a tick event when $\mathsf{trans}(E) = \mathsf{trans}(E')$, we use the weaker condition $\mathsf{trans}(E/\tau) = \mathsf{trans}(E'/\tau)$ where $X/\tau$ is $X$ from which all the tick events have been removed. This new criterion leads to the prefix given on the right of the figure 5 in which the only remaining tick event denotes that time *must* pass. All the situations in which time only *may* pass have been hidden. Choosing one or the other criterion depends if one wants to always know when ticks are possible or not. But, in both cases, we achieved our goal which was to remove the sensitivity to constants.

# 5    Conclusion

We defined a process algebra with multiway communication and timing feature through clocks directly handled. This model, the *Causal Time Calculus* (*CTC*), was provided with a structural operational semantics as well as with a consistent denotational semantics in terms of labelled coloured Petri nets. These nets use the so called *causal time* approach to the modelling of time which was shown in a previous paper [3] having the potentiality for efficient verification but suffering from a sensitivity to the constants compared to clocks. An important contribution of this paper was to show how to remove this weakness.

As an extension of the *Petri Box Calculus* (*PBC*) [2], CTC is similar to the approach in [11] where the author extends PBC with time using *time Petri nets* [15] for the denotational semantics. A similar result is also obtained in [14] where *timed Petri nets* [19] are used. It should be noted that, in both cases, the model checking of the underlying models is known to be much less efficient than that of standard Petri nets. This makes an importance difference with CTC for which the efficiency of the verification was a major concern. Moreover, we introduced time through explicit clocks directly handled by the processes which is known to be useful for modelling timed systems (this is indeed the scheme used in timed automata). Among process algebras not related to PBC, we should distinguish ARTS [6, chap. 5] which has been designed in order to denote timed automata while CTC denotes Petri nets; ARTS thus provides continuous time while CTC uses discrete time. Another difference is that the operational semantics in ARTS is used to give a translation from terms to automata while in CTC it is independent of Petri nets (even if both semantics are consistent). It finally appears that both algebras may be complementary as they denote objects on which model checking can be performed efficiently. Which one to use in which case is still a topic for future research. Concerning the other process algebra with time (for instance those based on CCS, see [5]), it may be remarked that most of them also use ticks to model the passing of time. However, they generally consider an interleaving semantics of parallelism while CTC considers true concurrency and most of these algebras do not provide multiway as in CTC.

Several extensions to the model presented here can be envisaged, in particular: actions with parameters, allowing to exchange data during handshakes; buffered communication, allowing to model program variables; and guards, allowing to specify conditions under which an atomic process may be executed. Incorporating these features should be straightforward since they are already defined in several extensions of PBC (in particular in [4]). Another extension would be to allow the maximum values of clocks to be changed dynamically. This must be addressed carefully in order to guaranty that either a finite prefix of the unfoldings can always be found or methods dedicated to infinite state spaces can be used.

Last but nor least, an in-depth study of the unfolding simplification proposed here appears necessary in order to know exactly what is its influence on the properties which can be verified: which one are preserved and which one are hidden. One way to reach this goal is to define a timed temporal logic in order to specify properties which could then be verified automatically. The more complete this logic will be, the more we will know about the properties preserved by our unfolding simplification.

## Acknowledgement

I am very grateful to Victor Khomenko for his advice about Petri nets unfoldings.

## References

[1] R. Alur and D. Dill. A theory of timed automata. Theoretical Computer Science 126(2). Elsevier, 1994. · 260, 261

[2] E. Best, R. Devillers and J. Hall. The Petri Box Calculus: a new causal algebra with multilabel communication. Advance in Petri nets 1992, LNCS 609. Springer, 1992. 260, 270

[3] C. Bui Thanh, H. Klaudel and F. Pommereau. Petri nets with causal time for system verification. MTCS'02, Electronic Notes in Theoretical Computer Sciences 68.5. Elsevier, 2002. 260, 261, 270

[4] C. Bui Thanh, H. Klaudel and F. Pommereau. Box Calculus with Coloured Buffers. LACL Technical report, 2002. Available at http://www.univ-paris12.fr/lacl. 270

[5] F. Corradini, D. D'Ortenzio and P. Inverardi. On the relationship among four timed process algebras. Fundamenta Informaticae 34. IOS Press, 1999. 270

[6] P. R. D'Argenio. Algebras and automata for real-time systems. PhD. Thesis, Department of Computer Science, University of Twente, 1999. 270

[7] R. Durchholz. Causality, time, and deadlines. Data & Knowledge Engineering 6. North-Holland, 1991. 262

[8] J. Esparza. Model checking using net unfoldings. Science of Computer Programming 23. Elsevier, 1994. 260, 261, 266

[9] V. Khomenko, M. Koutny et W. Vogler. Canonical prefixes of Petri net unfoldings. CAV'02, LNCS 2404. Springer, 2002. 267

[10] V. Khomenko and M. Koutny. Branching processes of high-level Petri nets. TACAS'03, LNCS 2619. Springer, 2003. 266

[11] M. Koutny. A compositional model of time Petri nets. ICATPN'00, LNCS 1825. Springer, 2000. 270

[12] K. G. Larsen, P. Pettersson et W. Yi. UPPAAL in a nutshell. International Journal on Software Tools and Technology Transfer 1(1-2). Springer, 1997. 260

[13] M. Mäkelä. MARIA: modular reachability analyser for algebraic system nets. Online manual, http://www.tcs.hut.fi/maria, 1999. 260

[14] O. Marroquín Alonzo and D. de Frutos Escrig. Extending the Petri Box Calculus with time. ICATPN'01, LNCS 2075. Springer, 2001. 270

[15] P. M. Merlin and D. J. Farber. Recoverability of communication protocols—implications of a theoretical study. IEEE Transaction on Communication 24. IEEE Society, 1976. 270

[16] R. Milner. Communication and concurrency. Prentice Hall, 1989. 261

[17] G. D. Plotkin. A Structural approach to Operational Semantics. Technical Report FN-19, Computer Science Department, University of Aarhus, 1981. 260

[18] F. Pommereau. Causal Time Calculus. LACL Technical report, 2002. Available
at http://www.univ-paris12.fr/lacl. 261

[19] C. Ramchandani. Analysis of asynchronous concurrent systems using Petri nets.
PhD. Thesis, project MAC, MAC-TR 120. MIT, 1974. 270

[20] S. Yovine. Kronos: A verification tool for real-time systems. International Journal
of Software Tools for Technology Transfer, 1(1/2). Springer, 1997. 260

# *ELSE*: A New Symbolic State Generator for Timed Automata

Sarah Zennou, Manuel Yguel, and Peter Niebert

Laboratoire d'Informatique Fondamentale de Marseille
Université de Provence – CMI, 39, rue Joliot-Curie / F-13453 Marseille Cedex 13
{zennou,niebert}@cmi.univ-mrs.fr

**Abstract.** We present ELSE, a new state generator for timed automata. ELSE is based on VERIMAG's IF-2.0 specification language and is designed to be used with state exploration tools like CADP. In particular, it compiles IF-2.0 specifications to C programs that link with CADP. It thus concentrates on the generation of comparatively small state spaces and integrates into existing tool chains. The emphasis of the ELSE development is on fundamentally different data structures and algorithms, notably on the level of zones. Rather than representing possible values of clocks at a given symbolic state, event zones represent in an abstract way the timing constraints of past and future events.

## 1  Introduction

Timed automata [AD94] are a powerful tool for the modeling and analysis of timed systems. They extend classical automata by *clocks*, continuous variables "measuring" the flow of time. A state of a timed automaton is thus a combination of its discrete control locations and the *clock values* taken from the real domain. While the resulting state spaces are infinite, *clock constraints* have been introduced to abstract the state spaces to a finite set of equivalence classes, thus yielding a finite (although often huge) symbolic state graph on which reachability and some other verification problems can be resolved.

While the theory, algorithms and tools like IF/Kronos [BFG+99] and Uppaal [LPY95] for timed automata represent a considerable achievement, they suffer for various reasons from combinatory explosion which still limits their applicability in practice. A great effort has been invested into optimization of representations of clock constraints, e.g. [DY96, BLP+99].

*ELSE*, developed at the Laboratoire d'Informatique Fondamentale in Marseille, is a new state generator – engine of algorithmic analysis – for timed automata incorporating alternative semantics that may allow certain partial order reduction approaches. *ELSE* is designed to be compatible with IF [BFG+99], notably the IF-2.0 specification language, for which it implements a new semantics allowing state space reduction with respect to parallelism while preserving reachability properties: The components of a parallel system (like networks of communicating transition systems) can sometimes progress independently, sometimes

K.G. Larsen and P. Niebert (Eds.): FORMATS 2003, LNCS 2791, pp. 273–280, 2004.
© Springer-Verlag Berlin Heidelberg 2004

interact. Basic verification techniques rely on an interleaving approach, where *global states* are *tuples of local states*. The resulting global transition systems can have a very redundant structure (which is responsible for the state explosion) including so called *diamonds*, pairs of commuting transitions that can be executed in either order leading to the same state. *Partial order reduction techniques* [Pel93, God96] together with their tools (e.g. SPIN) give an answer to this phenomenon in reducing the search space based on this redundancy for discrete systems.

**Partial Order Semantics for Timed Automata.** A natural question is the applicability of partial order reductions to networks of timed automata. However, as has been observed by several authors, the standard interleaving semantics combined with symbolic states via clock constraints results in transitions that do *not commute*. Several kinds of answers have been given to this problem: Adapt *persistent set method* [YS97]; Define a local time semantics [BJLY98, Min99]. In [DT98], the authors starts adapting the notion of equivalence classes of transition sequences in timed automata.

This approach followed by *ELSE* is formally defined in [LNZ03]. Basically, the chosen semantics relaxes constraints between independent components (automata and clocks) so that diamonds are almost preserved. The termination and bound on the number of symbolic states, a problem known from [BJLY98, Min99], is ensured by a symbolic state equivalence relation, which allows us to explore just one state of an equivalence class. The equivalence is closely related to well known abstractions on clock constraints in classical timed automata, but the use is radically different: Rather than modifying/abstracting states during exploration, states with an already explored equivalent state are *cut*. This guarantees that The price for this guarantee, that our symbolic transition systems are not bigger than those resulting from the classical approach, but that compared to [BJLY98, Min99], we do not preserve full commutation and partial order reductions cannot be applied naively.

But partial order reduction is not the only concern about the alternative semantics. As we will indicate, classical semantics suffers from a state splitting phenomenon (equivalent discrete paths lead to incomparable symbolic states), which we can avoid with the alternative semantics.

## 2     Scope and Architecture of *ELSE*

The generator *ELSE* is a tool to automatically translate a description of a network of timed automata in the IF syntax to C code providing a data structure for the symbolic states as defined in [LNZ03], and a mechanism to compute symbolic transitions from a given symbolic state. These elements can then be used by tools for exploring a symbolic reachability graph, CADP in particular. *ELSE* thus remains just one, yet a crucial component in a tool chain: Specifications may either directly be written in IF or obtained by (existing) translations to IF from various commonly used specification languages. *ELSE* provides a state generator; CADP may be used to explore it for reachability or more complex verification algorithms.

To achieve this, *ELSE* is composed of:

- A compiler, *else2c*, which generates the symbolic transition systems. It translates the description of a set of timed automata into C functions which compute them. The description must be done in the IF syntax;
- A library computing operations on clock constraints, called *elsezone*. This one contains all functions which are needed by functions generated by the compiler.

To summarize, the following figure shows how *ELSE* works: Given a network of timed automata which are described in IF (file sys.els), the compiler *else2c* generates C functions (file sys.c) computing the corresponding symbolic transition system. The generated functions may call functions on clock constraints of the clock constraint library (file elsezone.c).

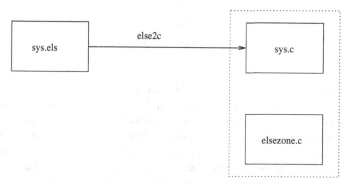

**Internals of *else2c*.** After the syntactic verification of the system description, the generation of the C code used as entry by a graph explorer is done in two main steps:

First, the C representation of (symbolic) states of the reachability graph is created. The generated data structure, a static record with some dynamic attachments, represents in a hierarchical manner the hierarchy of the system structure, as much as is possible in C, e.g. the automata are represented as subrecords, etc. Clock constraint representations in our setting do not have a fixed size and are kept apart.

Then the compiler computes a function for the interface, which, given a symbolic state of the reachability graph, computes and returns implicitly the list of its successors. A specific function is generated for the initial state creation.

## 3   Event Zones and the Library *elsezone*

For this section, we assume some familiarity with zones as used in classical timed automata tools.

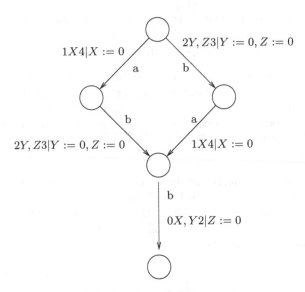

**Fig. 1.** Example automaton

The library *elsezone* contains all the functions computing the operations on constraints, called *event zones*, as defined in [LNZ03]. For illustration, let us consider a part of a timed automaton, transitions name $a$, $b$ and $c$, where we assume $a$ and $c$ to belong to process number 0 whereas $b$ and $c$ are transitions of process number 0. In other words, transitions $a$ and $b$ are of independent origin (and address different clocks here), whereas transition $c$ is a common transition, dependent on both $a$ and $b$ (and the clocks it addresses). Such a situation is depicted in Figure 1.

Zones of classical timed automata represent, "*clock zones*", sets of clock values symbolically by differences bounds matrices. Matrices as in Figure 2 represent for pairs of variables $(x, y)$ constraints for the difference $x - y$. The passage from one symbolic state to another here consists of several steps (letting the time advance, i.e. relax upper bounds of clock variables; intersection with transition bound; resetting clocks (to zero) by coupling them to the variable corresponding to 0). The nature of these steps implies that symbolic paths executing $a$ first and then $b$ or the other way around lead to incomparable clock zones and these differences may be propagated. This is one additional source of state explosion in timed automata.

In contrast to this classical approach, we have taken a philosophical shift for *ELSE*: Event zones represent constraints for the *occurrence times* of certain events. A constraint "$Y3$" guarding a transition $b$ can indeed be read in two ways: (a) at the moment of occurrence of the transition, clock $Y$ must have a value inferior to 3. (b) the difference of occurrence times of the transition in question and the last preceeding reset of clock $Y$ is 3.

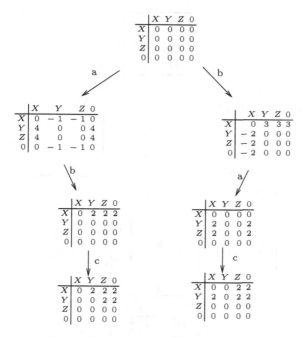

**Fig. 2.** Symbolic states with classical zones

Event zones (described in more detail in [LNZ03]) thus consist of a matrix of constraints on events and a list of "pointers" from clocks (and from components) to indicate, which was the last event of a reset of clock $X$, etc. The latter pointers are needed as it is via them that constraints link future events with the events already present in the event zone: either from clock constraints or due to causality (an event $f$ causally depending on a preceeding event $e$ must occur *later*). After a reset of clock $X$, no preceeding reset of $X$ can be linked to by a clock constraint on $C$ in a future transition. Since no more reference to this event is possible, we can remove it like "garbage collection". Events that are not referenced by pointers may be projected. As a consequence, the dimension of the constraints matrix is variable (may grow and shrink again). But its growth is bounded to $n+m$ dimensions (number of clocks plus number of processes). In case of a fully sequential system, this bound is equal to $n+1$ which corresponds to the dimension of classical zones (one dimension per clock and one dimension for "0"). The initial event zone has dimension one, consisting of a hypothetical "start event" (which occurs at time 0 and where all clocks are reset to 0).

As a transition occurs,(1) add a new event, (2) recompute constraints with respect to its links to previous events (this is an incremental Floyd-Warshall algorithm that is at worst of quadratic complexity), (3) change references, (4) garbage collection of events (lines and columns) in the matrix that lack a reference.

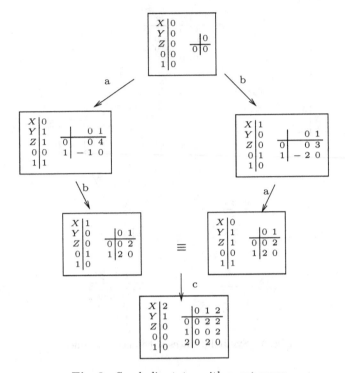

**Fig. 3.** Symbolic states with event zones

The event based view has as consequence that the independent events of transitions $a$ and $b$ indeed are not directly linked by constraints and are in particularly not affected by the order of occurrence: The paths $ab$ and $ba$ lead to the same event zone (up to renaming of events). This equivalence is easily detected by an appropriate hash function. This hash function interprets the event zone as constraints on clocks and processes (via the pointers) and thus cannot distinguish (renaming equivalent) event zones.

It is interesting to note that the event zone(s) reached by $ab$ and $ba$, when read as constraints on the clocks, actually *contains* the clock zones of both executions $ab$ and $ba$ in the classical timed automaton.

## Bounds Abstraction

The bounds on the dimension of the event zones are not sufficient to guarantee termination. In classical timed automata, arguments related to bisimulation equivalence of concrete states are used to justify semantics preserving modifications of zones (widening to infinity of bounds that are beyond a certain threshold).

This widening is incompatible with clock zones (it does not preserve reachability), but in [LNZ03], a sophisticated argument is given on how a closely related abstraction allows to explore only one of several clock zones that exceed certain thresholds in the same sense. This abstraction is used before hashing. A rather interesting difference between the classical approaches and ELSE is that computation on zones in the latter is always precise while preserving the same worst case bounds for the size of the explored state space.

# 4    Status of Development and Future Work

The development of *ELSE* is recent, up to now the invested effort is about 12 person months. The complete translation chain is running, i.e. syntactic analysis, semantic analysis and code generation, and there exists a prototype implementation of the elsezone library. However, coverage of the IF-2.0 language is very partial, we add code when we need it for modelling. The main current objective is to improve efficiency of the zone library.

We have begun experimenting with the prototype, in particular exploring artificial academic examples where the event zone approach seems superior to classical zones. On some example series, exponential savings in running *ELSE* against itself with the two semantics have been achieved.

# References

[AD94]     R. Alur and D. Dill, A theory of timed automata, Theoretical Computer Science **126(2)** (1994), 183–235. 273

[BFG+99]   Marius Bozga, Jean-Claude Fernandez, Lucian Ghirvu, Susanne Graf, Jean-Pierre Krimm, and Laurent Mounier, IF: An intermediate representation and validation environment for timed asynchronous systems, World Congress on Formal Methods (1), 1999, pp. 307–327. 273

[BJLY98]   J. Bengtsson, B. Jonsson, J. Lilius, and W. Yi, Partial order reductions for timed systems, Proceedings, Ninth International Conference on Concurrency Theory, LNCS, vol. 1466, Springer-Verlag, 1998, pp. 485–500. 274

[BLP+99]   G. Behrmann, K. Larsen, J. Pearson, C. Weise, W. Yi, and J. Lind-Nielsen, Efficient timed reachability analysis using clock difference diagrams, International Conference on Computer Aided Verification (CAV), LNCS, vol. 1633, 1999, pp. 341–353. 273

[DT98]     D. D'Souza and P. S. Thiagarajan, Distributed interval automata: A subclass of timed automata, 1998, Internal Report TCS-98-3. 274

[DY96]     C. Daws and S. Yovine, Reducing the number of clock variables of timed automata, IEE Real-Time Systems Symposium, December 1996, pp. 73–81. 273

[God96]    P. Godefroid, Partial-order methods for the verification of concurrent systems: an approach to the state-explosion problem, Lecture Notes in Computer Science, vol. 1032, Springer-Verlag Inc., New York, NY, USA, 1996. 274

[LNZ03]    Denis Lugiez, Peter Niebert, and Sarah Zennou, Clocked mazurkiewicz traces for partial order reductions of timed automata, 2003, draft, available at http://www.cmi.univ-mrs.fr/~niebert/docs/clockedmazu.pdf. 274, 276, 277, 278

[LPY95]    K. Larsen, P. Pettersson, and W. Yi, Model-checking for real-time systems, Fundamentals of Computation Theory, LNCS, August 1995, Invited talk, pp. 62–88.  273

[Min99]    Marius Minea, Partial order reduction for verification of timed systems, Ph.D. thesis, Carnegie Mellon University, 1999.  274

[Pel93]    D. Peled, All from one, one for all: On model checking using representatives, International Conference on Computer Aided Verification (CAV), LNCS, vol. 697, 1993, pp. 409–423.  274

[YS97]     Tomohiro Yoneda and Bernd-Holger Schlingloff, Efficient verification of parallel real-time systems, Formal Methods in System Design 11 (1997), no. 2, 197–215.  274

# Author Index